D1465757

Springer Series in
OPTICAL SCIENCES
85

founded by H.K.V. Lotsch

Springer
Berlin
Heidelberg
New York
Hong Kong
London
Milan
Paris
Tokyo

Physics and Astronomy

ONLINE LIBRARY

http://www.springer.de/phys/

Springer Series in
OPTICAL SCIENCES

The Springer Series in Optical Sciences, under the leadership of Editor-in-Chief *William T. Rhodes*, Georgia Institute of Technology, USA, and Georgia Tech Lorraine, France, provides an expanding selection of research monographs in all major areas of optics: lasers and quantum optics, ultrafast phenomena, optical spectroscopy techniques, optoelectronics, quantum information, information optics, applied laser technology, industrial applications, and other topics of contemporary interest.
With this broad coverage of topics, the series is of use to all research scientists and engineers who need up-to-date reference books.

The editors encourage prospective authors to correspond with them in advance of submitting a manuscript. Submission of manuscripts should be made to the Editor-in-Chief or one of the Editors. See also http://www.springer.de/phys/books/optical_science/

Daniel Mittleman   (Ed.)

Sensing
with Terahertz Radiation

With 207 Figures and 14 Tables

 Springer

Professor Daniel Mittleman
Rice University
Electrical and Computer Engineering Department MS-366
6100 Main Street
Houston, TX 77005
USA
E-mail: daniel@rice.edu

ISSN 0342-4111

ISBN 3-540-43110-1 Springer-Verlag Berlin Heidelberg New York

Cataloging-in-Publication Data applied for

Bibliographic information published by Die Deutsche Bibliothek.
Die Deutsche Bibliothek lists this publication in the Deutsche Nationalbibliothek;
detailed bibliographic data is available in the Internet at <http://dnb.ddb.de>.

Springer-Verlag Berlin Heidelberg New York
a member of BertelsmannSpringer Science+Business Media GmbH

http://www.springer.de

Data conversion by EDV-Beratung F. Herweg, Hirschberg
Cover concept by eStudio Calamar Steinen using a background picture from The Optics Project. Courtesy of
John T. Foley, Professor, Department of Physics and Astronomy, Mississippi State University, USA.
Cover production: *design & production* GmbH, Heidelberg

Printed on acid-free paper   SPIN 10780709   56/3141/di   5 4 3 2 1 0

Preface

One aspect of the field of THz radiation is the marriage of microwave and optical techniques. By its very nature, THz radiation bridges the gap between the microwave and optical regimes. The former can be characterized by the fact that most devices are comparable in size to the wavelength of the radiation. As a result, the propagation of energy in these devices is generally in the form of single-mode or low-order-mode guided waves. In contrast, the optical and infrared ranges are generally characterized by beams containing many modes, with dimensions much larger than the wavelength. Of course, there are exceptions to these rules, notably the single-mode propagation of optical radiation in fibers. Nonetheless, the general description holds true. Because of these fundamental differences, it is natural that the techniques used in their implementation are quite distinct. Much of the research in the THz field has been based on the melding of these disparate ideas.

Historically, techniques borrowed from the microwave and millimeter-wave communities dominated the progress in THz science and technology for many years. With continuing progress in intense optical sources, a plethora of new generation and detection schemes have been developed. Improvements have come not only in the radiation intensity, detection sensitivity, pulse duration, and bandwidth of the instruments. but also in the size, ease of use, and cost as well. These have had a dramatic impact on this field of research, since the difficulties are no longer rooted primarily in the production and sensing of the radiation. As a result, researchers are able to concentrate more on what one can do with THz radiation, and less on simply how to produce it. It is now possible to envision many applications of THz radiation which were previously impractical.

The purposes of this volume are twofold. First, the various different methods of accessing the THz range discussed here should serve as convincing evidence that there have been qualitative and significant improvements over older, more conventional techniques. It should be clear that these improvements enable practical "real-world" applications of THz technology, in a manner which would not have been possible before. Significantly, it should be clear from the descriptions provided here that these applications are not purely speculative; at least some of them can be realistically achieved. Second, the demonstrations and feasibility tests described here should serve as compelling evidence of the need for these new technologies. Owing to the unique characteristics of THz radiation and its interaction with materials, these devices have substantial advantages over other, competing technologies in a number of different areas. The work described here makes a compelling case that not only can THz radiation provide solutions to a host of problems, it can in some cases provide the best solution.

The introduction, by Grischkowsky and Mittleman, provides a historical overview of some of the recent developments in THz science. Owing to length restrictions, it is impossible to cover all of the excellent work in this rapidly growing area, but the authors hope to provide a reasonable cross section which illustrates the breadth of the area and the many exciting new advances. Equally importantly, the introduction provides a comprehensive bibliography for the interested reader to learn more about any of these techniques.

The chapter by De Lucia deals with the subject of gas-phase spectroscopy in the THz range. This was the genesis of far-infrared spectroscopy, and as such provides a valuable historical perspective. However, this field has continued to see rapid progress in the development of experimental tools for THz spectroscopy. These include electron beam and solid-state oscillators, as well as harmonic mixers and optically pumped gas lasers. The chapter by De Lucia provides a detailed bibliography of work in many of these areas, with particular focus on the most recent developments in gas sensing and spectroscopy. In addition, novel methods involving nonlinear electromagnetic propagation can be used as the basis for all-electronic sources of THz radiation, which can be extremely compact. The chapter by van der Weide describes this technology, and details some of the useful sensing and ranging applications being explored.

The combination of the time-domain spectroscopy technique with THz beams has some powerful advantages over traditional continuous-wave spectroscopy. Numerous applications of this spectroscopic tool are described in the chapters by Mittleman, Cheville et al., and Koch in this volume. A related technique for THz generation is photomixing. Recent advances in this field are described in the chapter by Duffy et al. Free-space electro-optic sensing is described in detail in the chapter by Jiang and Zhang.

The adoption of a new technology of this type presents something of a "chicken and egg" problem: there is no incentive to develop new technologies until an application has been demonstrated, but there is no motivation to explore new applications unless the implementation of the technology is feasible. A recognition of this paradox in the mid-1990s spurred many in the optics community to action. In several different cases, inspiring successes have been achieved in the development of useful new tools and techniques. Terahertz time-domain spectroscopy represents one of these ongoing stories. Many of the other techniques described here are also beginning to have an impact outside the scientific laboratories in which they were developed. The topic of sensing with terahertz radiation is already far too broad to be covered in a single volume, but we hope that this work provides a useful roadmap for some of the most exciting new developments.

Houston
August 2002   Daniel Mittleman

Contents

Bio-medical Applications of THz Imaging

**Electronic Sources and Detectors for Wideband Sensing
in the Terahertz Regime**

List of Contributors

R. Alan Cheville
Oklahoma State University
Electrical & Computer Engineering
202 Engineering South
Stillwater, OK 74078, USA
Email: kridnix@thzsun.ecen.
  okstate.edu
URL: elec-engr.okstate.edu/
  thzlab/

Frank C. De Lucia
Department of Physics
174 W. 18th Ave.
Ohio State University
Columbus, OH 43210, USA
Email: fcd@mps.ohio-state.edu
URL: www.physics.ohio-state.edu/
  graduate_brochure/people/
  delucia.html

Sean M. Duffy
MIT Lincoln Laboratory
244 Wood Street
Lexington, MA 02420-9108, USA
Email: sduffy@ll.mit.edu

Daniel R. Grischkowsky
Oklahoma State University
Electrical & Computer Engineering
202 Engineering South
Stillwater, OK 74078, USA
Email: grischd@master.ceat.
  okstate.edu
URL: elec-engr.okstate.edu/
  thzlab/

Zhiping Jiang
Department of Physics
Rensselaer Polytechnic Institute
Troy, NY 12180, USA
Email: jiangz@rpi.edu
URL: www.rpi.edu/~zhangxc/

Martin Koch
TU Braunschweig
Institut für Hochfrequenztechnik
Schleinitzstr. 22
38106 Braunschweig, Germany
Email: Martin.Koch@tu-bs.de
URL: www.tu-bs.de/institute/
  hochfrequenz/allgemein/
  mitarbeiter/koch.html

Roger McGowan
Oklahoma State University
Electrical & Computer Engineering
202 Engineering South
Stillwater, OK 74078, USA
Email: rmcgowan@thzsun.ecen.
  okstate.edu
URL: elec-engr.okstate.edu/
  thzlab/

K. Alexander McIntosh
MIT Lincoln Laboratory
244 Wood Street
Lexington, MA 02420-9108, USA
Email: alex@ll.mit.edu

Daniel Mittleman
Rice University
Electrical & Computer Engineering
Dept., MS-366
PO Box 1892
Houston. TX 77251-1892. USA
Email: daniel@rice.edu
URL: www-ece.rice.edu/
  ~daniel/Mittleman.html

Matthew T. Reiten
School of Electrical
and Computer Engineering
202 Engineering South
Stillwater, OK 74078, USA

Daniel W. van der Weide
University of Wisconsin
Dept. of Electrical
& Computer Engineering
1415 Engineering Dr
Madison, WI 53706, USA
Email: danvdw@engr.wisc.edu
URL: www.engr.wisc.edu/ece/
faculty/vanderweide_daniel.html

Simon Verghese
MIT Lincoln Laboratory
244 Wood Street
Lexington, MA 02420-9108, USA
Email: simonv@ll.mit.edu

Xi-Cheng Zhang
Rensselaer Polytechnic Institute
Department of Physics.
Applied Physics and Astronomy
Troy. NY 12180. USA
Email: zhangxc@rpi.edu
URL: www.rpi.edu/~zhangxc/

Introduction

Daniel R. Grischkowsky and Daniel Mittleman

> "It is appropriate to view the submillimeter wave region as a transition region lying between the millimeter wave and the infrared portions of the electromagnetic spectrum and posssessing as yet no hallmark of its own."
> - B. Senitzky and A.A. Oliner, 1970 [Senitzky, 1970]

This statement is the opening remark from the *Proceedings of the Symposium on Submillimeter Waves*, which was one in a continuing series of meetings held at the Polytechnic Institute of Brooklyn since the early 1950s. That volume covers a diverse array of topics, involving spectroscopy, applications, and technologies of the submillimeter region of the spectrum, and as such provides a useful snapshot of the state of the art in 1970. In some ways, the sentiments expressed in the above quotation remain true even today, some 30 years later. As of 1999, there was not a single commercial product which relied on submillimeter wave radiation as an enabling aspect of its functionality. However, it is also true that there has been much activity and progress in this area, and it seems inevitable that the submillimeter region is about to develop the hallmark it lacked in 1970. Indeed, as if to dramatize this point, the first commercial THz spectrometer became available in early 2000. In this volume, we shall provide an updated look at the status of the field of submillimeter-wave science and technology, and attempt to offer some educated guesses at the future prospects.

A first issue which must be faced is that of terminology. In 1970, the most commonly used terminology for the spectral range in question was "submillimeter waves". This restricts the discussion to frequencies above 300 GHz, corresponding to a free-space wavelength of 1 mm. Others have used "far infrared" to describe this spectral range, although this term is hazardous because its high-frequency limit is ambiguous. Our present interests are better defined by technological limits. There exist many methods for the generation and detection of radiation up to several tens of GHz. Frequencies at or below this range are generally compatible with electronic technologies, and are usually grouped together with microwaves. In contrast, frequencies above 100 GHz become increasingly less compatible with these established technologies, and are therefore often dealt with by other means. A distinct array of techniques is required, and a different community of scientists and engineers is often involved. On the high-frequency end of the range, one encounters the

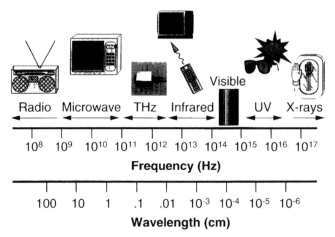

Fig. 1. The electromagnetic spectrum. The icon symbolizing the THz portion of the spectrum is a fiber-pigtailed Auston switch for use as an emitter in terahertz time-domain spectroscopy. Photo courtesy of Picometrix, Inc.

limits of optical techniques, including both coherent and incoherent sources. Several researchers have demonstrated that, with great care, one can push the limits of incoherent spectrometers down to perhaps 500 GHz, but the vast majority of infrared spectroscopists have limited their studies to frequencies above a few THz. Similarly, while numerous different laser systems can be used to provide coherent radiation in the range below 1 THz, few of these have attained a widespread degree of use. The most common far-infrared laser remains the CO_2 laser, operating at 10.6 microns ($\sim$30 THz). Our interests focus primarily on the spectral range which lies outside of the range of conventional technologies, from $\sim$0.1 THz to a few tens of THz (see Fig. 1). As such, the term "terahertz radiation" seems the most appropriately descriptive.

One aspect of the field which was recognized in 1970 as crucial to its development was the marriage of microwave and optical techniques. By its very nature, THz radiation bridges the gap between the microwave and optical regimes. The former can be characterized by the fact that most devices are comparable in size to the wavelength of the radiation. As a result, the propagation of energy in these devices is generally in the form of single-mode or low-order-mode guided waves. In contrast, the optical and infrared ranges are generally characterized by beams containing many modes, with dimensions much larger than the wavelength. Of course, there are exceptions to these rules, notably the single-mode propagation of optical radiation in fibers. Nonetheless, the general description holds true. Because of these fundamental differences, it is natural that the techniques used in the two regimes are quite distinct. Much of the research in the THz field has been based on the melding of these disparate ideas.

Historically, techniques borrowed from the microwave and millimeter-wave communities dominated the progress in THz science and technology for many years. Indeed, it is revealing to browse through the abstracts from the 1970 meeting, and note the very few papers that made use of "quasi-optical" methods. This balance began to shift during the 1970s, with early work in the generation of THz radiation using intense lasers, by nonlinear frequency conversion [Yang, 1971]. With continuing progress in intense optical sources, a plethora of new generation and detection schemes have been developed. Improvements have come not only in the radiation intensity, detection sensitivity, pulse duration, and bandwidth of the instruments, but also in the size, ease of use, and cost as well. These have had a dramatic impact on this field of research, since the difficulties are no longer rooted primarily in the production and sensing of the radiation. As a result, researchers are able to concentrate more on what one can do with THz radiation, and less on simply how to produce it. It is now possible to envision many applications of THz radiation which were previously impractical.

Senitzky and Oliner noted that the development of technologies for the THz spectral range was hampered by two important factors: the difficulty involved in the generation and detection of this radiation, and a lack of a perceived need. Unlike, for example, microwave technology, whose development was motivated by the need for radar systems, no equivalently compelling application had been identified. Since 1970, however, the first of these two factors has in some measure been overcome. So, it is now appropriate to consider the latter of the two problems. It is interesting to note that Senitzky and Oliner's article and others in the same volume list a number of proposed uses for THz radiation. Some of these are still of interest today, such as point-to-point communication and imaging. Other possibilities have been explored which were not anticipated in 1970. These exciting new prospects are among the primary motivating factors for this book.

The purposes of this volume are twofold. First, the various different methods of accessing the THz range discussed here should serve as convincing evidence that there have been qualitative and significant improvements over older, more conventional techniques. It should be clear that these improvements enable practical "real-world" applications of THz technology, in a manner which would not have been possible before. Significantly, it should be clear from the descriptions provided here that these applications are not purely speculative; at least some of them can be realistically achieved. Second, the demonstrations and feasibility tests described here should serve as compelling evidence of the need for these new technologies. Owing to the unique characteristics of THz radiation and its interaction with materials, these devices have substantial advantages over other, competing technologies in a number of different areas. The work described here makes a compelling case that not only can THz radiation provide solutions to a host of problems, it can in some cases provide the best solution.

In this introduction, we provide a historical overview of some of the recent developments in THz science. Owing to length restrictions, it is impossible to cover all of the excellent work in this rapidly growing area, but we hope to provide a reasonable cross section which illustrates the breadth of the area and the many exciting new advances. Equally importantly, we provide a comprehensive bibliography for the interested reader to learn more about any of these techniques. In order to emphasize the historical progress in this field, this bibliography is arranged chronologically. Together, this introduction and bibliography should be useful background material for the remaining chapters in this volume, which provide a great deal of additional depth in some of the areas described here.

The chapter in this volume by De Lucia deals with the subject of gas phase spectroscopy in the THz range. This was the genesis of far-infrared spectroscopy, and as such provides a valuable historical perspective. However, this field has continued to see rapid progress in the development of experimental tools for THz spectroscopy. These include electron beam and solid-state oscillators, as well as harmonic mixers and optically pumped gas lasers. The chapter by De Lucia provides a detailed bibliography of work in many of these areas, with particular focus on the most recent developments in gas sensing and spectroscopy. In addition, novel methods involving nonlinear electromagnetic propagation can be used as the basis for all-electronic sources of THz radiation, which can be extremely compact. The chapter by van der Weide describes this technology, and details some of the useful sensing and ranging applications being explored.

The development of far-infrared optoelectronic techniques was presaged by a series of pioneering experiments by Y.R. Shen in the early 1970s, which represented the first demonstration of far-infrared generation by rectification of short optical pulses [Yang, 1971; Shen, 1976]. Subsequently, a number of groups recognized the utility of ultrafast photoconductive switching for the manipulation of radiation at microwave, and higher, frequencies [Mourou, 1981; Auston, 1983; Mourou, 1984]. The concepts of terahertz time-domain spectroscopy (THz-TDS) were introduced by the pioneering series of experiments performed by Auston's group at Bell Laboratories [Auston, 1985; Cheung, 1985; Cheung, 1986]. As their source of ultrashort far-infrared pulses, they generated an electromagnetic shock wave via optical rectification in a nonlinear dielectric crystal with an ultrashort laser pulse [Auston, 1983]. By measuring the change in the time dependence of the shock wave as a function of propagation distance, far-infrared spectroscopic measurements of the non-linear dielectric were obtained. In addition, measurements of the change in the time dependence after a reflection at the surface of the dielectric permitted measurements of other materials brought into contact with the dielectric surface [Nuss, 1987].

Subsequently this powerful new technique, involving either free or guided-wave propagation of short electromagnetic pulses with THz bandwidths, has proven itself by the characterization of many different types of samples and systems using a variety of sources and receivers. With this technique two electromagnetic pulse shapes are measured, that of the input pulse and of the propagated pulse, which has changed shape owing to its passage through the sample under study. It is important to realize that the actual field amplitudes are measured, thereby enabling the determination of the relative phase of the corresponding components of the amplitude Fourier spectrum. Consequently, via a numerical complex Fourier analysis of the input and propagated pulses, knowledge of the complex Fresnel transmission coefficients through the sample, and knowledge of the effect of the sample on the transfer function between the transmitter and receiver, the frequency-dependent absorption and dispersion of the sample can be obtained. The useful frequency range of the method is determined by the initial pulse duration and the time resolution of the detection process. Therefore, with each reduction in the width of the generated electromagnetic pulse and/or the time resolution of detection, there is a corresponding increase in the available frequency range.

An early demonstration of guided-wave THz-TDS was the pioneering optoelectronic experiment of Cooper, which characterized the dispersion of a microstrip transmission line [Cooper, 1985]. In other work, the "sliding contact" generation of subpicosecond electrical pulses propagating as a single TEM mode on coplanar transmission lines was introduced [Ketchen, 1986; Grischkowsky, 1988]. Besides their many technical applications, these single-mode pulses can be used for THz-TDS as well [Grischkowsky, 1987; Sprik, 1987]. This is due to two important features. First, the useful bandwidth of the pulses extends up to 1 THz, and second, the pulses propagate as a single mode of excitation on the transmission line. This latter feature is due to the micron-sized dimensions of the coplanar line and to the "sliding contact" method of excitation, which matches the TEM mode of the transmission line. Thus, as the ultrashort electrical pulse propagates down the transmission line, its pulse shape changes only because of the frequency-dependent electrical and magnetic properties of the transmission line, e.g. the metal of the line, the dielectric substrate, and radiation processes. Consequently, via THz-TDS analyses of the input and propagated pulses, the absorption and dispersion of the line can be obtained.

For more general TDS applications, it is possible to bring other materials into contact with the transmission line and thereby intersect the field lines of the propagating pulse. Consequently, Fourier analysis of the propagated pulse characterizes these materials. As an example of this approach, when a powder of erbium–iron garnet was applied to a transmission line, strong far-infrared magnetic resonances were observed at low temperatures [Gallagher, 1987]. Because the electrical pulse propagates as a guided wave along the transmission line, the electric and magnetic fields are strongly localized at the surface. This feature makes TDS with transmission lines especially appropriate to studying surface excitations and thin-film substrates and superstrates, because it is possible to obtain long path lengths with small quantities of material.

Although electro-optic Cherenkov radiation was used in the early THz-TDS demonstrations, it was several years before a method to couple this radiation out of the electro-optic material was demonstrated [Hu, 1990]. Subsequently, several other sources of freely propagating electromagnetic pulses have been demonstrated [DeFonzo, 1987a; DeFonzo, 1987b; Fattinger, 1988; Smith, 1988; Fattinger, 1989a; van Exter, 1989b]. The spectral content of these sources extends from low frequencies up to the THz frequency range. One of the most useful of these sources is based on an optical approach whereby a transient Hertzian dipole is tightly coupled to a dielectric collimating lens; in the initial demonstrations, beams of single-cycle 0.5 THz pulses were produced [Smith, 1988; Fattinger, 1989a]. With the addition of paraboloidal focusing and collimating mirrors [van Exter, 1989b; van Exter, 1990c], the resulting system has high brightness and extremely high collection efficiency. Another method of generation, which has gained in popularity recently, is optical rectification [Hu, 1990; Xu, 1992]. This technique has the advantage that it is capable of producing extremely broadband radiation, limited only by the duration of the optical pulse [Bonvalet, 1995; Wu, 1997b].

The combination of the TDS technique with THz beams has some powerful advantages over traditional continuous-wave spectroscopy. First, the coherent detection of the far-infrared radiation is extremely sensitive. Although the energy per THz pulse was very low (0.1 fJ) in the initial characterization of this system, the 100 MHz repetition rate and the coherent detection allowed the determination of the electric field of the propagated pulse with a signal-to-noise ratio of about 10^4 for an integration time of 125 ms [van Exter, 1990c]. In terms of average power, this sensitivity exceeds that of conventional liquid-helium-cooled bolometers by more than 1000 times. Subsequent improvements have pushed the limits of these techniques significantly, with signal-to-noise ratios in excess of 10^6, and average powers in excess of 1 μW reported [Cai, 1997]. Secondly, because of the gated and coherent detection, the thermal background which plagues traditional measurements in this frequency range is observationally absent. It should be clear that the frequency resolution of this technique is similar to that of Fourier transform spectroscopy, as they are both based on a scanning delay line. Numerous applications of this spectroscopic tool are described in the chapters by Mittleman, Cheville et al., and Koch in this volume.

A related technique for THz generation is photomixing. Here, two stabilized single-frequency visible lasers are mixed on a fast photoconductive material, such as low-temperature-grown GaAs. A current is driven in the photoconductor at a frequency corresponding to the beat note between the two lasers, generating radiation in much the same way as in an Auston switch [Brown, 1993]. This technique affords a number of advantages over other methods for THz generation, including wide tunability, and narrow linewidths limited by the stabilization of the visible lasers used. Much of the motivation for the development of these technologies has arisen from the need for local oscillators for the heterodyne detection of extremely faint astrophysical sources of THz radiation. In this case, a narrow (but tunable) linewidth and high power are required. Recent advances in this field are described in the chapter by Duffy et al. in this volume.

In the mid-1990s, the field took another tremendous stride forward with the demonstration of free-space electro-optic sensing [Wu, 1995; Nahata, 1996a; Wu, 1996a; Nahata, 1996c; Winnewisser, 1997]. This detection scheme relies on the polarization rotation of a short optical probe pulse as it copropagates through a medium with a THz pulse. Because the technique is non-resonant, the response time is limited by the duration of the sampling pulse, which can be shorter than 10 fs. A direct advantage of this technique, therefore, is the extremely broad detection bandwidths that can be achieved [Wu, 1997a; Wu, 1997b; Leitenstorfer, 1999]. With these detectors, ultrabroadband spectroscopy, covering nearly the entire range from below 100 GHz to more than 50 THz, is now feasible. This technique is described in detail in the chapter by Jiang and Zhang in this volume.

As noted more than 30 years ago [Senitzky, 1970], the adoption of a new technology of this type presents something of a "chicken and egg" problem: there is no incentive to develop new technologies until an application has been demonstrated, but there is no motivation to explore new applications unless the implementation of the technology is feasible. A recognition of this paradox in the mid 1990s spurred many in the optics community to action. In several different cases, inspiring successes have been achieved in the development of useful new tools and techniques. Terahertz time-domain spectroscopy represents one of these ongoing stories. Many of the other techniques described here are also beginning to have an impact outside of the scientific laboratories in which they were developed. The topic of sensing with terahertz radiation is already far too broad to be covered in a single volume, but we hope that this work provides a useful roadmap for some of the most exciting new developments.

Chronological Bibliography

Nicolson, A. M., Ross, G. F., "Measurement of the intrinsic properties of materials by time-domain techniques", *IEEE Trans. Instrum. Meas.* **19**, 377 (1970).

Senitzky, B., Oliner, A. A., "Submillimeter waves – a transition region", in *Proceedings of the Symposium on Submillimeter Waves*, ed. by J. Fox (Polytechnic Press of the Polytechnic Institute of Brooklyn, New York, 1970), vol. 20.

Yang, K. H., Richards, P. L., Shen, Y. R., "Generation of far-infrared radiation by picosecond light pulses in LiNbO$_3$", *Appl. Phys. Lett.* **19**, 320 (1971).

Shen, Y. R., "Far-infrared generation by optical mixing", *Prog. Quantum Electron.* **4**, 207 (1976).

Mourou, G., Stancampiano, C. V., Blumenthal, D., "Picosecond microwave pulse generation", *Appl. Phys. Lett.* **38**, 470 (1981).

Auston, D. H., "Subpicosecond electro-optic shockwaves". *Appl. Phys. Lett.* **43**, 713 (1983).

Auston, D. H., Cheung, K. P., Smith, P. R., "Picosecond photoconducting Hertzian dipoles", *Appl. Phys. Lett.* **45**, 284 (1984a).
Auston, D. H., Cheung, K. P., Valdmanis, J. A., Kleinman, D. A., "Cherenkov radiation from femtosecond optical pulses in electro-optic media". *Phys. Rev. Lett.* **53**, 1555 (1984b).
Mourou, G. A., Meyer, K. E., "Subpicosecond electro-optic sampling using coplanar strip transmission lines". *Appl. Phys. Lett.* **45**, 492 (1984).

Auston, D. H., Cheung, K. P., "Coherent time-domain far-infrared spectroscopy", *J. Opt. Soc. Am. B* **2**, 606 (1985).
Cheung, K. P., Auston, D. H., "Excitation of coherent phonon polaritons with femtosecond optical pulses", *Phys. Rev. Lett.* **55**, 2152 (1985).
Cooper, D. E., "Picosecond optoelectronic measurement of microstrip dispersion". *Appl. Phys. Lett.* **47**, 33 (1985).

Cheung, K. P., Auston, D. H., "A novel technique for measuring far-infrared absorption and dispersion", *Infrared Phys.* **26**, 23 (1986).
Ketchen, M. B., Grischkowsky, D., Chen, T. C., Chi, C.-C., Duling, I. N., Halas, N. J., Halbout, J.-M., Kash, J. A., Li, G. P., "Generation of subpicosecond electrical pulses on coplanar transmission lines". *Appl. Phys. Lett.* **48**, 751 (1986).

DeFonzo, A. P., Jarwala, M., Lutz, C., "Transient response of planar integrated optoelectronic antennas", *Appl. Phys. Lett.* **50**, 1155 (1987a).
DeFonzo, A. P., Lutz, C. R., "Optoelectronic transmission and reception of ultra-short electrical pulses". *Appl. Phys. Lett.* **51**, 212 (1987b).
Gallagher, W. J., Chi, C.-C., Duling, I. N., III, Grischkowsky, D., Halas, N. J., Ketchen, M. B., Kleinsasser, A. W., "Subpicosecond optoelectronic study of resistive and superconductive transmission lines", *Appl. Phys. Lett.* **50**, 350 (1987).
Grischkowsky, D., Duling, I. N., III, Chen, J. C., Chi, C.-C., "Electromagnetic shock waves from transmission lines", *Phys. Rev. Lett.* **59**, 1663 (1987).
Nuss, M. C., Auston, D. H., Capasso, F., "Direct subpicosecond measurement of carrier mobility of photoexcited electrons in gallium arsenide", *Phys. Rev. Lett.* **58**, 2355 (1987).
Sprik, R., Duling, I. N., III, Chi, C. C., Grischkowsky, D., "Far infrared spectroscopy with subpicosecond electrical pulses on transmission lines". *Appl. Phys. Lett.* **51**, 548 (1987).

Auston, D. H., Nuss, M. C., "Electro-optical generation and detection of femtosecond electrical transients". *IEEE J. Quant. Elec.* **24**, 184 (1988).

Dykaar, D. R., Sobolewski, R., Chwalek, J. M., Whitaker, J. F., Hsiang, T. Y., Mourou, G. A., Lathrop, D. K., Russek, S. E., Buhrman, R. A., "High-frequency characterization of thin-film Y–Ba–Cu oxide superconducting transmission lines", *Appl. Phys. Lett.* **52**, 1444 (1988).

Fattinger, C., Grischkowsky, D., "Point source terahertz optics", *Appl. Phys. Lett.* **53**, 1480 (1988).

Freeman, J. L., Bloom, D. M., Jeffries, S. R., Auld, B. A., "Accuracy of electro-optic measurements of coplanar waveguide transmission lines", *Appl. Phys. Lett.* **53**, 7 (1988).

Grischkowsky, D., Ketchen, M. B., Chi, C.-C., Duling, I. N., III, Halas, N. J., Halbout, J.-M., May, P. G., "Capacitance free generation and detection of sub-picosecond electrical pulses on coplanar transmission lines", *IEEE J. Quant. Elec.* **24**, 221 (1988).

Smith, P. R., Auston, D. H., Nuss, M. C., "Subpicosecond photoconducting dipole antennas", *IEEE J. Quantum Electron.* **24**, 255 (1988).

Fattinger, C., Grischkowsky, D., "Terahertz beams", *Appl. Phys. Lett.* **54**, 490 (1989a).

Fattinger, C., Grischkowsky, D., "Observation of electromagnetic shock waves from propagating surface-dipole distributions", *Phys. Rev. Lett.* **62**, 2961 (1989b).

Knox, W. H., Henry, J. E., Groosen, K. W., Li, K. D., Tell, B., Miller, D. A. B., Chemla, D. S, Gossard, A. C., English, J., Schmitt-Rink, S., "Femtosecond excitonic optoelectronics", *IEEE J. Quantum Electron.* **25**, 2586 (1989).

Lutz, C., DeFonzo, A. P., "Far-field characteristics of optically pulsed millimeter wave antennas", *Appl. Phys. Lett.* **54**, 2186 (1989).

Nuss, M. C., Mankiewich, P. M., Howard, R. E., Straughn, B. L., Harvey, T. E., Brandle, C. D., Berkstresser, G. W., Goossen, K. W., Smith, P. R., "Propagation of terahertz bandwidth electrical pulses on $YBa_2Cu_3O_{7-\delta}$ transmission lines on lanthanum aluminate", *Appl. Phys. Lett.* **54**, 2265 (1989a).

Nuss, M. C., Goossen, K. W., "Investigation of high-temperature superconductors with terahertz bandwidth electrical pulses", *IEEE J. Quantum Electron.* **25**, 2596 (1989b).

Pastol, Y., Arjavalingam, G., Halbout, J.-M., Kopcsay, G. V., "Coherent broadband microwave spectroscopy using picosecond optoelectronic antennas", *Appl. Phys. Lett.* **54**, 307 (1989a).

Pastol, Y., Arjavalingam, G., Kopcsay, G. V., Halbout, J.-M., "Dielectric properties of uniaxial crystals measured with optoelectronically generated microwave transient radiation", *Appl. Phys. Lett.* **55**, 2277 (1989b).

Pastol, Y., Arjavalingam, G., Halbout, J.-M., Kopcsay, G. V., "Absorption and dispersion of low-loss dielectrics measured with microwave transient radiation", *Electron. Lett.* **25**, 523 (1989c).

van Exter, M., Fattinger, C., Grischkowsky, D., "Terahertz time-domain spectroscopy of water vapor", *Opt. Lett.* **14**, 1128 (1989a).

van Exter, M., Fattinger, C., Grischkowsky, D., "High-brightness terahertz beams characterized with an ultrafast detector", *Appl. Phys. Lett.* **55**, 337 (1989b).

Arjavalingam, G., Pastol, Y., Halbout, J.-M., Kopcsay, G. V., "Broad-band microwave measurements with transient radiation from optoelectronically pulsed antennas", *IEEE Trans. Microwave Theory Tech.* **38**, 615 (1990).

Grischkowsky, D., Keiding, S. R., "THz time-domain spectroscopy of high T_c substrates", *Appl. Phys. Lett.* **57**, 1055 (1990a).

Grischkowsky, D., Keiding, S., van Exter, M., Fattinger, C., "Far-infrared time-domain spectroscopy with terahertz beams of dielectrics and semiconductors", *J. Opt. Soc. Am. B* **7**, 2006 (1990b).

Hu, B. B., Zhang, X.-C., Auston, D. H., Smith, P. R., "Free-space radiation from electro-optic crystals", *Appl. Phys. Lett.* **56**, 506 (1990).

Keiding, S. R., Grischkowsky, D., "Measurements of the phase shift and reshaping of terahertz pulses due to total internal reflection", *Opt. Lett.* **15**, 48 (1990).

van Exter, M., Grischkowsky, D., "Optical and electrical properties of doped silicon from 0.1 to 2 THz", *Appl. Phys. Lett.* **56**, 1694 (1990a).

van Exter, M., Grischkowsky, D., "Carrier dynamics of electrons and holes in moderately doped silicon", *Phys. Rev. B* **41**, 12140 (1990b).

van Exter, M., Grischkowsky, D., "Characterization of an optoelectronic terahertz beam system", *IEEE Trans. Microwave Theory Tech.* **38**, 1684 (1990c).

Chwalek, J. M., Whitaker, J. F., Mourou, G. A., "Submillimetre wave response of superconducting YBa$_2$Cu$_3$O$_{7-x}$ using coherent time-domain spectroscopy", *Elec. Lett.* **27**, 447 (1991).

Frankel, M. Y., Gupta, S., Valdmanis, J. A., Mourou, G. A., "Terahertz attenuation and dispersion characteristics of coplanar transmission lines", *IEEE Trans. Microwave Theory Tech.* **39**, 910 (1991).

Greene, B. I., Federici, J. F., Dykaar, D. R., Jones, R. R., Bucksbaum, P. H., "Interferometric characterization of 160 fs far-infrared light pulses", *Appl. Phys. Lett.* **59**, 893 (1991a).

Greene, B. I., Federici, J. F., Dykaar, D. R., Levi, A. F. J., Pfeiffer, L., "Picosecond pump and probe spectroscopy utilizing freely propagating terahertz radiation", *Opt. Lett.* **16**, 48 (1991b).

Harde, H., Keiding, S., Grischkowsky, D., "THz commensurate echoes: periodic rephasing of molecular transitions in free-induction decay", *Phys. Rev. Lett.* **66**, 1834 (1991a).

Harde, H., Grischkowsky, D., "Coherent transients excited by subpicosecond pulses of terahertz radiation", *J. Opt. Soc. Am. B* **8**, 1642 (1991b).

Hu, B. B., Zhang, X.-C., Auston, D. H., "Terahertz radiation induced by subband-gap femtosecond optical excitation of GaAs", *Phys. Rev. Lett.* **67**, 2709 (1991).

Nuss, M. C., Goossen, K. W., Mankiewich, P. M., O'Malley, M. L., "Terahertz surface impedance of thin YBa2Cu3O7 superconducting films", *Appl. Phys. Lett.* **58**, 2561 (1991a).

Nuss, M. C., Mankiewich, P. M., O'Malley, M. L., Westerwick, E. H., "Dynamic conductivity and 'coherence peak' in YBa$_2$Cu$_3$O$_7$ superconductors", *Phys. Rev. Lett.* **66**, 3305 (1991b).

Nuss, M. C., Goossen, K. W., Gordon, J. P., Mankiewich, P. M., O'Malley, M. L., Bhushan, M., "Terahertz time-domain measurement of the conductivity and superconducting band gap in niobium", *J. Appl. Phys.* **70**, 2238 (1991c).

Robertson, W. M., Arjavalingam, G., Kopcsay, G. V., Halbout, J.-M., "Microwave reflection measurements on doped semiconductors with picosecond transient radiation", *IEEE Microwave Guided Wave Lett.* **1**, 371 (1991).

Roskos, H., Nuss, M. C., Goossen, K. W., Kisker, D. W., White, A. E., Short, K. T., Jacobson, D. C., Poate, J. M., "Propagation of picosecond electrical pulses on a silicon-based microstrip line with buried cobalt silicide ground plane", *Appl. Phys. Lett.* **58**, 2604 (1991).

Xu, L., Zhang, X.-C., Auston, D. H., Jalali, B., "Terahertz radiation from large aperture Si p–i–n diodes", *Appl. Phys. Lett.* **59**, 3357 (1991).

Bakker, H. J., Hunsche, S., Kurz, H., "Observation of THz phonon-polariton beats in LiTaO$_3$", *Phys. Rev. Lett.* **69**, 2823 (1992).

Chuang, S. L., Schmitt-Rink, S., Greene, B. I., Saeta, P. N., Levi, A. F. J., "Optical rectification at semiconductor surfaces", *Phys. Rev. Lett.* **68**, 102 (1992).

Darrow, J. T., Zhang, X.-C., Auston, D. H., Morse, J. D., "Saturation properties of large-aperture photoconducting antennas", *IEEE J. Quantum Electron.* **28**, 1607 (1992).

Federici, J. F., Greene, B. I., Saeta, P. N., Dykaar, D. R., Sharifi, F., Dynes, R. C., "Direct picosecond measurement of photoinduced Cooper-pair breaking in lead", *Phys. Rev. B* **46**, 153 (1992).

Froberg, N. M., Hu, B. B., Zhang, X.-C., Auston, D. H., "Terahertz radiation from a photoconducting antenna array", *IEEE J. Quantum Electron.* **28**, 2291 (1992).

Greene, B. I., Saeta, P. N., Dykaar, D. R., Schmitt-Rink, S., Chuang, S. L., "Far-infrared light generation at semiconductor surfaces and its spectroscopic applications", *IEEE J. Quantum Electron.* **28**, 2302 (1992).

Grischkowsky, D., "An ultrafast optoelectronic THz beam system: applications to time-domain spectroscopy", *Opt. Photon. News* **3**, 21 (1992).

Katzenellenbogen, N., Grischkowsky, D., "Electrical characterization to 4 THz of N- and P-type GaAs using THz time-domain spectroscopy", *Appl. Phys. Lett.* **61**, 840 (1992).

Konishi, Y., Kamegawa, M., Case, M., Yu, R., Rodwell, M. J. W., York, R. A., Rutledge, D. B., "Picosecond electrical spectroscopy using monolithic GaAs circuits", *Appl. Phys. Lett.* **61**, 2829 (1992).

Pedersen, J. E., Balslev, I., Hvam, J. M., Keiding, S. R., "Terahertz pulses from semiconductor-air interfaces", *Appl. Phys. Lett.* **61**, 1372 (1992a).

Pedersen, J. E., Keiding, S. R., "THz time-domain spectroscopy of nonpolar liquids", *IEEE J. Quantum Electron.* **28**, 2518 (1992b).

Planken, P. C. M., Nuss, M. C., Knox, W. H., Miller, D. A. B., Goossen, K. W., "THz pulses from the creation of polarized electron–hole pairs in biased quantum wells", *Appl. Phys. Lett.* **61**, 2009 (1992a).

Planken, P. C. M., Nuss, M. C., Brener, I., Goossen, K. W., Luo, M. S. C., Chuang, S. L., Pfeiffer, L., "Terahertz emission in single quantum wells after coherent optical excitation of light hole and heavy hole excitons", *Phys. Rev. Lett.* **69**, 3800 (1992b).

Ralph, S. E., Grischkowsky, D., "THz spectroscopy and source characterization by optoelectronic interferometry", *Appl. Phys. Lett.* **60**, 1070 (1992).

Ree, M., Chen, K.-J., Kirby, D. P., Katzenellenbogen, N., Grischkowsky, D., "Anisotropic properties of high temperature polyimide thin films: dielectric and thermal expansion behavior", *J. Appl. Phys.* **72**, 2014 (1992).

Robertson, W. M., Arjavalingam, G., Meade, R. D., Brommer, K. D., Rappe, A. M., Joannopoulos, J. D., "Measurement of photonic band structure in a two-dimensional periodic dielectric array", *Phys. Rev. Lett.* **68**, 2023 (1992).

Roskos, H. G., Nuss, M. C., Shah, J., Leo, K., Miller, D. A. B., Fox, A. M., Schmitt-Rink, S., Kohler, K., "Coherent submillimeter-wave emission from charge oscillations in a double-well potential", *Phys. Rev. Lett.* **68**, 2216 (1992).

Saeta, P. N., Federici, J. F., Greene. B. I.. Dykaar. D. R.. "Intervalley scattering in GaAs and InP probed by pulsed far-infrared transmission spectroscopy". *Appl. Phys. Lett.* **60**. 1477 (1992).

Sha, W.. Norris. T. B.. Burm. J. W.. Woodard. D.. Schaff. W. J.. "New coherent detector for terahertz radiation based on excitonic electroabsorption". *Appl. Phys. Lett.* **61**, 1763 (1992).

Xu. L., Zhang. X.-C.. Auston, D. H.. "Terahertz beam generation by femtosecond optical pulses in electro-optic materials". *Appl. Phys. Lett.* **61**, 1784 (1992).

Brown, E. R., Smith, F. W.. McIntosh. K. A.. "Coherent millimeter-wave generation by heterodyne conversion in low-temperature-grown GaAs photo-conductors". *J. Appl. Phys.* **73**, 1480 (1993).

Carin. L.. Agi. K.. Kralj, D., Leung, K. M.. Garetz, B. A.. "Characterization of layered dielectrics with short electromagnetic pulses", *IEEE J. Quantum Electron.* **29**. 2141 (1993a).

Carin. L., Agi, K.. "Ultra-wideband transient microwave scattering measurements using optoelectronically switched antennas". *IEEE Trans. Microwave Theory Tech.* **41**, 250 (1993b).

Dykaar, D. R., Kopf. R. F., Keil, U. D.. Laskowski. E. J.. Zydzik. G. J.. "Electro-optic sampling using an aluminum gallium arsenide probe". *Appl. Phys. Lett.* **62**. 1733 (1993).

Federici. J. F.. Greene. B. I.. Saeta, P. N.. Dykaar, D. R., Sharifi, F., Dynes. R. C., "Cooper pair breaking in lead measured by pulsed terahertz spectroscopy". *IEEE Trans. Appl. Supercond.* **3**, 1461 (1993).

Hamster, H.. Sullivan, A.. Gordon. S.. White, W.. Falcone, R. W., "Subpicosecond electromagnetic pulses from intense laser–plasma interaction". *Phys. Rev. Lett.* **71**. 2725 (1993).

Jones. R. R., You, D., Bucksbaum, P.. "Ionization of Rydberg atoms by subpicosecond half-cycle electromagnetic pulses". *Phys. Rev. Lett.* **70**, 1236 (1993).

Kralj. D.. Carin, L.. "Short-pulse scattering measurements from dielectric spheres using photoconductively switched antennas". *Appl. Phys. Lett.* **62**. 1301 (1993).

Kuznetsov, A. V.. Stanton. C. J., "Ultrafast optical generation of carriers in a dc electric field: Transient localization and photocurrent". *Phys. Rev. B* **48**. 10828 (1993).

Lin. S. Y.. Arjavalingam. G.. "Tunneling of electromagnetic waves in two-dimensional photonic crystals". *Opt. Lett.* **18**. 1666 (1993).

Pedersen, J. E., Lyssenko, V. G.. Hvam. J. M.. Jepsen, P. U.. Keiding, S. R.. Sorensen. C. B.. Lindelof. P. E.. "Ultrafast local field dynamics in photoconductive THz antennas". *Appl. Phys. Lett.* **62**, 1265 (1993).

Planken. P. C. M.. Brener. I.. Nuss, M. C.. Luo, M. S. C.. Chuang. S. L.. "Coherent control of terahertz charge oscillations in a coupled quantum well using phase-locked optical pulses". *Phys. Rev. B* **48**, 4903 (1993).

Robertson. W. M.. Arjavalingam. G.. Meade. R. D., Brommer, K. D., Rappe. A. M., Joannopoulos, J. D., "Observation of surface photons on periodic dielectric arrays", *Opt. Lett.* **18**, 528 (1993a).

Robertson, W. M., Arjavalingam, G., Meade, R. D., Brommer. K. D., Rappe, A. M.. Joannopoulos, J. D., "Measurement of the photon dispersion relation in two-dimensional ordered dielectric arrays", *J. Opt. Soc. Am. B* **10**, 322 (1993b).

Walecki, W. J., Some, D., Kozlov, V. G., Nurmikko, A. V., "Terahertz electromagnetic transients as probes of a two-dimensional electron gas", *Appl. Phys. Lett.* **63**, 1809 (1993).

Xu, L., Hu, B. B., Xin, W., Auston, D. H., Morse, J. D., "Hot electron dynamics study by terahertz radiation from large aperture GaAs p-i-n diodes", *Appl. Phys. Lett.* **62**, 3507 (1993).

You, D., Jones, R. R., Bucksbaum, P. H., Dykaar, D. R., "Generation of high-power sub-single-cycle 500-fs electromagnetic pulses", *Opt. Lett.* **18**, 290 (1993).

Zhang, X.-C., Jin, Y., Hewitt, T. D., Sangsiri, T., Kingsley, L. E., Weiner, M., "Magnetic switching of THz beams", *Appl. Phys. Lett.* **62**, 2003 (1993).

Brener, I., Planken, P. C. M., Nuss, M. C., Luo, M. S. C., Chuang, S. L., Pfeiffer, L., Leaird, D. E., Weiner, A. M., "Coherent control of terahertz emission and carrier populations in semiconductor heterostructures", *J. Opt. Soc. Am. B* **11**, 2457 (1994).

Brorson, S. D., Zhang, J., Keiding, S. R., "Ultrafast carrier trapping and slow recombination in ion-bombarded silicon on sapphire measured via THz spectroscopy", *Appl. Phys. Lett.* **64**, 2385 (1994).

Buhleier, R., Brorson, S. D., Trofimov, I. E., White, J. O., Habermeier, H. U., Kuhl, J., "Anomalous behavior of the complex conductivity of $Y_{1-x}Pr_xBa_2Cu_3O_7$ observed with THz spectroscopy", *Phys. Rev. B* **50**, 9672 (1994).

Frankel, M. Y., Voelker, R. H., Hilfiker. J. N., "Coplanar transmission lines on thin substrates for high-speed low-loss propagation", *IEEE Trans. Microwave Theory Tech.* **42**, 396 (1994).

Groeneveld. R. H. M., Grischkowsky, D., "Picosecond time-resolved far-infrared experiments on carriers and excitons in GaAs–AlGaAs multiple quantum wells", *J. Opt. Soc. Am. B* **11**, 2502 (1994).

Hamster, H., Sullivan, A., Gordon, S., Falcone, R. W., "Short-pulse terahertz radiation from high-intensity laser-produced plasmas", *Phys. Rev. E* **49**, 671 (1994).

Harde, H., Katzenellenbogen, N., Grischkowsky, D., "Terahertz coherent transients from methyl chloride vapor", *J. Opt. Soc. Am. B* **11**, 1018 (1994).

Hu, B. B., Weling, A. S., Auston, D. H.. Kuznetsov, A. V., Stanton, C. J., "dc-electric-field dependence of THz radiation induced by femtosecond optical excitation of bulk GaAs", *Phys. Rev. B* **49**, 2234 (1994).

Jaekel, C., Waschke, C., Roskos. H. G., Kurz, H., Prusseit, W., Kinder, H., "Surface resistance and penetration depth on $YBa_2Cu_3O_{7-\delta}$ thin films on silicon at ultrahigh frequencies", *Appl. Phys. Lett.* **64**, 3326 (1994).

Karadi, C., Jauhar, S., Kouwenhoven, L. P., Wald, K., Orenstein, J., McEuen, P. L., Nagamune, Y., Sakaki, H., "Dynamic response of a quantum point contact", *J. Opt. Soc. Am. B* **11**, 2566 (1994).

Kralj, D., Carin, L., "Wideband dispersion measurements of water in reflection and transmission", *IEEE Trans. Microwave Theory Tech.* **42**, 553 (1994).

Lin, S. Y., Arjavalingam, G., "Photonic bound states in two-dimensional photonic crystals probed by coherent microwave transient spectroscopy", *J. Opt. Soc. Am. B* **11**, 2124 (1994).

Nuss, M. C., Planken, P. C. M., Brener, I., Roskos, H. G., Luo, M. S. C., Chuang, S. L., "Terahertz electromagnetic radiation from quantum wells", *Appl. Phys. B* **58**, 249 (1994).

Ozbay, E., Michel, E., Tuttle, G., Biswas. R., Ho. K. M., Bostak, J., Bloom, D. M., "Terahertz spectroscopy of three-dimensional photonic band-gap crystals". *Opt. Lett.* **19**. 1155 (1994).

Planken, P. C. M., Brener, I., Nuss, M. C., Luo. M. S. C., Chuang, S. L., Pfeiffer, L. N., "THz radiation from coherent population changes in quantum wells". *Phys. Rev. B* **49**. 4668 (1994).

Rahman, A., Kralj, D., Carin. L., Melloch, M. R., Woodall. J. M., "Photoconductively switched antennas for measuring target resonances". *Appl. Phys. Lett.* **64**, 2178 (1994).

Ralph, S. E., Perkowitz, S., Katzenellenbogen. N., Grischkowsky, D., "Terahertz spectroscopy of optically thick multilayered semiconductor structures". *J. Opt. Soc. Am. B* **11**, 2528 (1994).

Some, D., Nurmikko, A. V., "Real-time electron cyclotron oscillations observed by terahertz techniques in semiconductor heterostructures", *Appl. Phys. Lett.* **65**, 3377 (1994).

Son, J.-H., Norris, T. B., Whitaker. J. F., "Terahertz electromagnetic pulses as probes for transient velocity overshoot in GaAs and Si", *J. Opt. Soc. Am. B* **11**, 2519 (1994).

Spielman, S., Parks. B., Orenstein. J., Nemeth. D. T., Ludwig, F., Clarke, J., Merchant, P., Lew, D. J., "Observation of the quasiparticle Hall effect in superconducting $YBa_2Cu_3O_{7-\delta}$", *Phys. Rev. Lett.* **73**, 1537 (1994).

Verghese, S., Karadi, C., Mears, C. A., Orenstein, J. O., Richards, P. L., Barfknecht, A. T., "Picosecond response of the quasiparticle current in superconducting tunnel junctions", *Physica B* **194–196**, 133 (1994).

Weling, A. S., Hu, B. B., Froberg, N. M., Auston. D. H., "Generation of tunable narrow-band THz radiation from large aperture photoconducting antennas". *Appl. Phys. Lett.* **64**, 137 (1994).

White, J. O., Buhleier, R., Brorson. S. D., Trofimov, I. E., Habermeier, H.-U., Kuhl, J., "Complex conductivity of $Y_{1-x}Pr_xBa_2Cu_3O_7$ thin films measured by coherent terahertz spectroscopy". *Physica C* **235–240**, 2025 (1994).

Bonvalet, A., Joffre, M., Martin, J. L., Migus. A., "Generation of ultrabroadband femtosecond pulses in the mid-infrared by optical rectification of 15 fs light pulses at 100 MHz repetition rate", *Appl. Phys. Lett.* **67**, 2907 (1995).

Bulzacchelli, J. F., Lee, H.-S., Stawiasz. K. G., Alexandrou, S., Ketchen. M. B., "Picosecond optoelectronic study of superconducting microstrip transmission lines", *IEEE Trans. Appl. Supercond.* **5**, 2839 (1995).

Cheville, R. A., Grischkowsky, D., "Far-infrared terahertz time-domain spectroscopy of flames", *Opt. Lett.* **20**, 1646 (1995a).

Cheville, R. A., Grischkowsky, D., "Time-domain terahertz impulse ranging studies", *Appl. Phys. Lett.* **67**, 1960 (1995b).

Gao, F., Whitaker, J. F., Liu, Y., Uher, C., Platt, C. E., Klein, M. V., "Terahertz transmission of a $Ba_{1-x}K_xBiO_3$ film probed by coherent time-domain spectroscopy", *Phys. Rev. B* **52**, 3607 (1995).

Harde, H., Katzenellenbogen, N., Grischkowsky, D., "Line-shape transition of collision broadened lines", *Phys. Rev. Lett.* **74**, 1307 (1995).

Howells, S. C., Schlie, L. A., "Temperature dependence of terahertz pulses produced by difference-frequency mixing in InSb". *Appl. Phys. Lett.* **67**, 3688 (1995).

Hu, B. B., de Souza, E. A., Knox, W. H., Cunningham, J. E., Nuss, M. C., Kuznetsov, A. V., Chuang, S. L., "Identifying the distinct phases of carrier transport in semiconductors with 10 fs resolution", *Phys. Rev. Lett.* **74**, 1689 (1995a).

Hu, B. B., Nuss, M. C., "Imaging with terahertz waves", *Opt. Lett.* **20**, 1716 (1995b).

Jepsen, P., Keiding, S. R., "Radiation patterns from lens-coupled terahertz antennas", *Opt. Lett.* **20**, 807 (1995).

Li, M., Sun, F. G., Wagoner, G. A., Alexander, M., Zhang, X.-C., "Measurement and analysis of terahertz radiation from bulk semiconductors", *Appl. Phys. Lett.* **67**, 25 (1995).

Liu, Y., Whitaker, J. F., Uher, C., Hou, S. Y., Phillips, J. M., "Pulsed terahertz-beam spectroscopy as a probe of the thermal and quantum response of $YBa_2Cu_3O_{7-\delta}$ superfluid", *Appl. Phys. Lett.* **67**, 3022 (1995).

Mcintosh, K. A., Brown, E. R., Nichols, K. B., Mcmahon, O. B., Dinatale, W. F., Lyszczarz, T. M., "Terahertz photomixing with diode lasers in low-temperature-grown GaAs", *Appl. Phys. Lett.* **67**, 3844 (1995).

Nahata, A., Auston, D. H., Wu, C., Yardley, J. T., "Generation of terahertz radiation from a poled polymer", *Appl. Phys. Lett.* **67**, 1358 (1995).

Parks, B., Spielman, S., Orenstein, J., Nemeth, D. T., Ludwig, F., Clarke, J., Merchant, P., Lew, D. J., "Phase-sensitive measurements of vortex dynamics in the terahertz domain", *Phys. Rev. Lett.* **74**, 3265 (1995).

Sun, F. G., Wagoner, G. A., Zhang, X.-C., "Measurement of free-space terahertz pulses via long-lifetime photoconductors", *Appl. Phys. Lett.* **67**, 1656 (1995).

Thrane, L., Jacobsen, R. H., Jepsen, P. U., Keiding, S. R., "THz reflection spectroscopy of liquid water", *Chem. Phys. Lett.* **240**, 330 (1995).

White, J. O., Ludwig, C., Kuhl, J., "Response of grating pairs to single-cycle electromagnetic pulses", *J. Opt. Soc. Am. B* **12**, 1687 (1995).

Wu, Q., Zhang, X.-C., "Free-space electro-optic sampling of terahertz beams", *Appl. Phys. Lett.* **67**, 3523 (1995).

Bonvalet, A., Nagle, J., Berger, V., Migus, A., Martin, J.-L., Joffre, M., "Femtosecond infrared emission resulting from coherent charge oscillations in quantum wells", *Phys. Rev. Lett.* **76**, 4392 (1996).

Brener, I., Dykaar, D., Frommer, A., Pfeiffer, L. N., Lopata, J., Wynn, J., West, K., Nuss, M. C., "Terahertz emission from electric field singularities in biased semiconductors", *Opt. Lett.* **21**, 1924 (1996).

Brorson, S. D., Buhleier, R., Trofimov, L. E., White, J. O., Ludwig, C., Balakirev, F. E., Habermeier, H.-U., Kuhl, J., "Electrodynamics of high-temperature superconductors investigated with coherent terahertz pulse spectroscopy", *J. Opt. Soc. Am. B* **13**, 1979 (1996).

Budiarto, E., Margolies, J., Jeong, S., Son, J., Bokor, J., "High-intensity terahertz pulses at 1-kHz repetition rate", *IEEE J. Quantum Electron.* **32**, 1839 (1996).

Chansungsan, C., "Coherent terahertz radiative dynamics of excitons from ultrafast optical excitation of single semiconductor quantum wells", *J. Opt. Soc. Am. B* **13**, 2792 (1996).

Dekorsy, T., Auer, H., Bakker, H. J., Roskos, H. G., Kurz, H., "THz electromagnetic emission by coherent infrared-active phonons", *Phys. Rev. B* **53**, 4005 (1996).

Duvillaret, L., Garet, F., Coutaz, J.-L., "A reliable method for extraction of material parameters in terahertz time-domain spectroscopy", *IEEE J. Sel. Top. Quantum Electron.* **2**, 739 (1996).

Flanders, B. N.. Cheville, R. A.. Grischkowsky. D.. Scherer. N. F.. "Pulsed terahertz transmission spectroscopy of liquid $CHCl_3$. CCl_4, and their mixtures", *J. Phys. Chem.* **100**. 11824 (1996).

Hangyo. M.. Tomozawa. S.. Murakami. Y.. Tonouchi. M.. Tani. M.. Wang. Z.. Sakai. K., Nakashima. S.. "Terahertz radiation from superconducting $YBa_2Cu_3O_{7-\delta}$ thin films excited by femtosecond optical pulses". *Appl. Phys. Lett.* **69**. 2122 (1996).

Heyman. J. N.. Unterrainer. K.. Craig. K., Williams. J.. Sherwin. M. S.. Campman. K.. Hopkins. P. F.. Gossard. A. C.. Murdin, B. N.. Langerak, C. J. G. M.. "Far-infrared pump-probe measurements of the intersubband lifetime in an AlGaAs/GaAs coupled quantum well". *Appl. Phys. Lett.* **68**, 3019 (1996).

Hirakawa. K. Wilke, I.. Yamanaka. K.. Roskos. H. G.. Vosseburger. M., Wolter. F., Waschke. C.. Kurz. H.. Grayson. M.. Tsui. D. C.. "Coherent submillimeter-wave emission from non-equilibrium two-dimensional free carrier plasmas in AlGaAs/GaAs heterojunctions". *Surf. Sci.* **361/362**. 368 (1996).

Howells. S. C.. Schlie. L. A.. "Transient terahertz reflection spectroscopy of undoped InSb from 0.1 to 1.1 THz". *Appl. Phys. Lett.* **69**. 550 (1996).

Jacobsen. R. H.. Birkelund. K.. Holst. T.. Jepsen. P. U.. Keiding. S. R.. "Interpretation of photocurrent correlation measurements used for ultrafast photoconductive switch characterization". *J. Appl. Phys.* **79**. 2649 (1996a).

Jacobsen. R. H.. Mittleman. D. M.. Nuss. M. C.. "Chemical recognition of gases and gas mixtures with terahertz waves". *Opt. Lett.* **21**. 2011 (1996b).

Jaekel. C.. Roskos. H. G.. Kurz. H.. "Emission of picosecond electromagnetic pulses from optically excited superconducting bridges". *Phys. Rev. B* **54**. 6889 (1996).

Jepsen. P. U.. Jacobsen. R. H.. Keiding. S. R.. "Generation and detection of terahertz pulses from biased semiconductor antennas". *J. Opt. Soc. Am. B* **13**. 2424 (1996).

Joffre. M.. Bonvalet, A.. Migus. A.. Martin. J.-L.. "Femtosecond diffracting Fourier-transform infrared interferometer". *Opt. Lett.* **21**. 964 (1996).

Kindt. J. T., Schmuttenmaer. C. A.. "Far infrared dielectric properties of polar liquids probed by femtosecond THz pulse spectroscopy". *J. Phys. Chem.* **100**. 10373 (1996).

Liu. Y., Park, S.-G.. Weiner. A. M.. "Terahertz waveform synthesis via optical pulse shaping". *IEEE J. Sel. Top. Quantum Electron.* **2**. 709 (1996).

Ludwig. C.. Kuhl. J.. "Studies of the temporal and spectral shape of terahertz pulses generated from photoconducting switches". *Appl. Phys. Lett.* **69**, 1194 (1996a).

Ludwig. C.. Jiang, Q.. Kuhl. J.. Zegenhagen. J.. "Electrodynamic properties of oxygen reduced $YBa_2Cu_3O_{7-x}$ thin films in the THz frequency range". *Physica C* **269**. 249 (1996b).

Martini. R.. Klose, G.. Roskos. H. G.. Kurz. H.. Grahn. H. T.. Hey. R.. "Superradiant emission from Bloch oscillations in semiconductor superlattices", *Phys. Rev. B* **54**. 14325 (1996).

Mittleman, D. M.. Jacobsen. R. H.. Nuss. M. C.. "T-ray imaging". *IEEE J. Sel. Top. Quantum Electron.* **2**, 679 (1996).

Nahata. A.. Weling. A. S.. Heinz. T. F.. "A wideband coherent terahertz spectroscopy system using optical rectification and electro-optic sampling". *Appl. Phys. Lett.* **69**, 2321 (1996a).

Nahata. A.. Heinz, T. F.. "Reshaping of freely propagating terahertz pulses by diffraction". *IEEE J. Sel. Top. Quantum Electron.* **2**. 701 (1996b).

Nahata, A., Auston, D. H., Heinz, T. F., Wu, C., "Coherent detection of freely propagating terahertz radiation by electro-optic sampling", *Appl. Phys. Lett.* **68**, 150 (1996c).

Nuss, M. C., "Chemistry is right for T-ray imaging", *IEEE Circuits Devices*, March, p. 25 (1996).

Pine, A. S., Suenram, R. D., Brown, E. R., McIntosh, K. A., "A terahertz photomixing spectrometer: Application to SiO_2 self broadening", *J. Mol. Spectrosc.* **175**, 37 (1996).

Ralph, S. E., Chen, Y., Woodall, J., McInturff, D., "Subpicosecond photoconductivity of $In_{0.53}Ga_{0.47}As$: Intervalley scattering rates observed via THz spectroscopy", *Phys. Rev. B* **54**, 5568 (1996).

Raman, C., Conover, C. W. S., Sukenik, C. I., Bucksbaum, P. H., "Ionization of Rydberg wave packets by subpicosecond half-cycle electromagnetic pulses", *Phys. Rev. Lett.* **76**, 2436 (1996).

Rodriguez, G., Taylor, A. J., "Screening of the bias field in terahertz generation from photoconductors", *Opt. Lett.* **21**, 1046 (1996).

Son, J.-H., Jeong, S., Bokor, J., "Noncontact probing of metal–oxide–semiconductor inversion layer mobility", *Appl. Phys. Lett.* **69**, 1779 (1996).

Tonouchi, M., Tani, M., Wang, Z., Sakai, K., Tomozawa, S., Hangyo, M., Murakami, Y., Nakashima, S., "Ultrashort electromagnetic pulse radiation from YBCO thin films excited by femtosecond optical pulse", *Jp. J. Appl. Phys. Pt. 1* bf 35, 2624 (1996b).

Tonouchi, M., Tani, M., Wang, Z., Sakai, K., Wada, N., Hangyo, M., "Terahertz emission study of femtosecond time-transient nonequilibrium state in optically excited $YBa_2Cu_3O_{7-\delta}$ thin films", *Japan J. Appl. Phys.* **35**, 1578 (1996a).

Weling, A. S., Auston, D. H., "Novel sources and detectors for coherent tunable narrow-band terahertz radiation in free space", *J. Opt. Soc. Am. B* **13**, 2783 (1996).

Wu, Q., Zhang, X.-C., "Ultrafast electro-optic field sensors", *Appl. Phys. Lett.* **68**, 1604 (1996a).

Zielbauer, J., Wegener, M., "Ultrafast optical pump THz-probe spectroscopy on silicon", *Appl. Phys. Lett.* **68**, 1223 (1996).

Wu, Q., Sun, F. G., Campbell, P., Zhang, X.-C., "Dynamic range of an electro-optic field sensor and its imaging applications", *Appl. Phys. Lett.* **68**, 3224 (1996b).

Bensky, T. J., Haeffler, G., Jones, R. R., "Ionization of Na Rydberg atoms by subpicosecond quarter-cycle circularly polarized pulses", *Phys. Rev. Lett.* **79**, 2018 (1997).

Bromage, J., Radic, S., Agrawal, G. P., Stroud, C. R., Fauchet, P. M., Sobolewski, R., "Spatiotemporal shaping of terahertz pulses", *Opt. Lett.* **22**, 627 (1997).

Cai, Y., Brener, I., Lopata, J., Wynn, J., Pfeiffer, L., Federici, J., "Design and performance of singular electric field terahertz photoconducting antennas", *Appl. Phys. Lett.* **71**, 2076 (1997).

Cheville, R. A., McGowan, R. W., Grischkowsky, D., "Late-time target response measured with THz impulse ranging", *IEEE Trans. Antennas Propag.* **45**, 1518 (1997).

Haran, G., Wynne, K., Hochstrasser, R. M., "Femtosecond far-infrared pump–probe spectroscopy: a new tool for studying low-frequency vibrational dynamics in molecular condensed phases", *Chem. Phys. Lett.* **274**, 365 (1997).

Harde, H., Cheville, R. A., Grischkowsky, D., "Collision-induced tunneling in methyl halides", *J. Opt. Soc. Am. B* **14**, 3282 (1997).

Haring Bolivar, P., Wolter, F., Muller, A., Roskos, H. G., Kurz, H., "Excitonic emission of THz radiation: Experimental evidence of the shortcomings of the Bloch equation method", *Phys. Rev. Lett.* **78**, 2232 (1997).

Huggard, P. G., Cluff, J. A., Shaw, C. J., Andrews, S. R., Linfield, E. H., Ritchie, D. A., "Coherent control of cyclotron emission from a semiconductor using sub-picosecond electric field transients", *Appl. Phys. Lett.* **71**, 2647 (1997).

Jeon, T.-I., Grischkowsky, D., "Nature of conduction in doped silicon", *Phys. Rev. Lett.* **78**, 1106 (1997).

Keiding, S. R., "Dipole correlation functions in liquid benzenes measured with THz time domain spectroscopy", *J. Phys. Chem. A* **101**, 3646 (1997).

Kersting, R., Unterrainer, K., Strasser, G., Kauffmann, H. F., Gornik, E., "Few-cycle THz emission from cold plasma oscillations", *Phys. Rev. Lett.* **79**, 3038 (1997).

Lu, Z. G., Campbell, P., Zhang, X.-C., "Free-space electro-optic sampling with a high-repetition rate regenerative amplified laser", *Appl. Phys. Lett.* **71**, 593 (1997).

McElroy, R., Wynne, K., "Ultrafast dipole solvation measured in the far infrared", *Phys. Rev. Lett.* **79**, 3078 (1997).

McGowan, R. W., Grischkowsky, D., Misewich, J. A., "Demonstrated low radiative loss of a quadrupole ultrashort electrical pulse propagated on a three strip coplanar transmission line", *Appl. Phys. Lett.* **71**, 2842 (1997).

Messner, C., Sailer, M., Kostner, H., Hopfel, R. A., "Coherent generation of tunable narrow-band THz radiation by optical rectification of femtosecond pulse trains", *Appl. Phys. B* **64**, 619 (1997).

Mittleman, D., Cunningham, J., Nuss, M. C., Geva, M., "Noncontact semiconductor wafer characterization with the terahertz Hall effect", *Appl. Phys. Lett.* **71**, 16 (1997a).

Mittleman, D. M., Nuss, M. C., Colvin, V. L., "Terahertz spectroscopy of water in inverse micelles", *Chem. Phys. Lett.* **275**, 332 (1997b).

Mittleman, D. M., Hunsche, S., Boivin, L., Nuss, M. C., "T-ray tomography", *Opt. Lett.* **22**, 904 (1997c).

Prabhu, S. S., Ralph, S. E., Melloch, M. R., Harmon, E. S., "Carrier dynamics of low-temperature-grown GaAs observed via THz spectroscopy", *Appl. Phys. Lett.* **70**, 2419 (1997).

Riordan, J., Sun, F. G., Lu, Z. G., Zhang, X.-C., "Free-space transient magneto-optic sampling", *Appl. Phys. Lett.* **71**, 1452 (1997).

Ronne, C., Thrane, L., Astrand, P.-O., Wallqvist, A., Mikkelsen, K. V., Keiding, S. R., "Investigation of the temperature dependence of dielectric relaxation in liquid water by THz reflection spectroscopy and molecular dynamics simulation", *J. Chem. Phys.* **107**, 5319 (1997).

Tani, M., Matsuura, S., Sakai, K., Nakashima, S., "Emission characteristics of photoconductive antennas based on low-temperature-grown GaAs and semi-insulating GaAs", *Appl. Opt.* **36**, 7853 (1997a).

Tani, M., Sakai, K., Mimura, H., "Ultrafast photoconductive detectors based on semi-insulating GaAs and InP", *Jp. J. Appl. Phys.* **36**, 1175 (1997b).

Taylor, A. J., Rodriguez, G., Some, D., "Ultrafast field dynamics in large-aperture photoconductors", *Opt. Lett.* **22**, 715 (1997).

Tonouchi, M., Tani, M., Wang, Z., Sakai, K., Hangyo, M., Wada, N., Murakami, Y., "Enhanced THz radiation from YBCO thin film bow-tie antennas with hyper-hemispherical MgO lens", *IEEE Trans. Appl. Supercond.* **7**, 2913 (1997a).

Tonouchi, M., Tani, M., Wang, Z., Sakai, K., Wada, N., Hangyo, M., "Novel tera-hertz radiation from flux-trapped $YBa_2Cu_3O_{7-\delta}$ thin films excited by femtosec-ond laser pulses", *Jp. J. Appl. Phys. Pt. 2* **36**, 93 (1997b).

Vrijen, R. B., Lankhuijzen, G. M., Noordam, L. D., "Delayed electron emission in the ionization of Rydberg atoms with half-cycle THz pulses", *Phys. Rev. Lett.* **79**, 617 (1997).

Winnewisser, C., Jepsen, P. U., Schall, M., Schyja, V., Helm, H., "Electro-optic detection of THz radiation in $LiTaO_3$, $LiNbO_3$, and ZnTe", *Appl. Phys. Lett.* **70**, 3069 (1997).

Wu, Q., Zhang, X.-C., "7 terahertz broadband GaP electro-optic sensor", *Appl. Phys. Lett.* **70**, 1784 (1997a).

Wu, Q., Zhang, X.-C., "Free-space electro-optic sampling of mid-infrared pulses", *Appl. Phys. Lett.* **71**, 1285 (1997b).

You, D., Bucksbaum, P., "Propagation of half-cycle far-infrared pulses", *J. Opt. Soc. Am. B* **14**, 1651 (1997).

Bromage, J., Radic, S., Agrawal, G. P., Stroud, C. R., Fauchet, P. M., Sobolewsky, R., "Spatiotemporal shaping of half-cycle terahertz pulses by diffraction through conductive apertures of finite thickness", *J. Opt. Soc. Am. B* **15**, 1399 (1998).

Cai, Y., Brener, I., Lopata, J., Wynn, J., Pfeiffer, L., Stark, J. B., Wu, Q., Zhang, X.-C., Federici, J. F., "Coherent terahertz radiation detection: direct comparison between free-space electro-optic sampling and antenna detection", *Appl. Phys. Lett.* **73**, 444 (1998).

Cheville, R. A., Grischkowsky, D., "Observation of pure rotational absorption spec-tra in the ν_2 band of hot H_2O in flames", *Opt. Lett.* **23**, 531 (1998a).

Cheville, R. A., McGowan, R. W., Grischkowsky, D., "Time resolved measurements which isolate the mechanisms responsible for terahertz glory scattering from dielectric spheres", *Phys. Rev. Lett.* **80**, 269 (1998b).

Feng, S., Winful, H. G., Hellwarth, R. W., "Gouy shift and temporal reshaping of focused single-cycle electromagnetic pulses", *Opt. Lett.* **23**, 385 (1998).

Flanders, B. N., Arnett, D. C., Scherer, N. F., "Optical pump–terahertz probe spectroscopy utilizing a cavity-dumped oscillator-driven terahertz spectrome-ter", *IEEE J. Sel. Top. Quantum Electron.* **4**, 353 (1998).

Hunsche, S., Mittleman, D. M., Koch, M., Nuss, M. C., "New dimensions in T-ray imaging", *IEEE Trans. Elec.* **E81C**, 269 (1998a).

Hunsche, S., Koch, M., Brener, I., Nuss, M. C., "THz near-field imaging", *Opt. Commun.* **150**, 22 (1998b).

Jeon, T.-I., Grischkowsky, D., "Observation of a Cole–Davidson type complex con-ductivity in the limit of very low carrier densities in doped silicon", *Appl. Phys. Lett.* **72**, 2259 (1998a).

Jeon, T.-I., Grischkowsky, D., "Characterization of optically dense, doped semi-conductors by reflection THz time domain spectroscopy", *Appl. Phys. Lett.* **72**, 3032 (1998b).

Jiang, Z., Zhang, X.-C., "Single-shot spatiotemporal terahertz field imaging", *Opt. Lett.* **23**, 1114 (1998).

Kaplan, A., "Diffraction-induced transformation of near-cycle and subcycle pulses". *J. Opt. Soc. Am. B* **15**. 951 (1998).

Labbe-Lavigne. S., Barret. S., Garet. F.. Duvillaret. L., Coutaz. J.-L.. "Far-infrared dielectric constant of porous silicon layers measured by terahertz time-domain spectroscopy". *J. Appl. Phys.* **83**. 6007 (1998).

Lai, R. K., Hwang. J.-R., Norris, T. B.. Whitaker. J. F.. "A photoconductive miniature terahertz source". *Appl. Phys. Lett.* **72**. 3100 (1998).

Matsuura, S.. Tani. M.. Abe. H.. Sakai. K.. Ozeki. H.. Saito. S.. "High resolution THz spectroscopy by a compact radiation source based on photomixing with diode lasers in a photoconductive antenna". *J. Mol. Spectrosc.* **187**. 97 (1998).

Mittleman, D. M.. Jacobsen, R. H.. Neelamani, R.. Baraniuk. R. G.. Nuss. M. C.. "Gas sensing using terahertz time-domain spectroscopy". *Appl. Phys. B* **67**. 379 (1998).

Nuss. M. C.. Orenstein. J.. "Terahertz time-domain spectroscopy (THz-TDS)" in *Millimeter and Sub-Millimeter-Wave Spectroscopy of Solids*, ed. by G. Grüner (Springer, Berlin. Heidelberg, 1998). p. 7.

Park, S.-G.. Melloch, M. R.. Weiner. A. M., "Comparison of terahertz waveforms measured by electro-optic and photoconductive sampling". *Appl. Phys. Lett.* **73**, 3184 (1998).

Venables, D. S.. Schmuttenmaer, C. A.. "Far-infrared spectra and associated dynamics in acetonitrile–water mixtures measured with femtosecond THz pulse spectroscopy". *J. Chem. Phys.* **108**. 4935 (1998).

Verghese, S.. McIntosh, K. A.. Calawa. S., Dinatale. W. F.. Duerr. E. K.. Molvar. K. A.. "Generation and detection of coherent terahertz waves using two photomixers". *Appl. Phys. Lett.* **73**. 3824 (1998).

Winnewisser. C.. Lewen. F.. Helm. H.. "Transmission characteristics of dichroic filters measured by THz time-domain spectroscopy". *Appl. Phys. A* **66**. 593 (1998).

Bromage. J., Walmsley. I. A.. Stroud. C. R., "Dithered-edge sampling of terahertz pulses". *Appl. Phys. Lett.* **75**, 2181 (1999a).

Bromage. J., Walmsley. I. A.. Stroud. C. R.. "Direct measurement of a photoconductive receiver's temporal response by dithered edge sampling". *Opt. Lett.* **24**, 1771 (1999b).

Buot, F. A.. Krowne, C. M.. "Double-barrier THz source based on electrical excitation of electrons and holes". *J. Appl. Phys.* **86**, 5215 (1999).

Carey. J. J., Zawadzka. J.. Jaroszynski, D. A.. Wynne. K.. "Noncausal time response in frustrated total internal reflection?". *Phys. Rev. Lett.* **84**, 1431 (1999).

Chen, Q., Zhang. X.-C.. "Polarization modulation in optoelectronic generation and detection of terahertz beams", *Appl. Phys. Lett.* **74**, 3435 (1999).

Cheville, R. A., Grischkowsky, D., "Foreign and self-broadened rotational linewidths of high temperature water vapor". *J. Opt. Soc. Am. B* **16**. 317 (1999).

Cook, D. J., Chen, J. X.. Morlino, E. A.. Hochstrasser, R. M.. "Terahertz-field-induced second-harmonic generation measurements of liquid dynamics". *Chem. Phys. Lett.* **309**, 221 (1999).

Cote. D., Fraser. J. M., De Camp. M., Bucksbaum. P. H.. van Driel. H. M.. "THz emission from coherently controlled photocurrents in GaAs", *Appl. Phys. Lett.* **75**, 3959 (1999).

Duvillaret, L., Garet, F., Coutaz, J.-L., "Highly precise determination of optical constants and sample thickness in terahertz time-domain spectroscopy", *Appl. Opt.* **38**, 409 (1999).

Englert, C. R., Maurer, H., Birk, M., "Photon induced far infrared absorption in pure single crystal silicon", *Infrared Phys. Tech.* **40**, 447 (1999).

Feise, M. W., Citrin, D. S., "Semiclassical theory of terahertz multiple-harmonic generation in semiconductor superlattices", *Appl. Phys. Lett.* **75**, 3536 (1999).

Flanders, B. N., Shang, X., Scherer, N. F., Grischkowsky, D., "The pure rotational spectrum of solvated HCl: solute–bath interaction strength and dynamics", *J. Phys. Chem. A* **103**, 10054 (1999).

Gallot, G., Zhang, J., McGowan, R. W., Jeon, T.-I., Grischkowsky, D., "Measurements of the THz absorption and dispersion of ZnTe and their relevance to the electro-optic detection of THz radiation", *Appl. Phys. Lett.* **74**, 3450 (1999a).

Gallot, G., Grischkowsky, D., "Electro-optic detection of terahertz radiation", *J. Opt. Soc. Am. B* **16**, 1204 (1999b).

Garet, F., Duvillaret, L., Coutaz, J.-L., "Evidence of frequency dependent THz beam polarization in time-domain spectroscopy", *Proc. SPIE* **3617**, 30 (1999).

Gerecht, E., Musante, C. F., Zhuang, Y., Yngvesson, K. S., Gol'tsman, G. N., Voronov, B. M., Gershenzon, E. M., "NbN hot electron bolometric mixers – a new technology for low-noise THz receivers", *IEEE Trans. Microwave Theory Tech.* **47**, 2519 (1999).

Gu, P., Chang, F., Tani, M., Sakai, K., Pan, C. L., "Generation of coherent cw-terahertz radiation using a tunable dual-wavelength external cavity laser diode", *Jp. J. Appl. Phys. Pt. 2* **38**, 1246 (1999).

Hughes, S., Citrin, D., "Tunability in the terahertz regime: charge-carrier wavepacket manipulation in quantum wells", *Appl. Opt.* **38**, 7153 (1999).

Hunsche, S., Feng, S., Winful, H. G., Leitenstorfer, A., Nuss, M. C., Ippen, E. P., "Spatiotemporal focusing of single-cycle light pulses", *J. Opt. Soc. Am. A* **16**, 2025 (1999).

Jiang, Z., Sun, F. G., Zhang, X.-C., "Terahertz pulse measurement with an optical streak camera", *Opt. Lett.* **24**, 1245 (1999a).

Jiang, Z., Zhang, X.-C., "Terahertz imaging via electrooptic effect", *IEEE Trans. Microwave Theory Tech.* **47**, 2644 (1999b).

Jiang, Z., Zhang, X.-C., "2D measurement and spatio-temporal coupling of few-cycle THz pulses", *Opt. Express* **5**, 243 (1999c).

Kawamura, J., Jian, C., Miller, D., Kooi, J., Zmuidzinas, J., Bumble, B., LeDuc, H. G., Stern, J. A., "Low-noise submillimeter-wave NbTiN superconducting tunnel junction mixers", *Appl. Phys. Lett.* **75**, 4013 (1999).

Kindt, J. T., Schmuttenmaer, C. A., "Theory for determination of the low-frequency time-dependent response function in liquids using time-resolved terahertz pulse spectroscopy", *J. Chem. Phys.* **110**, 8589 (1999).

Kondo, T., Sakamoto, M., Tonouchi, M., Hangyo, M., "Terahertz radiation from (111) InAs surface using $1.55\,\mu m$ femtosecond laser pulses", *Japan J. Appl. Phys. Pt. 2* **38**, 1035 (1999).

Kume, E., Iguchi, I., Takahashi, H., "On-chip spectroscopic detection of terahertz radiation emitted from a quasiparticle-injected nonequilibrium superconductor using a high-T_c Josephson junction", *Appl. Phys. Lett.* **75**, 2809 (1999).

Kuzel, P., Khazan, M. A., Kroupa, J., "Spatiotemporal transformations of ultra-short terahertz pulses", *J. Opt. Soc. Am. B* **16**, 1795 (1999).

Leitenstorfer, A., Hunsche, S., Shah, J., Nuss, M. C., Knox, W. H., "Detectors and sources for ultrabroadband electro-optic sampling: experiment and theory". *Appl. Phys. Lett.* **74**, 1516 (1999).

Li, M., Cho, G. C., Lu, T.-M., Zhang, X.-C., Wang, S.-Q., Kennedy, J. T., "Time-domain dielectric constant measurement of thin film in GHz–THz frequency range near the Brewster angle". *Appl. Phys. Lett.* **74**, 2113 (1999).

Liu, T. A., Huang, K. F., Pan, C. L., Liu, Z. L., Ono, S., Ohtake, H., Sarukura, N., "THz radiation from intracavity saturable Bragg reflector in magnetic field with self-started mode-locking by strained saturable Bragg reflector". *Jp. J. Appl. Phys. Pt. 2* **38**, 1333 (1999).

McGowan, R. W., Gallot, G., Grischkowsky, D., "Propagation of ultrawideband short pulses of THz radiation through submillimeter-diameter circular waveguides". *Opt. Lett.* **24**, 1431 (1999).

Mittleman, D. M., Gupta, M., Neelamani, R., Baraniuk, R. G., Rudd, J. V., Koch, M., "Recent advances in terahertz imaging", *Appl. Phys. B* **68**, 1085 (1999).

Morikawa, O., Yamashita, M., Saijo, H., Morimoto, M., Tonouchi, M., Hangyo, M., "Vector imaging of supercurrent flow in $YBa_2Cu_3O_{7-\delta}$ thin films using terahertz radiation", *Appl. Phys. Lett.* **75**, 3387 (1999a).

Morikawa, O., Tonouchi, M., Hangyo, M., "Sub-THz spectroscopic system using a multimode laser diode and photoconductive antenna". *Appl. Phys. Lett.* **75**, 3772 (1999b).

Mouret, G., Chen, W., Boucher, D., Bocquet, R., Mounaix, P., Lippens, D., "Gas filter correlation instrument for air monitoring at submillimeter wavelengths". *Opt. Lett.* **24**, 351 (1999).

Nahata, A., Yardley, J. T., Heinz, T. F., "Free-space electro-optic detection of continuous-wave terahertz radiation", *Appl. Phys. Lett.* **75**, 2524 (1999).

Ohtake, H., Ono, S., Liu, Z., Sarukura, N., Ohta, M., Watanabe, K., Matsumoto, Y., "Enhanced THz radiation from femtosecond laser pulse irradiated InAs clean surface", *Jp. J. Appl. Phys. Pt. 2* **38**, 1186 (1999).

Park, S.-G., Weiner, A. M., Melloch, M. R., Siders, C. W., Siders, J. L. W., Taylor, A. J., "High-power narrow-band terahertz generation using large-aperture photoconductors", *IEEE J. Quantum Electron.* **35**, 1257 (1999).

Ronne, C., Astraand, P.-O., Keiding, S. R., "THz spectroscopy of liquid H_2O and D_2O", *Phys. Rev. Lett.* **82**, 2888 (1999).

Ruffin, A. B., Rudd, J. V., Whitaker, J. F., Feng, S., Winful, H. G., "Direct observation of the Gouy phase shift with single-cycle terahertz pulses", *Phys. Rev. Lett.* **83**, 3410 (1999).

Schall, M., Helm, H., Keiding, S. R., "Far infrared properties of electro-optic crystals measured by THz time-domain spectroscopy", *Int. J. Infrared Millim. Waves* **20**, 595 (1999).

Siders, C. W., Siders, J. L. W., Taylor, A. J., Park, S.-G., Melloch, M. R., Weiner, A. M., "Generation and characterization of terahertz pulse trains from biased large-aperture photoconductors", *Opt. Lett.* **24**, 241 (1999).

Telles, E. M., Zink, L. R., Evenson, K. M., "New FIR laser lines and frequency measurements of CD3OD", *Int. J. Infrared Millime. Waves* **20**, 1631 (1999).

Weling, A. S., Heinz, T. F., "Enhancement in the spectral irradiance of photoconducting terahertz emitters by chirped-pulse mixing", *J. Opt. Soc. Am. B* **16**, 1455 (1999).

Williams, B. S., Xu, B., Hu, Q., Melloch, M. R., "Narrow-linewidth terahertz intersubband emission from three-level systems", *Appl. Phys. Lett.* **75**, 2927 (1999).

Winnewisser, C., Lewen, F., Weinzierl, J., Helm, H., "Transmission features of frequency-selective components in the far infrared determined by terahertz time-domain spectroscopy", *Appl. Opt.* **38**, 3961 (1999).

Wynne, K., Jaroszynski, D. A., "Superluminal terahertz pulses", *Opt. Lett.* **24**, 25 (1999).

Averitt, R., Rodriguez, G., Siders, J. L. W., Trugman, S. A., Taylor, A. J., "Conductivity artifacts in optical-pump THz-probe measurements of $YBa_2Cu_3O_7$", *J. Opt. Soc. Am. B* **17**, 327 (2000).

Beard, M. C., Turner, G. M., Schmuttenmaer, C. A., "Transient photoconductivity in GaAs as measured by time-resolved terahertz spectroscopy", *Phys. Rev. B* **62**, 15764 (2000).

Bieler, M., Hein, G., K.Pierz, Siegner, U., Koch, M., "Spatial pattern formation of optically excited carriers in photoconductive switches", *Appl. Phys. Lett.* **77**, 1002 (2000).

Boucaud, P., Gill, K. S., Williams, J. B., Sherwin, M. S., Schoenfeld, W. V., Petroff, P. M., "Saturation of THz-frequency intraband absorption in InAs/GaAs quantum dot molecules", *Appl. Phys. Lett.* **77**, 510 (2000).

Bozzi, M., Perregrini, L., Weinzierl, J., Winnewisser, C., "Design, fabrication, and measurement of frequency-selective surfaces", *Opt. Eng.* **39**, 2263 (2000).

Bratschitsch, R., Muller, T., Kersting, R., Strasser, G., Unterrainer, K., "Coherent terahertz emission from optically pumped intersubband plasmons in parabolic quantum wells", *Appl. Phys. Lett.* **76**, 3501 (2000).

Brucherseifer, M., Nagel, M., Bolivar, P. H., Kurz, H., Bosserhoff, A., Buttner, R., "Label-free probing of the binding state of DNA by time-domain terahertz sensing", *Appl. Phys. Lett.* **77**, 4049 (2000).

Brundermann, E., Chamberlin, D. R., Haller, E. E., "High duty cycle and continuous terahertz emission from germanium", *Appl. Phys. Lett.* **76**, 2991 (2000).

Chen, Q., Jiang, Z., Xu, G., Zhang, X.-C., "Near-field terahertz imaging with a dynamic aperture", *Opt. Lett.* **25**, 1122 (2000a).

Chen, Q., Jiang, Z., Tani, M., Zhang, X.-C., "Electro-optic terahertz transceiver", *Electron. Lett.* **36**, 1298 (2000b).

Cho, G. C., Han, P. Y., Zhang, X.-C., Bakker, H. J., "Optical phonon dynamics of GaAs studied with time-resolved terahertz spectroscopy", *Opt. Lett.* **25**, 1609 (2000).

Ciesla, C. M., Arnone, D. D., Corchia, A., Crawley, D., Longbottom, C., Linfield, E. H., Pepper, M., "Biomedical applications of terahertz pulse imaging", *Proc. SPIE* **3934**, 73 (2000).

Cook, D. J., Hochstrasser, R. M., "Intense terahertz pulses by four-wave rectification in air", *Opt. Lett.* **25**, 1210 (2000).

Corchia, A., Ciesla, C. M., Arnone, D. D., Linfield, E. H., Simmons, M. Y., Pepper, M., "Crystallographic orientation dependence of bulk optical rectification", *J. Mod. Opt.* **47**, 1837 (2000).

Corson, J., Orenstein, J., Seongshik, O., O'Donnell, J., Eckstein, J. N., "Nodal quasiparticle lifetime in the superconducting state of $Bi_2Sr_2CaCu_2O_{8+\delta}$", *Phys. Rev. Lett.* **85**, 2569 (2000).

Dean, R. N., Nordine, P. C., Christodoulou, C. G., "3-D helical THz antennas", *Microwave Opt. Tech. Lett.* **24**, 106 (2000).

Digby, J. W., McIntosh, C. E., Parkhurst, G. M., Towlson, B. M., Hadjiloucas, S., Bouwen, J. W., Chamberlain, J. M., Pollard, R. D., Miles, R. E., Steenson, D. P., Karatzas, L. S., Cronin, N. J., Davies, S. R., "Fabrication and characterization of micromachined rectangular waveguide components for use at millimeter-wave and terahertz frequencies", *IEEE Trans. Microwave Theory Tech.* **48**, 1293 (2000).

Divin, Y. Y., Poppe, U., Volkov, O. Y., Pavlovskii, V. V., "Frequency-selective incoherent detection of terahertz radiation by high-T_c Josephson junctions", *Appl. Phys. Lett.* **76**, 2826 (2000).

Dodge, J. S., Weber, C. P., Corson, J., Orenstein, J., Schlesinger, Z., Reiner, J. W., Beasley, M. R., "Low frequency crossover of the fractional power-law conductivity in $SrRuO_3$", *Phys. Rev. Lett.* **85**, 4932 (2000).

Duvillaret, L., Garet, F., Coutaz, J.-L., "Influence of noise on the characterization of materials by terahertz time-domain spectroscopy", *J. Opt. Soc. Am. B* **17**, 452 (2000).

Feng, S., Winful, H. G., "Cavity phase engineering for stable enhanced terahertz pulse trains", *J. Opt. Soc. Am. A* **17**, 2096 (2000).

Gaidis, M. C., Pickett, H. M., Smith, C. D., Martin, S. C., Smith, R. P., Siegel, P. H., "A 2.5 THz receiver front end for spaceborne applications", *IEEE Trans. Microwave Theory Tech.* **48**, 733 (2000).

Gallot, G., Grischkowsky, D., "Terahertz waveguides", *J. Opt. Soc. Am. B* **17**, 851 (2000).

Ganzevles, W. F. M., Swart, L. R., Gao, J. R., de Korte, P. A. J., Klapwijk, T. M., "Direct response of twin-slot antenna-coupled hot-electron bolometer mixers designed for 2.5 THz radiation detection", *Appl. Phys. Lett.* **76**, 3304 (2000).

Gatesman, A. J., Waldman, J., Ji, M., Musante, C., Yagvesson, S., "An anti-reflection coating for silicon optics at terahertz frequencies", *IEEE Microwave Guided Wave Lett.* **10**, 264 (2000).

Gendriesch, R., Lewen, F., Winnewisser, G., Hahn, J., "Precision broadband spectroscopy near 2 THz: frequency-stabilized laser sideband spectrometer with backward-wave oscillators", *J. Molec. Spec.* **203**, 205 (2000).

Grischkowsky, D. R., "Optoelectronic characterization of transmission lines and waveguides by terahertz time-domain spectroscopy", *IEEE J. Sel. Top. Quantum Electron.* **6**, 1122 (2000).

Gu, P., Tani, M., Sakai, K., Yang, T. R., "Detection of terahertz radiation from longitudinal optical phonon–plasmon coupling modes in InSb film using an ultrabroadband photoconductive antenna", *Appl. Phys. Lett.* **77**, 1798 (2000).

Gundlach, K. H., Schicke, M., "SIS and bolometer mixers for terahertz frequencies", *Supercond. Sci. Tech.* **13**, 171 (2000).

Guo, B., Wen, J. H., Zhang, H. C., Zhong, W. B., Lin, W. Z., "Terahertz dispersion and attenuation characteristics of optically excited coplanar strip lines on LT-GaAs", *J. Infrared Millim. Waves* **19**, 98 (2000).

Gurtler, A., Wirmewisser, C., Helm, H., Jepsen, P. U., "Terahertz pulse propagation in the near field and the far field", *J. Opt. Soc. Am. A* **17**, 74 (2000).

Han, P. Y., Cho, G. C., Zhang, X.-C., "Time-domain transillumination of biological tissues with terahertz pulses", *Opt. Lett.* **25**, 242 (2000a).

Han, P. Y., Tani, M., Pan, F., Zhang, X.-C., "Use of the organic crystal DAST for terahertz beam applications", *Opt. Lett.* **25**, 675 (2000b).

Han, P. Y., Huang, X. G., Zhang, X.-C., "Direct characterization of terahertz radiation from the dynamics of the semiconductor surface field", *Appl. Phys. Lett.* **77**, 2864 (2000c).

Harel, R., Brener, I., Pfeiffer, L. N., West, K. W., Vandenberg, J. M., Chu, S. G., Wynn, J. D., "Coherent terahertz radiation from cavity polaritons in GaAs/AlGaAs microcavities", *Phys. Stat. Sol. A* **178**, 365 (2000).

Hashimshony, D., Zigler, A., Papadopoulos, K., "Miniature photoconducting capacitor array as a source for tunable THz radiation", *Rev. Sci. Instrum.* **71** (2000).

Hegmann, F. A., Williams, J. B., Cole, B., Sherwin, M. S., Beeman, J. W., Haller, E. E., "Time-resolved photoresponse of a gallium-doped germanium photoconductor using a variable pulse-width terahertz source", *Appl. Phys. Lett.* **76**, 262 (2000).

Hermann, M., Tani, M., Sakai, K., "Display modes in time-resolved terahertz imaging", *Jp. J. Appl. Phys. Pt. 1* **39**, 6254 (2000).

Holzman, J. F., Vermeulen, F. E., Elezzabi, A. Y., "Ultrafast photoconductive self-switching of subpicosecond electrical pulses", *IEEE J. Quantum Electron.* **36**, 130 (2000a).

Holzman, T. E., Vermeulen, F. E., Elezzabi, A. Y., "Frozen wave generation of bandwidth tunable two-cycle THz radiation", *J. Opt. Soc. Am. B* **17**, 1457 (2000b).

Hovenier, J. N., Van Es, R. W., Klaassen, T. O., Wenckebach, W. T., Kratschmer, M., Klappenberger, F., Schomburg, E., Winnerl, S., Knippels, G. M. H., van der Meer, A. F. G., "Differential electronic gating: A method to measure the shape of short THz pulses with a poorly defined trigger signal", *Appl. Phys. Lett.* **77**, 1762 (2000a).

Hovenier, J., Carmen-Diez, M., Klaassen, T. O., Wenckebach, W. T., Muravjov, A. V., Pavlov, S. G., Shastin, V. N., "The p-Ge terahertz laser: properties under pulsed and mode-locked operation", *IEEE Trans. Microwave Theory Tech.* **48**, 670 (2000b).

Huber, R., Brodschelm, A., Tauser, F., Leitenstorfer, A., "Generation and field-resolved detection of femtosecond electromagnetic pulses tunable up to 41 THz", *Appl. Phys. Lett.* **76**, 3191 (2000).

Huggard, P. G., Cluff, J. A., Moore, G. P., Shaw, C. J., Andrews, S. R., Keiding, S. R., Linfield, E. H., Ritchie, D. A., "Drude conductivity of highly doped GaAs at terahertz frequencies", *J. Appl. Phys.* **87**, 2382 (2000a).

Huggard, P. G., Shaw, C. J., Andrews, S. R., Cluff, J. A., Grey, R., "Mechanism of THz emission from asymmetric double quantum wells", *Phys. Rev. Lett.* **84**, 1023 (2000b).

Iguchi, I., Kume, E., Takahashi, H., "Emitted spectra of electromagnetic waves from a tunnel-injected nonequilibrium high-T_c $YBa_2Cu_3O_{7-y}$ superconductor", *Phys. Rev. B* **62**, 5370 (2000).

Jamison, S. P., McGowan, R. W., Grischkowsky, D., "Single-mode waveguide propagation and reshaping of sub-ps terahertz pulses in sapphire fibers", *Appl. Phys. Lett.* **76**, 1987 (2000).

Jeon, T.-I., Grischkowsky, D., Mukherjee, A. K., Menon, R., "Electrical characterization of conducting polypyrrole by THz time-domain spectroscopy", *Appl. Phys. Lett.* **77**, 2452 (2000).

Jiang, Z., Li, M., Zhang, X.-C., "Dielectric constant measurement of thin films by differential time-domain spectroscopy", *Appl. Phys. Lett.* **76**, 3221 (2000a).

Jiang, Z., Xu, G., Zhang, X.-C., "Improvement of terahertz imaging with a dynamic subtraction technique", *Appl. Opt.* **39**, 2982 (2000b).

Jiang, Z., Zhang, X.-C., "Measurement of spatio-temporal terahertz field distribution by using chirped pulse technology", *IEEE J. Quantum Electron.* **36**, 1214 (2000c).

Kadow, C., Jackson, A. W., Gossard, A. C., Matsuura, S., Blake, G. A., "Self-assembled ErAs islands in GaAs for optical-heterodyne THz generation", *Appl. Phys. Lett.* **76**, 3510 (2000).

Kaestner, A., Volk, M., Ludwig, F., Schilling, M., Menzel, J., "YBa$_2$Cu$_3$O$_7$ Josephson junctions on LaAlO3 bicrystals for terahertz frequency applications", *Appl. Phys. Lett.* **77**, 3057 (2000).

Kawase, K., Hatanaka, T., Takahashi, H., Nakamura, K., Taniuchi, T., Ito, H., "Tunable terahertz-wave generation from DAST crystal by dual signal-wave parametric oscillation of periodically poled lithium niobate", *Opt. Lett.* **25**, 1714 (2000).

Kersting, R., Bratschitsch, R., Strasser, G., Unterrainer, K., Heyman, J. N., "Sampling a terahertz dipole transition with subcycle time resolution", *Opt. Lett.* **25**, 272 (2000a).

Kersting, R., Strasser, G., Unterrainer, K., "Terahertz phase modulator", *Electron. Lett.* **36**, 1156 (2000b).

Khmyrova, I., Ryzhii, V., "Resonant detection and frequency multiplication in barrier-injection heterostructure transistors", *Jp. J. Appl. Phys. Pt. 1* **39**, 4727 (2000).

Kida, N., Hangyo, M., Tonouchi, M., "Low-energy charge dynamics in La$_{0.7}$Ca$_{0.3}$MnO$_3$: THz time-domain spectroscopic studies", *Phys. Rev. B* **62**, 11965 (2000).

Kiwa, T., Kawashima, I., Nashima, S., Hangyo, M., Tonouchi, M., "Optical response in amorphous GaAs thin films prepared by pulsed laser deposition", *Jp. J. Appl. Phys. Pt. 1* **39**, 6304 (2000).

Koch, M., Bieler, M., Hein, G., Pierz, K., Siegner, U., "Photoconductive switches: the role of spatial effects in carrier dynamics", *Phys. Stat. Sol. B* **221**, 429 (2000).

Kono, S., Tani, M., Gu, P., Sakai, K., "Detection of up to 20 THz with a low-temperature-grown GaAs photoconductive antenna gated with 15 fs light pulses", *Appl. Phys. Lett.* **77**, 4104 (2000a).

Kono, S., Gu, P., Tani, M., Sakai, K., "Temperature dependence of terahertz radiation from n-type InSb and n-type InAs surfaces", *Appl. Phys. B* **71**, 901 (2000b).

Kral, P., Sipe, J. E., "Quantum kinetic theory of two-beam current injection in bulk semiconductors", *Phys. Rev. B* **61**, 5381 (2000).

Kuzel, P., Petzelt, J., "Time-resolved terahertz transmission spectroscopy of dielectrics", *Ferroelectrics* **239**, 79 (2000).

Lachaine, J. M., Hawton, M., Sipe, J. E., Dignam, M. M., "Asymmetry in the excitonic Wannier-Stark ladder: A mechanism for the stimulated emission of terahertz radiation", *Phys. Rev. B* **62**, 4829 (2000).

Lee, Y. S., Meade, T., Perlin, V., Winful, H., Norris, T. B., Galvanauskas, A., "Generation of narrow-band terahertz radiation via optical rectification of femtosecond pulses in periodically poled lithium niobate", *Appl. Phys. Lett.* **76**, 2505 (2000a).

Lee, Y. S., Meade, T., DeCamp, M., Norris, T. B., Galvanauskas, A., "Temperature dependence of narrow-band terahertz generation from periodically poled lithium niobate", *Appl. Phys. Lett.* **77**, 1244 (2000b).

Lee, Y. S., Meade, T., Naudeau, M. L., Norris, T. B., Galvanauskas, A., "Domain mapping of periodically poled lithium niobate via terahertz waveform analysis", *Appl. Phys. Lett.* **77**, 2488 (2000c).

Leitenstorfer, A., Hunsche, S., Shah, J., Nuss, M. C., Knox, W. H., "Femtosecond high-field transport in compound semiconductors", *Phys. Rev. B* **61**, 16642 (2000).

Libon, I. H., Baumgartner, S., Hempel, M., Hecker, N. E., Feldmann, J., Koch, M., Dawson, P., "An optically controllable terahertz filter", *Appl. Phys. Lett.* **76**, 2821 (2000).

Liu, W. Y., Steenson, D. P., "Investigation of subharmonic mixer based on a quantum barrier device", *IEEE Trans. Microwave Theory Tech.* **48**, 757 (2000a).

Liu, Z., Ono, S., Ohtake, H., Sarakura, N., Liu, T. A., Huang, K. F., Pan, C. L., "Efficient terahertz radiation generation from a bulk InAs mirror as an intracavity terahertz radiation emitter", *Jp. J. Appl. Phys. Pt. 2* **39**, 366 (2000b).

Loffler, T., Jacob, F., Roskos, H. G., "Generation of terahertz pulses by photoionization of electrically biased air", *Appl. Phys. Lett.* **77**, 453 (2000).

Lu, S. H., Li, J. L., Yu, J. S., Horng, S. F., Chi, C. C., "Observation of terahertz electric pulses generated by nearly filled-gap nonuniform illumination excitation", *Appl. Phys. Lett.* **77**, 3896 (2000).

MacDonald, M. E., Alexanian, A., York, R. A., Popovic, Z., Grossman, E. N., "Spectral transmittance of lossy printed resonant-grid terahertz bandpass filters", *IEEE Trans. Microwave Theory Tech.* **48**, 712 (2000).

McGowan, R. W., Cheville, R. A., Grischkowsky, D., "Direct observation of the Gouy phase shift in THz impulse ranging", *Appl. Phys. Lett.* **76**, 670 (2000a).

McGowan, R. W., Cheville, R. A., Grischkowsky, D., "Experimental study of the surface waves on a dielectric cylinder via terahertz impulse radar ranging", *IEEE Trans. Microwave Theory Tech.* **48**, 417 (2000b).

McLaughlin, R, Corchia, A. A., Johnston, M. B., Chen, Q., Ciesla, C. M., Amone, D. D., Pepper, M., Jones, G. A. C., Linfield, E. H., Davies, A. G., "Enhanced coherent terahertz emission from indium arsenide in the presence of a magnetic field", *Appl. Phys. Lett.* **76**, 2038 (2000a).

McLaughlin, R., Chen, Q., Corchia, A., Ciesla, C. M., Arnone, D. D., Zhang, X.-C., Jones, G. A. C., Lindfield, E. H., Pepper, M., "Enhanced coherent terahertz emission from indium arsenide", *J. Mod. Opt.* **47**, 1847 (2000b).

Maiwald, F., Lewen, F., Ahrens, V., Beaky, M., Grendriesch, R., Koroliev, A. N., Negirev, A. A., Paveljev, D. G., Vowinkel, B., Winnewisser, G., "Pure rotational spectrum of HCN in the terahertz region: use of a new planar Schottky diode multiplier", *J. Mol. Spectrosc.* **202**, 166 (2000).

Markelz, A. G., Roitberg, A., Heilweil, E. J., "Pulsed terahertz spectroscopy of DNA, bovine serum albumin and collagen between 0.1 and 2.0 THz", *Chem. Phys. Lett.* **320**, 42 (2000).

Matsuura, S., Chen, P., Blake, G. A., Pearson, J. C., Pickett, H. M., "A tunable cavity-locked diode laser source for terahertz photomixing", *IEEE Trans. Microwave Theory Tech.* **48**, 380 (2000).

Meinert, G., Banyan, L., Gartner, P., Haug, H., "Theory of THz emission from optically excited semiconductors in crossed electric and magnetic fields", *Phys. Rev. B* **62**, 5003 (2000).

Mendis, R., Grischkowsky, D., "Plastic ribbon THz waveguides", *J. Appl. Phys.* **88**, 4449 (2000).

Mickan, S., Abbott, D., Munch, J., Zhang, X.-C., Van Doorn, T., "Analysis of system trade-offs for terahertz imaging", *Microelectron. J.* **31**, 503 (2000).

Misochko, O. V., Gu, P., Sakai, K., "Coherent phonons in InSb and their properties from femtosecond pump-probe experiments", *Physica B* **293**, 33 (2000).

Mitrofanov, O., Brener, I., Wanke, M. C., Ruel, R. R., Wynn, J. D., Bruca, A. J., Federici, J., "Near-field microscope probe for far infrared time domain measurements", *Appl. Phys. Lett.* **77**, 591 (2000a).

Mitrofanov, O., Brener, I., Harel, R., Wynn, J. D., Pfeiffer, L. N., West, K. W., Federici, J., "Terahertz near-field microscopy based on a collection mode detector", *Appl. Phys. Lett.* **77**, 3496 (2000b).

Morikawa, O., Tonouchi, M., Hangyo, M., "A cross-correlation spectroscopy in subterahertz region using an incoherent light source", *Appl. Phys. Lett.* **76**, 1519 (2000a).

Morikawa, O., Yamashita, M., Tonouchi, M., Hangyo, M., "Vector map imaging of supercurrent distribution in high-T_C superconductive thin films", *Physica B* **284–288**, 2069 (2000b).

Ohtake, H., Ono, S., Sakai, M., Liu, Z., Tsukamoto, T., Sarakura, N., "Saturation of THz-radiation power from femtosecond-laser-irradiated InAs in a high magnetic field", *Appl. Phys. Lett.* **76**, 1398 (2000).

Ono, S., Tsukamoto, T., Sakai, M., Liu, Z., Ohtake, H., Sarakura, N., Nishizawa, S., Nakanishi, A., "Compact THz-radiation source consisting of a bulk semiconductor, a mode-locked fiber laser, and a 2 T permanent magnet", *Rev. Sci. Inst.* **71**, 554 (2000).

Piao, Z., Tani, M., Sakai, K., "Carrier dynamics and terahertz radiation in photoconductive antennas", *Jp. J. Appl. Phys. Pt. 1* **39**, 96 (2000).

Pimenov, A., Pronin, A. V., Loidl, A., Michelucci, U., Kampf, A. P., Krasnosvobodtsev, S. I., Nozdrin, V. S., Rainer, D., "Anisotropic conductivity of $Nd_{1.85}Ce_{0.15}CuO_{4-\delta}$ films at submillimeter wavelengths", *Phys. Rev. B* **62**, 9822 (2000).

Rahman, F., Thornton, T., "Superconducting quantum wells for the detection of submillimeter wave electromagnetic radiation", *Appl. Phys. Lett.* **77**, 432 (2000).

Reedijk, J. A., Martens, H. C. F., Smits, B. J. G., Brom, H. B., "Measurement of the complex dielectric constant down to helium temperatures. II. Quasioptical technique from 0.03 to 1 THz", *Rev. Sci. Inst.* **71**, 478 (2000).

Ronne, C., Jensby, K., Loughnan, B. J., Fourkas, J., Nielsen, O. F., Keiding, S. R., "Temperature dependence of the dielectric function of $C_6H_6(l)$ and $C_6H_5CH(l)$ measured with THz spectroscopy", *J. Chem. Phys.* **113**, 3749 (2000).

Roskos, H. G., "Overview on time-domain terahertz spectroscopy and its applications in atomic and semiconductor physics", *Phys. Scr. T* **86**, 51 (2000).

Roux, J. F., Coutaz, J.-L., Tedjini, S., "All-optical high-frequency characterization of optical devices for optomicrowave applications", *IEEE Photonics Tech. Lett.* **12**, 1031 (2000).

Rudd, J. V., Zimdars, D., Warmuth, M., "Compact fiber-pigtailed terahertz imaging system", *Proc. SPIE* **3934**, 27 (2000a).

Rudd, J. V., Johnson, J. L., Mittleman, D. M., "Quadrupole radiation from terahertz dipole antennas", *Opt. Lett.* **25**, 1556 (2000b).

Ryzhii, V., Khmyrova, I., Shur, M., "Resonant detection and frequency multiplication of terahertz radiation utilizing plasma waves in resonant-tunneling transistors", *J. Appl. Phys.* **88**, 2868 (2000).

Saitow, K. I., Nishikawa, K., Ohtake, H., Sarukura, N., Miyagi, H., Shimokawa, Y., Matsuo, H., Tominaga, K., "Supercritical-fluid cell with device of variable optical path length giving fringe-free terahertz spectra", *Rev. Sci. Instrum.* **71**, 4061 (2000).

Sassen, S., Witzigmann, B., Wolk, C.. Brugger, H., "Barrier height engineering on GaAs THz Schottky diodes by means of high–low doping, InGaAs- and InGaP-layers". *IEEE Trans. Elec. Dev.* **47**, 24 (2000).

Schall, M., Jepsen, P. U., "Photoexcited GaAs surfaces studied by transient terahertz time-domain spectroscopy", *Opt. Lett.* **25**, 13 (2000a).

Schall, M.. Jepsen, P. U., "Freeze-out difference-phonon modes in ZnTe and its applications in detection of THz pulses", *Appl. Phys. Lett.* **77**, 2801 (2000b).

Shan, J., Weling, A. S., Knoesel, E., Bartels, L., Bonn, M., Nahata, A., Reider, G. A.. Heinz, T. F., "Single-shot measurement of terahertz electromagnetic pulses by use of electro-optic sampling", *Opt. Lett.* **25**, 426 (2000).

Shikata, J., Kawase, K., Karino, K., Taniuchi, T., Ito, H., "Tunable terahertz-wave parametric oscillators using LiNbO$_3$ and MgO:LiNbO$_3$ crystals", *IEEE Trans. Microwave Theory Tech.* **48**, 653 (2000).

Shur, M. S.. Lu, J. Q., "Terahertz sources and detectors using two-dimensional electronic fluid in high electron-mobility transistors", *IEEE Trans. Microwave Theory Tech.* **48**, 750 (2000).

Siders, J. L. W.. Trugman, S. A., Garzon, F. H., Houlton, R. J., Taylor, A. J.. "Terahertz emission from YBa$_2$Cu$_3$O$_{7-\delta}$ thin films via bulk electric-quadrupole-magnetic-dipole optical rectification", *Phys. Rev. B* **61**, 13633 (2000).

Takahashi. H., Hosada. M., "Frequency domain spectroscopy of free-space terahertz radiation", *Appl. Phys. Lett.* **77**, 1085 (2000).

Tani, M.. Jiang, Z., Zhang, X.-C., "Photoconductive terahertz transceiver", *Electron. Lett.* **36**, 1298 (2000a).

Tani, M., Lee, K. S., Zhang, X.-C., "Detection of terahertz radiation with low-temperature-grown GaAs-based photoconductive antenna using $1.55\,\mu m$ probe". *Appl. Phys. Lett.* **77**, 1396 (2000b).

Taniuchi, T., Shikata. J., Ito, H., "Tunable terahertz-wave generation in DAST crystal with dual-wavelength KTP optical parametric oscillator", *Electron. Lett.* **36**, 1414 (2000).

Tomlinson. A. M., Chang, C. C., Stone, R. J.. Nicholas, R. J., Fox, A. M., Pate, M. A., Foxon, C. T., "Intersubband transitions in GaAs coupled-quantum-wells for use as a tunable detector at THz frequencies", *Appl. Phys. Lett.* **76**, 1579 (2000).

Tonouchi, M., Yamashita, M., Hangyo, M., "Terahertz radiation imaging of super-current distribution in vortex-penetrated YBa$_2$Cu$_3$O$_{7-\delta}$ thin film strips", *J. Appl. Phys.* **87**, 7366 (2000).

Ulrich, J., Zobl, R., Unterrainer, K., Strasser, G., Gornik, E., "Magnetic-field-enhanced quantum-cascade emission", *Appl. Phys. Lett.* **76**, 19 (2000a).

Ulrich, J., Zobl, R., Schrenk, W., Strasser, G., Unterrainer, K., Gornik, E., "Terahertz quantum cascade structures: Intra- versus interwell transition", *Appl. Phys. Lett.* **77**, 1928 (2000b).

van der Weide, D. W., Murakowski, J., Keilmann, F., "Gas-absorption spectroscopy with electronic terahertz techniques", *IEEE Trans. Microwave Theory Tech.* **48**, 740 (2000).

Venables, D. S., Chiu, A., Schmuttenmaer, C. A., "Structure and dynamics of non-aqueous mixtures of dipolar liquids. I. Infrared and far-infrared spectroscopy", *J. Chem. Phys.* **113**, 3243 (2000).

Walther, M., Jensby, K., Keiding, S. R., Takahashi, H., Ito, H., "Far-infrared properties of DAST", *Opt. Lett.* **25**, 911 (2000a).

Walther, M., Fischer, B., Schall, M., Helm, H., Jepsen, P. U., "Far-infrared vibrational spectra of all-trans, 9-cis and 13-cis retinal measured by THz time-domain spectroscopy", *Chem. Phys. Lett.* **332**, 389 (2000b).

Weiss, C., Wallenstein, R., Beigang, R., "Magnetic-field-enhanced generation of terahertz radiation in semiconductor surfaces", *Appl. Phys. Lett.* **77**, 4160 (2000).

Wilke, I., Khazan, M., Rieck, C. T., Kuzel, P., Kaiser, T., Jaekel, C., Kurz, H., "Terahertz surface resistance of high temperature superconducting thin films", *J. Appl. Phys.* **87**, 2984 (2000).

Winnerl, S., "GaAs/AlAs superlattices for detection of terahertz radiation", *Microelectron. J.* **31**, 389 (2000).

Winnewisser, C., Lewen, F. T., Schall, M., Walther, M., Helm, H., "Characterization and application of dichroic filters in the 0.1–3 THz region", *IEEE Trans. Microwave Theory Tech.* **48**, 744 (2000).

Wynne, K., Carey, J. J., Zawadzka, J., Jaroszynski, D. A., "Tunneling of single-cycle terahertz pulses through waveguides", *Opt. Commun.* **176**, 429 (2000).

Yamashita, M., Tonouchi, M., Hangyo, M., "Visualization of supercurrent distribution by THz radiation mapping", *Physica B* **284–288**, 2067 (2000).

Zharov, A. A., Dodin, E. P., Raspopin, A. S., "Compression of terahertz radiation in resonant systems with a quantum superlattice", *JETP Lett.* **72**, 453 (2000).

Ahn, J., Hutchinson, D. N., Rangan, C., Bucksbaum, P. H., "Quantum phase retrieval of a Rydberg wave packet using a half-cycle pulse", *Phys. Rev. Lett.* **86**, 1179 (2001).

Aoki, T., Takeda, M. W., Haus, J. W., Zhenyu-Yuan, Tani, M., Sakai, K., Kawai, N., Inoue, K., "Terahertz time-domain study of a pseudo-simple-cubic photonic lattice", *Phys. Rev. B* **64**, 1 (2001).

Averitt, R., Rodriguez, G., Lobad, A. I., Siders, J. L. W., Trugman, S. A., Taylor, A. J., "Nonequilibrium superconductivity and quasiparticle dynamics in YBa$_2$Cu$_3$O$_{7-\delta}$", *Phys. Rev. B* **63**, 140502 (2001).

Beard, M. C., Schmuttenmaer, C. A., "Using the finite-difference time-domain pulse propagation method to simulate time-resolved THz experiments", *J. Chem. Phys.* **114**, 2903 (2001).

Beard, M. C., Turner G. M., Schmuttenmaer, C. A., "Subpicosecond carrier dynamics in low-temperature grown GaAs as measured by time-resolved terahertz spectroscopy", *J. Appl. Phys.* **90**, 5915 (2001).

Boyd, J. E., Briskman, A., Colvin, V. L., Mittleman, D. M., "Direct observation of terahertz surface modes in nanometer-sized liquid water pools", *Phys. Rev. Lett.* **87**, 1 (2001).

Brucherseifer, M., Haring Bolivar, P., Klingenberg, H., Kurz, H., "Angle-dependent THz tomography characterization of thin ceramic oxide films for fuel cell applications", *Appl. Phys. B* **72**, 361 (2001).

Chen, Q., Tani, M., Jiang, Z., Zhang, X.-C., "Electro-optic transceivers for terahertz wave applications", *J. Opt. Soc. Am. B* **18**, 823 (2001).

Chen, Q., Zhang, X.-C., "Semiconductor dynamic aperture for near-field terahertz wave imaging", *IEEE J. Quantum Electron.* **7**, 608 (2001).

Chun-Zhang, Kwang-Su-Lee, Zhang, X.-C., Xing-Wei, Shen, Y. R., "Optical constants of ice Ih crystal at terahertz frequencies", *Appl. Phys. Lett.* **79**, 491 (2001).

Citrin, D., "Terahertz nonlinear optics with strained p-type quantum wells", *Opt. Lett.* **26**, 554 (2001).

Cole, B. E., Williams, J. B., King, B. T., Sherwin, M. S., Stanley, C. R., "Coherent manipulation of semiconductor quantum bits with terahertz radiation", *Nature* **410**, 60 (2001).

Colombelli, R., Straub, A., Capasso, F., Gmachl, C., Blakey, M. I., Sergent, A. M., Chu, S. N. G., Corchia, A., McLaughlin, R., Johnston, M. B., Whittaker, D. M., Arnone, D. D., Linfield, E. H., Davies, A. G., Pepper, M., "Effects of magnetic field and optical fluence on terahertz emission in gallium arsenide", *Phys. Rev. B* **64**, 1 (2001).

Demers, J. R., Goyette, T. M., Ferrio, K. B., Everitt, H. O., Guenther, B. D., De Lucia, F. C., "Spectral purity and sources of noise in femtosecond-demodulation terahertz sources driven by Ti:sapphire mode-locked lasers", *IEEE J. Quantum Electron.* **37**, 595 (2001).

Dorney, T. D., Baraniuk, R. G., Mittleman, D. M., "Material parameter estimation with terahertz time-domain spectroscopy", *J. Opt. Soc. Am. A* **18**, 1562 (2001).

Dorney, T. D., Johnson, J. L., Van-Rudd, J., Baraniuk, R. G., Symes, W. W., Mittleman, D. M., "Terahertz reflection imaging using Kirchhoff migration", *Opt. Lett.* **26**, 1513 (2001).

Dragoman, D., Dragoman, M., "Terahertz field characterization using Fabry–Perot-like cantilevers", *Appl. Phys. Lett.* **79**, 581 (2001).

Duffy, S. M., Verghese, S., McIntosh, A., Jackson, A., Gossard, A. C., Matsuura, S., "Accurate modeling of dual dipole and slot elements used with photomixers for coherent terahertz output power", *IEEE Trans. Microwave Theory Tech.* **49**, 1032 (2001).

Duvillaret, L., Garet, F., Roux, J.-F., Coutaz, J.-L., "Analytical modeling and optimization of terahertz time-domain spectroscopy experiments, using photoswitches as antennas", *IEEE J. Quantum Electron.* **7**, 615 (2001).

Ferguson, B., Abbott, D., "De-noising techniques for terahertz responses of biological samples", *Microelect. J.* **32**, 943 (2001).

Fifin, A., Stowe, M., Kersting, R., "Time-domain differentiation of terahertz pulses", *Opt. Lett.* **26**, 2008 (2001).

Gonzalo, R., Ederra, I., Mann, C. M., de Maagt, P., "Radiation properties of terahertz dipole antenna mounted on photonic crystal", *Elec. Lett.* **37**, 613 (2001).

Grebenev, V., Knoesel, E., Bartels, L., "Destructive interference of freely propagating terahertz pulses and its potential for high-resolution spectroscopy and optical computing", *Appl. Phys. Lett.* **79**, 145 (2001).

Graf, U., Heyminck, S., "Fourier gratings as submillimeter beam splitters", *IEEE Trans. Antennas Prop.* **49**, 542 (2001).

Han, P. Y., Tani, M., Usami, M., Kono, S., Kersting, R., Zhang, X.-C., "A direct comparison between terahertz time-domain spectroscopy and far-infrared Fourier transform spectroscopy", *J. Appl. Phys.* **89**, 2357 (2001).

Hangyo, M., Migita. M., Nakayama, K., "Magnetic field and temperature dependence of terahertz radiation from InAs surfaces excited by femtosecond laser pulses", *J. Appl. Phys.* **90**, 3409 (2001).

Harde. H., Zhao. J., Wolff, M., Cheville, R. A., Grischowsky, D., "THz time-domain spectroscopy on ammonia", *J. Phys. Chem. A* **105**. 6038 (2001).

Harrison, P., Soref. R. A., "Population-inversion and gain estimates for a semiconductor TASER", *IEEE J. Quantum Electron.* **37**, 153 (2001).

Hashimoto, H., Takahashi. H., Yamada. T., Kuroyanagi, K., Kobayashi. T., "Characteristics of the terahertz radiation from single crystals of N-substituted 2-methyl-4-nitroaniline", *J. Phys.* **13**, L529 (2001).

Hashimshony. D., Geltner, I., Cohen. G., Avitzour, Y., Zigler, A., Smith, C., "Characterization of the electrical properties and thickness of thin epitaxial semiconductor layers by THz reflection spectroscopy". *J. Appl. Phys.* **90**, 5778 (2001).

Hattori, T., Tukamoto, K., Nakatsuka, H., "Time-resolved study of intense terahertz pulses generated by a large-aperture photoconductive antenna", *Japan. J. Appl. Phys.* **40**, 4907 (2001).

Heeseok-Lee, Jongjoo-Lee, Joungho-Kim. "Picosecond-domain radiation pattern measurement using fiber-coupled photoconductive antenna", *IEEE J. Quantum Electron.* **7**, 667 (2001).

Heyman, J. N., Neocleous, P., Hebert, D., Crowell, P. A., Muller, T., Unterrainer, K., "Terahertz emission from GaAs and InAs in a magnetic field", *Phys. Rev. B* **64**, 1 (2001).

Huber, R., Tauser, F., Brodschelm, A., Bichler, M., Abstreiter, G., Leitenstorfer, A., "How many-particle interactions develop after ultrafast excitation of an electron–hole plasma, *Nature* **414**. 286 (2001).

Hughes, S., Citrin, D., "Broadband terahertz emission through exciton trapping in a semiconductor quantum well". *Opt. Lett.* **26**, 1 (2001).

Imai, K., Kawase, K., Ito. H., "A frequency-agile terahertz-wave parametric oscillator". *Optics Express* **8** (2001a).

Imai, K., Kawase. K., Shikata, J., Minamide, H., Ito, H., "Injection-seeded terahertz wave parametric oscillator", *Appl. Phys. Lett.* **78**, 1026 (2001b).

Jager, B. G. L., Wimmer, S., Lorke, A., Kotthaus, J. P., Wegscheider, W., Bichler, M., "Edge and bulk effects in Terahertz photoconductivity of an antidot superlattice", *Phys. Rev. B* **63**. 45315 (2001).

Jeon, T. I., Grischkowsky, D., Mukherjee, A. K., Menon. R., "Electrical and optical characterization of conducting poly-3-methylthiophene film by THz time-domain spectroscopy". *Appl. Phys. Lett.* **79**, 4142 (2001).

Jepsen. P. U., Schairer, W., Libon. I. H., Lemmer. U., Hecker, N. E., Birkholz, M., Lips, K., Schall, M., "Ultrafast carrier trapping in microcrystalline silicon observed in optical pump-terahertz probe measurements". *Appl. Phys. Lett.* **79**, 1291 (2001).

Johnson, J. L., Dorney, T. D., Mittleman, D. M., "Interferometric imaging with terahertz pulses", *IEEE J. Quantum Electron.* **7**, 592 (2001a).

Johnson, J. L., Dorney, T. D., Mittleman, D. M., "Enhanced depth resolution in terahertz imaging using phase-shift interferometry", *Appl. Phys. Lett.* **78**, 835 (2001b).

Jongjoo, Lee, Heeseok-Lee, Sungkyu-Yu, Joungho-Kim, "A micromachined photoconductive near-field probe for picosecond pulse propagation measurement on coplanar transmission lines", *IEEE J. Quantum Electron.* **7**, 674 (2001).

Kamba, S., Petzelt, J., Buixaderas, E., Haubrich, D., Vanek, P., Kuzel, P., Jawahar, I. N., Sebastian, M. T., Mohanan, P., "High frequency dielectric properties of $A_5B_4O_{15}$ microwave ceramics", *J. Appl. Phys.* **89**, 3900 (2001).

Kawase, K., Shikata, J., Imai, K., Ito, H., "Transform-limited, narrow-linewidth, terahertz-wave parametric generator", *Appl. Phys. Lett.* **78**, 2819 (2001a).

Kawase, K., Shikata, J., Minamide, H., Imai, K., Ito, H., "Arrayed silicon prism coupler for a terahertz-wave parametric oscillator", *Appl. Opt.* **40**, 1423 (2001b).

Kee, C. S., Lim, H., "Tunable complete photonic band gaps of two-dimensional photonic crystals with intrinsic semiconductor rods", *Phys. Rev. B* **64**, 1 (2001).

Keutsch, F. N., Brown, M. G., Petersen, P. B., Saykally, R. J., Geleijns, M., van der Avoird, A., "Terahertz vibration–rotation–tunneling spectroscopy of water clusters in the translational band region of liquid water", *J. Chem. Phys.* **114**, 3994 (2001).

Khazan, M., Meissner, R., Wilke, I., "Convertible transmission-reflection time-domain terahertz spectrometer", *Rev. Sci. Instrum.* **72**, 3427 (2001).

Kida, N., Tonouchi, M., "Terahertz radiation from magnetoresistive $Pr_{0.7}Ca_{0.3}MnO_3$ thin films", *Appl. Phys. Lett.* **78**, 4115 (2001).

Kitahara, H., Tsumura, N., Kondo, H., Takeda, M. W., Haus, J. W., Zhenyu-Yuan, Kawai, N., Sakoda, K., Inoue, K., "Terahertz wave dispersion in two-dimensional photonic crystals", *Phys. Rev. B* **64**, 1 (2001).

Kitoh, Y., Yamashita, M., Nagashima, T., Hangyo, M., "Terahertz beam profiler using optical transmission modulation in silicon", *Japan. J. Appl. Phys.* **40**, 1113 (2001).

Kiwa, T., Tonouchi, M., "Time-domain terahertz spectroscopy of (100) $(LaAlO_3)_{0.3}$-$(Sr_2AlTaO_6)_{0.7}$ substrate", *Jp. J. Appl. Phys. Pt. 2* **40**, 38 (2001).

Klappenberger, F., Ignatov, A. A., Winnerl, S., Schomburg, E., Wegscheider, W., Renk, K. F., Bichler, M., "Broadband semiconductor superlattice detector for THz radiation", *Appl. Phys. Lett.* **78**, 1673 (2001).

Kleine-Ostmann, T., Knobloch, P., Koch, M., Hoffmann, S., Breede, M., Hoffmann, M., Hein, G., Pierz, K., Sperling, M., Donhuijsen, K., "Continuous-wave THz imaging", *Electron. Lett.* **37**, 1461 (2001).

Knoesel, E., Bonn, M., Shan, J., Heinz, T. F., "Charge transport and carrier dynamics in liquids probed by THz time-domain spectroscopy", *Phys. Rev. Lett.* **86**, 340 (2001).

Kondo, T., Tonouchi, M., Hangyo, M., "Terahertz radiation from superconducting $YBa_2Cu_3O_{7-\delta}$ thin films with 1.55 micron excitation", *Jp. J. Appl. Phys. Pt. 1* **40**, 640 (2001).

Kono, S., Tani, M., Sakai, K., "Ultrabroadband photoconductive detection: comparison with free-space electro-optic sampling", *Appl. Phys. Lett.* **79**, 898 (2001).

Krishnamurthy, S., Reiten, M. T., Harmon, S. A., Cheville, R. A., "Characterization of thin polymer films using terahertz time-domain interferometry", *Appl. Phys. Lett.* **79**, 875 (2001).

Kumar, A. K. S., Kawasaki, M., Koinuma, H., "Crack-free and c-axis in-plane aligned superconducting $Y_{0.7}Ca_{0.3}Ba_2Cu_3O_7$-d films for terahertz radiating Josephson plasma devices", *Japan. J. Appl. Phys.* **40**, 6335 (2001).

Lampin, J. F., Desplanque, L., Mollot, F., "Detection of picosecond electrical pulses using the intrinsic Franz–Keldysh effect", *Appl. Phys. Lett.* **78**, 4103 (2001).

Lee, Y. S., Meade, T., Norris, T. B., Galvanauskas, A., "Tunable narrow-band terahertz generation from periodically poled lithium niobate", *Appl. Phys. Lett.* **78**, 3583 (2001).

Lin, G. R., Pan, C. L., "Characterization of optically excited terahertz radiation from arsenic-ion-implanted GaAs", *Appl. Phys. B* **72**, 151 (2001).

Liu, K. P. H., Hegmann, F. A., "Ultrafast carrier relaxation in radiation-damaged silicon on sapphire studied by optical-pump–terahertz-probe experiments", *Appl. Phys. Lett.* **78**, 3478 (2001).

Loffler, T., Bauer, T., Siebert, K. J., Roskos, H. G., Fitzgerald, A., Czasch, S., "Terahertz dark-field imaging of biomedical tissue", *Opt. Exp.* **9** (2001).

Lu, J. Q., Shur, M., "Terahertz detection by high-electron-mobility transistor: Enhancement by drain bias", *Appl. Phys. Lett.* **78**, 2587 (2001).

Maslov, A. V., Citrin, D. S., "Extraction of the frequency-domain optical response function of a periodically modulated medium with short optical pulses", *J. Opt. Soc. Am. B* **18**, 1563 (2001).

Matsushima, F., Matsunaga, M., Qian, G. Y., Ohtaki, Y., Wang, R. L., Takagi, K., "Frequency measurement of pure rotational transitions of D_2O from 0.5 to 5 THz", *J. Molec. Spec.* **206**, 41 (2001).

McClatchey, K., Reiten, M. T., Cheville, R. A., "Time resolved synthetic aperture terahertz impulse imaging", *Appl. Phys. Lett.* **79**, 4485 (2001).

Melnik, D. G., Gopalakrishnan, S., Miller, T. A., De Lucia, F. C., Belov, S., "Submillimeter wave vibration–rotation spectroscopy of Ar.CO and Ar.ND_3", *J. Chem. Phys.* **114**, 6100 (2001).

Mendis, R., Grischkowsky, D., "THz interconnect with low-loss and low-group velocity dispersion", *IEEE Microwave Lett.* **11**, 444 (2001a).

Mendis, R., Grischkowsky, D., "Undistorted guided-wave propagation of subpicosecond terahertz pulses", *Opt. Lett.* **26**, 846 (2001b).

Messner, C., Kostner, H., Hopfel, R. A., Unterrainer, K., "Time-resolved THz spectroscopy of proton-bombarded InP", *J. Opt. Soc. Am. B* **18**, 1369 (2001).

Migita, M., Hangyo, M., "Pump-power dependence of THz radiation from InAs surfaces under magnetic fields excited by ultrashort laser pulses", *Appl. Phys. Lett.* **79**, 3437 (2001).

Ming-Li, Fortin, J., Kim, J. Y., Fox, G., Chu, F., Davenport, T., Toh-Ming-Lu, Xi-Cheng-Zhang, "Dielectric constant measurement of thin films using goniometric terahertz time-domain spectroscopy", *IEEE J. Quantum Electron.* **7**, 624 (2001).

Mitrofanov, O., Harel, R., Lee, M., Pfeiffer, L. N., West, K., Wynn, J. D., Federici, J., "Study of single-cycle pulse propagation inside a terahertz near-field probe", *Appl. Phys. Lett.* **78**, 252 (2001a).

Mitrofanov, O., Lee, M., Hsu, J. W. P., Brener, I., Harel, R., Federici, J. F., Wynn, J. D., Pfeiffer, L. N., West, K. W., "Collection-mode near-field imaging with 0.5-THz pulses", *IEEE J. Quantum Electron.* **7**, 600 (2001b).

Mitrofanov, O., Lee, M., Hsu, J. W. P., Pfeiffer, L. N., West, K. W., Wynn, J. D., Federici, J. F., "Terahertz pulse propagation through small apertures", *Appl. Phys. Lett.* **79**, 907 (2001c).

Mochan, W. L., Brudny, V. L., Carey, J. J., Zawadzka, J., Jaroszynski, D. A., Wynne, K., "Comment on 'Noncausal time response in frustrated total internal reflection?' [and reply]", *Phys. Rev. Lett.* **87**, 119101/1 (2001).

Moto, A., Hangyo, M., Tonouchi, M., "Terahertz radiation imaging of vortex penetration into YBCO thin films with and without ordered arrays of antidots", *IEICE Trans. Electron.* **E84-C**, 67 (2001).

Nagashima, T., Hangyo, M., "Measurement of complex optical constants of a highly doped Si wafer using terahertz ellipsometry", *Appl. Phys. Lett.* **79**, 3917 (2001).

Nagel, M., Dekorsky, T., Brucherseifer, M., Haring Bolivar, P., Kurz, H., "Characterization of polypropylene thin-film microstrip lines at millimeter and submillimeter wavelengths", *Microwave Opt. Tech. Lett.* **29**, 97 (2001).

Nashima, S., Morikawa, O., Takata, K., Hangyo, M., "Measurement of optical properties of highly doped silicon by terahertz time domain reflection spectroscopy", *Appl. Phys. Lett.* **79**, 3923 (2001a).

Nashima, S., Morikawa, O., Takata, K., Hangyo, M., "Temperature dependence of optical and electronic properties of moderately doped silicon at terahertz frequencies", *J. Appl. Phys.* **90**, 837 (2001b).

Nekkanti, S., Sullivan, D., Citrin, D. S., "Simulation of spatiotemporal terahertz pulse shaping in 3-D using conductive apertures of finite thickness", *IEEE J. Quantum Electron.* **37**, 1226 (2001).

Nemec, H., Pashkin, A., Kuzel, P., Khazan, M., Schnull, S., Wilke, I, "Carrier dynamics in low-temperature grown GaAs studied by terahertz emission spectroscopy", *J. Appl. Phys.* **90**, 1303 (2001).

Nymand, T. M., Ronne, C., Keiding, S. R., "The temperature dependent dielectric function of liquid benzene: interpretation of THz spectroscopy data by molecular dynamics simulation", *J. Chem. Phys.* **114**, 5246 (2001).

O'Hara, J., Grischkowsky, D., "Quasi-optic terahertz imaging", *Opt. Lett.* **26**, 1918 (2001).

Ohtake, H., Suzuki, Y., Sarukura, N., Ono, S., Tsukamoto, T., Nakanishi, A., Nishizawa, S., Stock, M. L., Yoshida, M., Endert, H., "THz-radiation emitter and receiver system based on a 2 T permanent magnet, 1040 nm compact fiber laser and pyroelectric thermal receiver", *Japan. J. Appl. Phys.* **40**, 1223 (2001).

Ohtaki, Y., Matsushima, F., Odashima H., Takagi K., "Rotational spectra of XeH$^+$ and its isotopic species", *J. Mol. Spectrosc.* **210**, 271 (2001).

Ono, S., Tsukamoto, T., Kawahata, E., Yano, T., Ohtake, H., Sarakura, N., "Terahertz radiation from a shallow incidence-angle InAs emitter in a magnetic field irradiated with femtosecond laser pulses", *Appl. Opt.* **40**, 1369 (2001).

Ostapchuk, T., Petzelt, J., Zelezny, V., Kamba, S., Bovtun, V., Porokhonskyy, V., Pashkin, A., Kuzel, P., Glinchuk, M. D., Bykov, I. P., Gorshunov, B., Dressel, M., "Polar phonons and central mode in antiferroelectric PbZrO$_3$ ceramics", *J. Phys.: Condens. Matter.* **13**, 2677 (2001).

Parks, B., Loomis, J., Rumberger, E., Hendrickson, D. N., Christou, G., "Linewidth of single-photon transitions in Mn$_{12}$-acetate", *Phys. Rev. B* **64**, 1 (2001).

Pearce, J., Mittleman, D. M., "Propagation of single-cycle terahertz pulses in random media", *Opt. Lett.* **26**, 2002 (2001).

Planken, P. C. M., Nienhuys, H., Bakker, H. J., Wenckebach, T., "Measurement and calculation of the orientation dependence of terahertz pulse detection in ZnTe", *J. Opt. Soc. Am. B* **18**, 313 (2001).

Quema, A., Migita, F., Nashima, S., Hangyo, M., "Terahertz time-domain spectroscopic measurement of moderately doped silicon using InAs emitter under magnetic field", *Jp. J. Appl. Phys. Pt. 1* **40**, 867 (2001).

Rangan, C., Bucksbaum, P. H., "Optimally shaped terahertz pulses for phase retrieval in a Rydberg-atom data register", *Phys. Rev. A* **64**, 1 (2001a).

Reiten, M. T., Grischkowsky, D., Cheville, R. A., "Optical tunneling of single-cycle terahertz bandwidth pulses", *Phys. Rev. E* **64**, 1 (2001a).

Reiten. M. T., Grischkowsky. D., Cheville. R. A.. "Properties of surface waves determined via bistatic terahertz impulse ranging", *Appl. Phys. Lett.* **78**. 1146 (2001b).

Reiten. M. T., McClatchey. K.. Grischkowsky. D.. Cheville. R. A.. "Incidence-angle selection and spatial reshaping of terahertz pulses in optical tunneling". *Opt. Lett.* **26**. 1900 (2001c).

Roux. J. F.. Aquistapace. F.. Garet. F.. Duvillaret. L.. Coutaz. J.-L.. "High efficiency grating coupling of THz pulse radiation into dielectric waveguide". *Electron. Lett.* **37**. 1390 (2001).

Ruffin, A. B., Decker. J., Sanchez-Palencia. L.. Le Hors, L.. Whitaker. J. F., Norris, T. B., Rudd, J. V.. "Time reversal and object reconstruction with single-cycle pulses", *Opt. Lett.* **26**. 681 (2001).

Saitow. K.-I., Ohtake. H., Sarukura, N.. Nishikawa. K.. "Terahertz absorption spectra of supercritical CHF_3 to investigate local structure through rotational and hindered rotational motions". *Chem. Phys. Lett.* **341**. 86 (2001).

Schall. M.. Walther. M.. Jepsen. P. U.. "Fundamental and second-order phonon processes in CdTe and ZnTe", *Phys. Rev. B* **64**. 1 (2001).

Sekine. N.. Hirakawa. K., Vosseburger. M.. Bolivar P. H.. Kurz H.. "Crossover from coherent to incoherent excitation of two-dimensional plasmons in $GaAs/Al_xGa_{1-x}As$ single quantum wells by femtosecond laser pulses". *Phys. Rev. B* **64**. 1 (2001).

Shan. J., Weiss. C.. Wallenstein. R.. Beigang. R.. Heinz. T. F.. "Origin of magnetic field enhancement in the generation of terahertz radiation from semiconductor surfaces", *Opt. Lett.* **26**. 849 (2001).

Smye, S. W.. Chamberlain. J. M.. Fitzgerald. A. J.. Berry. E.. "The interaction between terahertz radiation and biological tissue". *Phys. Med. Biol.* **46**. R101 (2001).

Starikov. E.. Shiktorov, P.. Gruzinskis. V.. Reggiani. L.. Varani, L.. Vaissiere. J. C.. Zhao, J. H.. "Monte Carlo simulation of the generation of terahertz radiation in GaN". *J. Appl. Phys.* **89**. 1161 (2001).

Tonouchi. M.. "Magnetic flux quanta in $YBa_2Cu_3O_{7-\delta}$ thin-film loops controlled by femtosecond optical pulses". *Japan. J. Appl. Phys.* **40**. L542 (2001).

Thorsmelle. V. K.. Averitt, R. D.. Maley. M. P.. Bulaevskii. L. N.. Helm. C.. Taylor. A. J.. "C-axis Josephson plasma resonance observed in $Tl_2Ba_2CaCu_2O_8$ superconducting thin films by use of terahertz time-domain spectroscopy". *Opt. Lett.* **26**. 1292 (2001).

Tze-An-Liu, Kai-Feng-Huang. Ci-Ling-Pan. Shingo-Ono. Ohtake. H.. Sarukura. N.. "Generation of THz radiation from resonant absorption in strained multiple quantum wells in a magnetic field". *Japan. J. Appl. Phys.* **40**. 681 (2001).

Unterrainer. K.. Kersting. R.. Bratschitsch. R.. Muller. T.. Strasser. G.. Heyman. J. N.. "Few-cycle THz spectroscopy of semiconductor quantum structures". *Physica E* **9**, 76 (2001).

Van-Rudd. J.. Johnson, J. L.. Mittleman, D. M.. "Cross-polarized angular emission patterns from lens-coupled terahertz antennas", *J. Opt. Soc. Am. B* **18**. 1524 (2001).

Yamashita, M., Tonouchi, M.. Hangyo. M.. "Supercurrent distribution in YBCO strip lines under bias current and magnetic fields observed by THz radiation imaging". *Physica C* **355**, 217 (2001).

Yoneda, H., Tokuyama, K., Ueda, K.-I., Yamamoto, H., Baba, K., "High-power terahertz radiation emitter with a diamond photoconductive switch array", *Appl. Opt.* **40**, 6733 (2001).

Weiss, C., Torosyan, G., Avetisyan, Y., Beigang, R., "Generation of tunable narrowband surface-emitted terahertz radiation in periodically poled lithium niobate", *Opt. Lett.* **26**, 563 (2001a).

Weiss, C., Torosyan, G., Meyn, J. P., Wallenstein, R., Beigang, R., Avetisyan, Y., "Tuning characteristics of narrowband THz radiation generated via optical rectification in periodically poled lithium niobate", *Opt. Express* **8**, 497 (2001b).

Zhang, W., Zhang, J., Grischkowsky, D., "Quasioptic dielectric terahertz cavity: coupled through optical tunneling", *Appl. Phys. Lett.* **78**, 2425 (2001).

Burke, P. J., Eisenstein, J. P., Pfeiffer, L. N., West, K. W., "An all-cryogenic THz transmission spectrometer", *Rev. Sci. Instrum.* **73**, 130 (2002).

Herrmann, M., Tani, M., Sakai, K., Fukasawa, R., "Terahertz imaging of silicon wafers", *J. Appl. Phys.* **91**, 1247 (2002).

Ignatov, A. A., Klappenberger, F., Schomburg, E., Renk, K. F., "Detection of THz radiation with semiconductor superlattices at polar-optic phonon frequencies", *J. Appl. Phys.* **91**, 1281 (2002).

Johnston, M. B., Whittaker, D. M., Corchia, A., Davies, A. G., Linfield, E. H., "Theory of magnetic-field enhancement of surface-field terahertz emission", *J. Appl. Phys.* **91**, 2104 (2002).

Kawase, K., Minamide, H., Imai, K., Shikata, J., Ito, H., "Injection-seeded terahertz-wave parametric generator with wide tunability", *Appl. Phys. Lett.* **80**, 195 (2002).

Kohler, R., Tredicucci, A., Beltram, F., Beere, H. E., Linfield, E. H., Davies, A. G., Ritchie, D. A., "High-intensity interminiband terahertz emission from chirped superlattices", *Appl. Phys. Lett.* **80**, 1867 (2002).

Loffler, T., Roskos H. G., "Gas-pressure dependence of terahertz-pulse generation in a laser-generated nitrogen plasma", *J. Appl. Phys.* **91**, 2611 (2002).

Mitrofanov, O., Mark-Lee, Pfeiffer, L. N., West, K. W., "Effect of chirp in diffraction of short electromagnetic pulses through subwavelength apertures", *Appl. Phys. Lett.* **80**, 1319 (2002).

Nagel, M., Bolivar, P. H., Brucherseifer, M., Kurz, H., Bosserhoff, A., Buttner, R., "Integrated THz technology for label-free genetic diagnostics", *Appl. Phys. Lett.* **80**, 154 (2002).

Ryzhii, V., Khmyrova, I., Shur, M., "Terahertz photomixing in quantum well structures using resonant excitation of plasma oscillations", *J. Appl. Phys.* **91**, 1875 (2002).

Shikata, J. I., Kawase, K., Taniuch, T., Ito, H., "Fourier-transform spectrometer with a terahertz-wave parametric generator", *Japan. J. Appl. Phys.* **41**, 134 (2002).

Sinyukov, A. M., Hayden, L. M., "Generation and detection of terahertz radiation with multilayered electro-optic polymer films", *Opt. Lett.* **27**, 55 (2002).

Tani, A., Watanabe, M., Sakai, K., "Photoconductive twin dipole antennas for THz transceiver", *Electron. Lett.* **38**, 5 (2002).

Van-Rudd, J., Mittleman, D. M., "Influence of substrate-lens design in terahertz time-domain spectroscopy", *J. Opt. Soc. Am. B* **19**, 319 (2002).

West, K. W., Pfeiffer, L. N., "Terahertz electroluminescence from superlattice quantum cascade structures", *J. Appl. Phys.* **91**, 3526 (2002).

Williams, G. P., "Far-IR/THz radiation from the Jefferson Laboratory, energy recovered linac, free electron laser", *Rev. Sci. Instrum.* **73**, 1461 (2002).

Spectroscopy
in the Terahertz Spectral Region

Frank C. De Lucia

Abstract. The principal applications of the THz region of the electromagnetic spectrum have been molecular spectral analysis and the astronomical and atmospheric remote sensing which have grown out of these laboratory activities. In this chapter we discuss the physical basis for the interctions between matter and radiation in the THz and the theoretical underpinnings of molecular rotational spectroscopy. Laboratory techniques, with an emphasis on the development of sources that are appropriate for spectroscopy, are presented, along with representative results. Finally, recently developed systems for atmospheric remote sensing, astrophysical studies, and analytical chemistry are presented.

1 Introduction

The genesis of microwave, and ultimately THz, spectroscopy, was the wartime development of microwave radar. However, this development was greatly aided by a fortuitous (for microwave spectroscopy at least!) accident that placed a previously unknown transition of water (the 6_{16}–5_{23} at 22 GHz) in the middle of the spectral region that was being developed at the end of the war as the next new radar band. Not only did this make vast quantities of sophisticated equipment that ordinarily would have been beyond the means of university researchers immediately available, but it also established the "relevance" of the field.

By 1948 the field was mature enough that an article in Reviews of Modern Physics [1] appeared, and by 1952 another review entitled "Microwave Spectroscopy above 60 kMC" reported work to above 125 GHz [2]. Technology advanced rapidly, and by 1954 the submillimeter threshold at 300 GHz had been passed [3]. This drive toward ever higher frequencies was aided by the rapidly increasing absorption strengths of the spectra of many of the most important small fundamental molecules (e. g. H_2O, O_2, CO, HCl, and O_3) which lay there.

A strong argument can be made that this scientific focus led spectroscopists toward the development of practical and robust "THz" technologies, including the crossed waveguide harmonic generator [4], electronic frequency control systems [5], quasi-optical propagation, and the exploitation of sensitive detectors to complement the harmonic-generation sources [6].

While microwave spectroscopy had a number of other streams, in the end this drive to higher frequency to observe the most fundamental species provided the enduring legacy for the field. Because these small species were not only scientifically fundamental but also pervasive in many physical and chemical systems, the strong interactions in the THz region have led to a number of important applications. In later sections we shall discuss some of them.

At appropriate places in this chapter, links are included to particularly useful web sites, which can be consulted for additional information. Since web addresses are prone to change, we shall attempt to keep an updated list on our web site http://www.physics.ohio-state.edu/ uwave/front.html.

1.1   Radiation and Matter

At a fundamental level, what distinguishes one spectral region from another is how radiation interacts with matter. While the underlying electromagnetic wave theory simply scales with frequency and wavelength, we associate dramatically dissimilar phenomena with different regions of the spectrum, ranging all the way from audio signals that drive classical vibrations we can hear and feel to cosmic rays.

So what is it that distinguishes the THz spectral region: what kinds of science, technology, and applications have arisen, and what kinds of scientific and technological applications can we foresee? One THz corresponds to an energy of 0.004 eV, a temperature of 50 K, a wavelength of 0.3 mm, and, in common spectroscopic terms, 33.3 cm^{-1}. If the commonly used bounds extending from perhaps 0.1 THz (near where a large proportion of current THz work is done) to 10 THz are adopted, the corresponding temperature scale ranges from 5 K to 500 K.

We shall argue below that the relative size of $h\nu$ and kT is important. Thus, these wide definitions of the THz regime include both the $h\nu/kT \gg 1$ and $h\nu/kT \ll 1$ limits, with a correspondingly broad range of phenomena. Likewise, this definition of the THz regime includes wavelengths from 3 mm to 30 μm. Thus, size considerations (whose scales are ordinarily set by the wavelength) lead to low-order-mode (i.e. microwave) devices in the longer-wavelength portions of this region and to high-order-mode (i.e. laser) devices at the shorter wavelengths. For now, we shall simply note here that jumping what is sometimes referred to as the "gap in the electromagnetic spectrum" is not equivalent to filling it.

1.2   What Phenomena Fall in this Energy Range?

To a reasonable approximation the interactions of THz radiation with matter can be divided into three categories: interactions with low-pressure gases, interactions with gases near atmospheric pressure, and interactions with liquids and solids. This separation is according to the Q of the resonances, with phenomena in the first category having a Q of $\sim 10^6$, in the second a Q of $\sim 10^2$, and in the last very low Q, or more often continuum interactions. Most of the successful applications of this spectral region fall into the first category and are dependent on the high Q for their success.

For an isolated system its linewidth $\Delta\nu$ is related to its relaxation time $\Delta\tau$ by $\Delta\nu \sim 1/\Delta\tau$. For gases, this is a reasonable approximation, and the concept of pressure broadening results from simple kinetic theory, with $\Delta\nu_{pb}$ typically $\sim 10\,\text{MHz/Torr}$. At low pressure, Doppler broadening with $\Delta\nu/\nu \sim 10^{-6}$ (1 MHz at 1 THz) sets a lower bound. However, in solids and liquids the resonant systems are not isolated or are often collective, and the linewidths are much broader.

Because of the six orders of magnitude difference in the linewidths of THz phenomena, the appropriate technology for the respective scientific studies varies widely, as does the basic physics of the phenomena involved. Thus, linewidth provides a useful and convenient classification for both the sciences associated with the THz phenomena and the appropriate technology for their study. We shall focus primarily on the high-Q/gas phase phenomena in this chapter. However, since one of the purposes of this volume is to provide cross-fertilization between the several THz communities, we shall attempt to build each category out of discussions from first principles, with extensive references and links for further exploration. We shall also explore some of the scientific and technological interface regions.

1.3   Gases

Figure 1 shows the strength of interactions as a function of frequency and molecular mass [7]. These absorptions result from the interaction of the rotation of the molecule with the radiation and exist only in gases. This figure shows that the strengths of the interactions increase as ν^2–ν^3. As a result, THz interactions are 10^3–10^6 times more intense than interactions in the conventional microwave (MW) region. Beyond a mass-dependent peak in the THz region, the interaction strength falls exponentially towards the infrared. *This very sharp peak in the THz region is one of the most important features of the spectral region and is closely related to many of the current and potential applications.*

Figure 2 shows the atmospheric propagation at sea level for a reasonably moist standard atmosphere [8]. This is both a spectroscopic problem of considerable interest and a factor of significant technological importance for THz systems. It is important to note that the vertical axis is a logarithmic plot of a logarithmic quantity and seemingly small differences are in fact very large. In this figure the rapid increase in absorption as we move out of the MW into the THz region (due primarily to water and oxygen) can be seen.

Figure 2 also shows the atmospheric absorption as a function of altitude. *This is a very important figure because it shows how rapidly atmospheric transmission changes as a function of altitude.* For example, in win-

Calculated MW Spectra for $K=0$, $B=A$

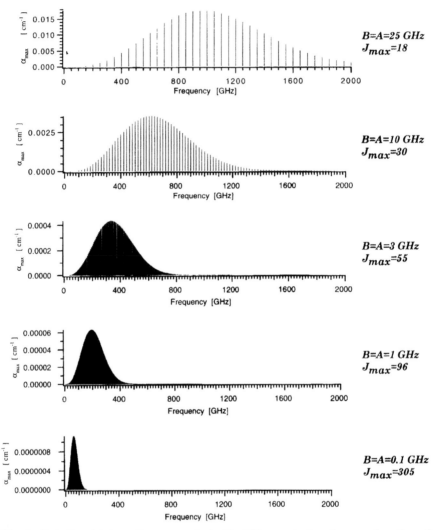

Fig. 1. Rotational interaction strengths in the THz spectral region as a function of molecular size (i. e. the molecular moments of inertia) and frequency

dows around 500 GHz, at sea level the attenuation is $\sim 100 \, \mathrm{dB/km}$ (virtually opaque), whereas at 16 km the attenuation is $\sim 0.01 \, \mathrm{dB/km}$ (so small as to be difficult to detect).

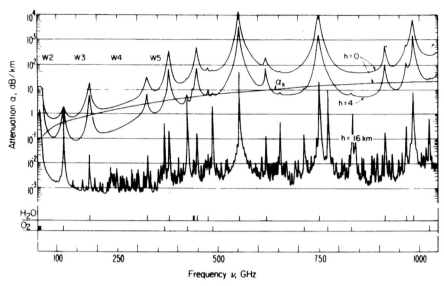

Fig. 2. Atmospheric propagation as a function of altitude in the THz spectral region. Note the rapid change with altitude, due to both the narrowing of the lines and the drying of the atmosphere with altitude

1.4   Liquids and Solids

All of the interactions discussed above are based on the interaction of molecular rotation with THz radiation. In liquids and solids there is no rotation and consequently there are no rotational spectra. However, for large molecules, collective motions are possible which result in energy-level spacings that correspond to THz frequencies. Solids and liquids are strongly interacting and as a result resonances are much broader, ordinarily leading to "continuum" spectra in the THz region. However, evidence of specific features in biological substances has been reported. While these are difficult measurements and further work needs to be done, if these reports are confirmed and the application can be developed, this is potentially very important. Although this is a speculation, it would appear to us that such spectra are most likely at relatively high THz frequencies. Since these resonances would be relatively broad, this could turn out to be an important application for some of the new laser source concepts being proposed. These operate best at high THz frequencies and initially may be rather broad.

1.5   Applications and Impact of Terahertz Spectroscopy

High-resolution THz spectroscopy has had a major impact on many important fields of science and technology. The earliest studies in this region were of species such as H_2O, O_2, NO, CH_3F. and OCS and served both to establish spectroscopic methodologies and to provide basic information about molecular structure and interactions [4,9,10]. A general understanding of the basic spectroscopic properties of these and the other small, fundamental species has been established [11 14]. Because these small, fundamental species have intrinsically interesting collisional properties, their dynamical properties have been studied as well. These studies have ranged from investigations of pressure broadening near room temperature (which are fundamental to the deconvolution of atmospheric remote sensing data) [15–18] to basic studies of the quantum nature of molecular collisions at low temperature [19–21].

Because the strength of the interactions between electromagnetic radiation and molecular rotation peaks sharply in the THz region, this spectral region has also been well suited for the study of reactive species such as free radicals [22 24] and ions [25 27], as well as weakly bound complexes [28,29]. Laboratory studies of molecular lasers and the collision-induced rotational and vibrational processes which are central to their operation have also been important [30 32].

A variety of spectroscopically based remote-sensing applications has grown out of this more basic work. Of these, two have become of major importance. The first is the study of the chemical processes in the upper atmosphere which are important in ozone formation and destruction [33–35]. Also, the vast majority of the over 100 molecular species which have been identified and studied in the interstellar medium have been observed by means of millimeter/submillimeter "radio" astronomy [36,37]. We shall discuss each of these in more detail below.

Because of these and other applications (e. g. the modeling of atmospheric propagation), the spectroscopic properties of virtually all of the important atmospheric and astronomical species have been collected into databases. These databases have become the standard for many applications and play an important role in the development of the spectral region. The *Submillimeter, Millimeter, and Microwave Spectral Line Catalog* (http://spec.jpl.nasa.gov/) has been maintained by the Jet Propulsion Laboratory for many years [13]. Likewise, the *HITRAN Molecular Spectroscopic Database* (http://www.hitran.com/) has been maintained by the US Air Force [14]. While the latter began primarily as an infrared database, the growth in both infrared and submillimeter experimental technologies has been such that for many molecular species the best spectral database results from a weighted fit of infrared and microwave data to a theoretical model.

Although most of the spectroscopic work in this spectral region has been referred to historically as millimeter and submillimeter spectroscopy, in this chapter we shall for the most part use the term THz, in keeping with the title of this book. An interesting study of the relationships among the communities that work in this spectral region can be done by using an Internet search engine to explore "THz" and "submillimeter" Boolean combined with "spectroscopy".

2   Theoretical Underpinnings

A number of excellent books and texts discuss the theoretical underpinnings of rotational spectra in detail. These include the early classic texts of Gordy, Smith, and Trambarulo [38] and Townes and Schawlow [39] as well as more recent texts by Carrington [40], Kroto [41], and Gordy and Cook [42]. Additionally, more specialized and concise reviews have appeared [43,44]. Here we shall briefly discuss the underlying physics and its implications so that we can discuss the general character of THz spectroscopy and its applications. The interested reader is referred to the aforementioned texts for methods of detailed calculation.

The physics that underlies spectroscopy in the THz region is very favorable. Briefly stated, THz spectroscopy is orders of magnitude more sensitive than spectroscopy in the adjacent microwave region, provides orders of magnitude greater resolution than does infrared spectroscopy (especially for experimental systems of comparable size and complexity) because of smaller Doppler widths, and provides absolute quantitative analysis traceable to fundamental theory.

Figure 3 shows an example of a portion of a spectrum of a molecule of moderate size and spectral complexity, nitric acid (HNO3), which can be used to illustrate the nature of the submillimeter rotational spectra observed with the fast scan submillimeter spectroscopy technique (FASSST) system described in Sect. 3. Because a single FASSST scan contains $\sim 10^6$ frequency resolution elements, it is not possible to graphically display a complete, full-band spectrum. Consequently, Fig. 3 shows a series of blowups in both frequency and sensitivity to provide a perspective. The $\sim 45\,\mathrm{GHz}$ scan shown at the top was recorded in a single $\sim 1\,\mathrm{s}$ scan, with the $300\,\mathrm{MHz}$ segment shown at the bottom recorded in $\sim 0.01\,\mathrm{s}$.

2.1   Absorption Strengths

If the absorption coefficient of the THz power P is defined by

$$\alpha = -\left(\frac{1}{P}\right)\left(\frac{\Delta P}{\Delta x}\right) \tag{1}$$

in the microwave limit (where $h\nu \ll kT$), the peak absorption coefficient between two rotational levels m and n is

$$\alpha_{mn} = \frac{8\pi^2 N F_m \nu^2}{3ckT(\Delta\nu)}\mu_{mn}^2\,, \tag{2}$$

where NF_m is the number of molecules per unit volume in state m, ν is the rotational transition frequency, $\Delta\nu$ is the linewidth, and μ_{mn} is the dipole matrix element [42]. The transition moment μ_{mn} contains contributions from the components of the permanent dipole moment along each of the principal

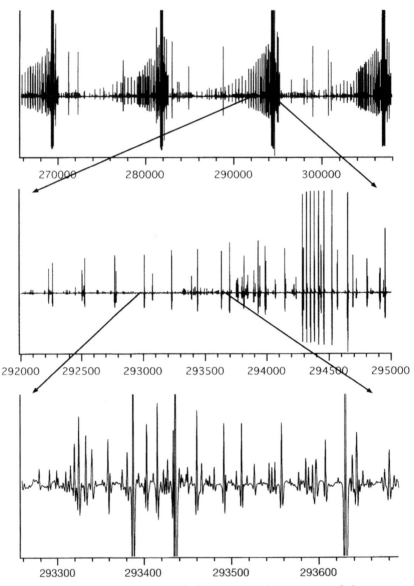

Fig. 3. A series of blowups in both frequency and sensitivity of the spectrum of HNO$_3$. The $\sim$ 45 GHz scan shown at the top was recorded in a single $\sim$ 1 s scan, with the 300 MHz segment shown at the bottom recorded in $\sim$ 0.01 s

axes of the moment-of-inertia tensor, which determine the rotational selection rules.

For optimum sensitivity, the gas pressure is adjusted in proportion to frequency so that the Doppler and pressure broadening contributions to the linewidth are equal and $N/\Delta\nu$ is independent of frequency. Because of degeneracy and rotational-partition-function effects, F_m is often proportional to ν until the declining Boltzmann population causes it to fall exponentially. These factors typically give rise to absorption coefficients that rise as ν^3 to reach a maximum at some optimum frequency in the THz range, before declining exponentially. This effect can be seen in Fig. 1 above.

2.2   Energy Levels and Transitions Frequencies

The rotational energy levels of a molecule result from the quantization of its rotational kinetic energy

$$E_{\rm r} = \frac{1}{2}\left(\frac{P_x^2}{I_x} + \frac{P_y^2}{I_y} + \frac{P_z^2}{I_z}\right), \tag{3}$$

where the P_j are the components of the molecular angular momentum and the I_j are the components of the principal moments of inertia. All real molecules have additional effects (e. g. centrifugal distortion, perturbations, and internal rotations) which significantly complicate the spectroscopic problem [42], but which have minimal impact on the overall character of the spectra.

The fundamental underpinnings of the high specificity of THz spectroscopy are (1) that the rotational degree of freedom is unique in that many levels are thermally populated, and (2) that the strong fundamental rotational transitions which arise from these levels are not associated with functional groups which may be constituents of many similar molecules. Rather, they depend upon the global moment-of-inertia tensor I of the molecule. Since THz spectroscopy is sensitive to changes in each of the I_j of $< 1/10^7$, each molecule has a unique signature if even a few of the lines of its rotational structure can be detected and measured.

2.3   The Character of Rotational Spectra

For the most general case of an asymmetric rotor, none of the I_j are equal, and the quantized rotational energy levels become

$$E = \frac{1}{2}(B+C)J(J+1) + \left[A - \frac{1}{2}(B+C)\right]W_{J_\tau}(b_p), \tag{4}$$

where A, B, and C are constants inversely proportional to the I_j's, and $W_{J_\tau}(b_p)$ is a complicated function related to the degree of asymmetry of the molecule and angular-momentum projection quantum numbers [42]. However, in the limit of a prolate symmetric top ($B = C$), $W_{J_\tau}(b_p) = K^2$, and it is possible to use this relation along with the dipole selection rule $J \rightarrow J+1$, $\Delta K = 0$ to illustrate the general character of rotational spectra.

Figure 1 above plots the absorption coefficients at $300\,\mathrm{K}$ for the $K = 0$ component of each $J \rightarrow J + 1$ transition for symmetric tops of various B

values. For rigid symmetric tops, the $K \neq 0$ components are degenerate with the $K = 0$ transitions, but for real molecules they are slightly (a few MHz) displaced owing to centrifugal distortion effects. In this symmetric-top limit, it is straightforward to show that the values of J and ν for which the rotational lines are strongest are given by

$$J_{\mathrm{opt}} \sim 5(T/B)^{1/2}, \tag{5}$$

$$\nu_{\mathrm{opt}} \sim 2B + 11(BT)^{1/2}. \tag{6}$$

where T is the temperature in kelvin and B a rotational constant in GHz.

For asymmetric tops the $K \neq 0$ components are more widely spread according to their asymmetry and (4), but on average have similar transition strengths. Additionally, because of the lower symmetry of asymmetric rotors, components of the dipole moment can exist along each of the three axes, and a new set of selection rules and spectra will appear for each. Although this can lead to a spectrum of great complexity, the character of the spectra of symmetric tops is representative of the average spectral density, distribution in frequency, and strength of lines. Although the complete spectral assignment of asymmetric rotors can be a complex task, from many perspectives it is unnecessary because either a reference spectrum can be recorded and archived or any of several spectral catalogs can be consulted [13,14].

Figure 1 also shows that as B decreases, there are more thermally populated transitions, and that the lines are more closely spaced and weaker. Calculations show that the average spacing of rotational lines becomes equal to the Doppler width for molecules whose rotational constants are $\sim 0.1\,\mathrm{GHz}$. For molecules whose atoms have mass distributions similar to those of HNO_3, the rotational constants are $\sim 0.1\,\mathrm{GHz}$ for molecules of mass $\sim 1000\,\mathrm{amu}$.

Figure 4 also illustrates another important feature of rotational spectra in the THz region: the contributions from pure rotational transitions in excited vibrational states. Because methyl formate ($HCOOCH_3$) has larger rotational constants (A, B, C) than does nitric acid (HNO_3), it has a sparser pure rotational spectrum. However, because it has low-lying ($< kT$) torsional and vibrational states, many of these contribute spectra of intensity similar to that of the ground state and contribute to the much denser spectrum shown in the figure. It is likely that molecules of $\sim 1000\,\mathrm{amu}$ will have many low-lying, thermally populated states, thereby leading to a more congested spectrum than in the simple estimate above. On the other hand, molecules of $\sim 1000\,\mathrm{amu}$ made of heavier atoms than those of HNO_3 will have larger rotational constants and less dense spectra. This large-molecule limit is an interesting and as yet unexplored topic in rotational spectroscopy, which will ultimately place a limit on the mass and size of molecules that have highly specific rotational signatures.

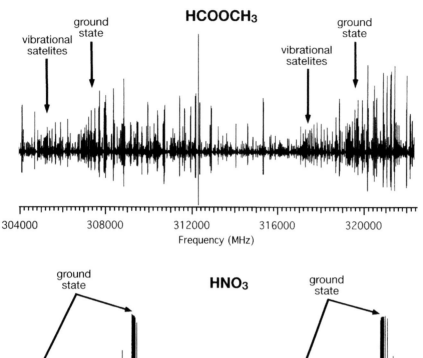

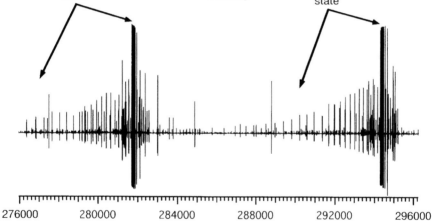

Fig. 4. Excited vibrational states in $HCOOCH_3$ and HNO_3

2.4 Rotation–Vibration Spectra

Although rotational spectra are unique to molecules, molecules also have spectra associated with their electronic, vibrational, and nuclear degrees of freedom. It is fortuitous that for most species the characteristic energy associated with each of these is separated from its neighbor by perhaps two orders of magnitude. Quantum mechanically, this allows the molecular wavefunction to be separated into parts and each solved separately in the context of an effective Hamiltonian. Classically, a similar picture evolves, with the much higher speed of the more energetic motions providing average potentials for their less energetic neighbors. In both pictures, the rapid electronic motion provides an average electrostatic potential in which the nuclei vibrate, the average positions of the vibrating nuclei provide the moments of rotational inertia, etc.

This large separation in energy also leads to a relation between each degree of freedom and a portion of the electromagnetic spectrum: between the electronic degree of freedom and the optical portion, the vibrational and the infrared, the rotational and the microwave, and the nuclear hyperfine interactions and the radio. As a result, rather independent communities have grown up around each of these combinations. However, while still rather distinct communities, the rotational (microwave) and vibrational (infrared) are being brought together by technological advances. In the early days, the resolution of infrared spectrometers was too low to resolve the rotational structure of any except the lightest species (e. g. H_2O), and microwave techniques did not have the frequency coverage to measure significant rotational spectra except in much heavier molecules. However, now FTIR and diode laser techniques can resolve the Doppler limit ($\sim 100\,MHz$) and THz technologies have very wide spectral coverage.

As a result, it is now common for "infrared" analyses to contain measured microwave data in large weighted least-squares analyses. Nitric acid provides a useful specific example. For this species, infrared techniques have been used to study the rotational structure of all of its fundamental vibrations. THz techniques have now been used to study not only the molecules which reside in the ground vibrational state, but also the rotational structure of molecules in many of the excited vibrational states. While the thermal population in the excited states is reduced by the Boltzmann factor (e. g. the ν_5 vibrational state lies at $\sim 900\,cm^{-1}$ and has a thermal population of $\sim 1\,\%$), the resolution and sensitivity of THz spectroscopy makes the study of these small populations reasonably straightforward. For example, Fig. 5 shows THz spectra of a number of these transitions, which were recorded with a FASSST spectrometer with $\sim 10^{-6}\,s$ of integration time. For molecules such as HNO_3 for which it is possible to record THz spectra over a large portion of the thermally populated rotational spectra of the molecule, weighted infrared–THz fits are completely dominated by the higher-accuracy THz data (the weighting ratio is typically 10^6). In fact, it has been shown that an analysis of the THz data can predict (except for the central vibrational frequency itself!) the entire infrared spectra to unprecedented accuracy. While this is useful for species such as HNO_3 (especially in highly congested and perturbed regions of its spectrum), its real promise lies in somewhat heavier species (e. g. $ClONO_2$) whose spectra are unresolved or marginally resolved in the Doppler limit in the infrared. The detailed ro-vibration structure of these species is of considerable interest for the recovery of atmospheric remote-sensing information. This requires accurate modeling of the pressure and temperature variation of the recovered profiles. These can only be obtained by a detailed

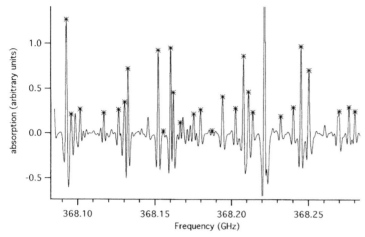

Fig. 5. Spectrum of HNO_3, showing the absorptions in the excited ν_5 vibration states marked with *crosses*

line model, which can be synthesized from analyses of the THz rotational structure in both the ground and the upper vibrational state of the infrared band of interest.

3 Spectroscopic Techniques and Results

As is well known, the THz spectral region is by far the least explored portion of the electromagnetic spectrum, largely because of the difficulty of generating and detecting radiation at these frequencies. In fact, Townes has pointed out that his original motivation for the development of the maser/laser was his desire to make a "molecular generator" to overcome this problem [45]. Although thermal, or nearly thermal [46], photons have always existed in the THz, the problem is one of generating "appropriate" radiation, not only in terms of power, but also spectral purity. Since other chapters in this volume address the technical issues of the generation and detection of radiation in more detail, we shall focus here on those properties important for the development of spectroscopic applications and for the development of useful spectroscopic systems.

Because there is an intimate interplay between the development of appropriate THz sources and the spectroscopic studies themselves, we shall also use this section to discuss laboratory spectroscopic results. These will include studies of the basic spectroscopy of small, fundamental species, the production and study of free radicals and ions, and the study of collisional processes.

Figure 6 shows an overview of the power available from representative sources as a function of frequency [47]. While at first it might seem that these power levels are so low as to preclude scientific work over much of the

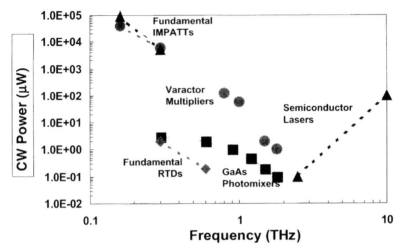

Fig. 6. Power available from solid-state sources as a function of frequency

region, this is far from the case. Most spectroscopic measurements are linear. and as a result it is possible to trade detector sensitivity for source power.

The initial measurements in this spectral region were made by the use of point contact detectors [4], which in their more modern microelectronic fabricated form [48] have been rechristened as "Schottky barrier" diodes and are still an important tool. The next important step in the development of spectroscopic systems in this region was the introduction of cryogenic detectors [6], especially the InSb hot-electron bolometer because of its speed ($\tau <$ 10^{-6} s) and sensitivity ($\sim 10^{-12}$ W/Hz$^{-1/2}$). Additionally, the much higher reliability of the cryogenic detectors made work in this spectral region much less of a "black art" and made THz science more accessible to nonspecialists.

In most configurations, the sensitivity of the InSb detectors begins to decrease above ~ 300 GHz and the slower ($\sim 10^{-3}$ s) Si and Ge bolometers become more advantageous in many systems [49]. Additionally, bolometers rapidly gain sensitivity with decreasing temperature and cooling them to ^{3}He temperatures (~ 300 mK) can increase their noise-equivalent power (NEP) to $\sim 10^{-15}$ W/Hz$^{1/2}$ [50]. In this limit these detectors approach the sensitivity limit set by the fluctuations in the black-body background [51–54].

Heterodyne detectors can be even more sensitive and in many cases can approach the quantum limit. However, for laboratory spectroscopy they in some sense beg the question of available power because of their requirements for local-oscillator power. However, for remote-sensing applications they have been developed to a very high degree, and we shall discuss them in this context in Sect. 4 below.

A very large majority of THz studies have taken advantage of the high Q's ($\sim 10^6$) associated with spectral lines of low-pressure gases. Consequently, correspondingly high spectral purity has been a requirement for most THz laboratory spectroscopic systems, as well as for their corresponding field applications.

Approaches to the high-Q source problem have included a series of advances in nonlinear frequency multiplication and cooled-detector development [4,6,50,55], the extension of fundamental electron beam oscillators [56,57] and fundamental solid-state oscillators to higher frequency [58,59], optical heterodyne down conversion [60–64], the production of microwave sidebands on FIR laser sources [65–69], and the demodulation of femtosecond laser pulse trains [70,71]. In the following sections we shall discuss those techniques which have been the most widely used, to illustrate results typical of high-resolution spectroscopy and its applications in the THz region. Because virtually all of the systems we shall discuss make use of the cryogenic detectors discussed above, we shall organize this section according to source technology.

3.1 Harmonic Generation

Harmonic generation was the basis for the initial explorations of the THz spectral region and for many applications continues to be the technology of choice. In combination with sensitive detector technology, it provides broad spectral coverage, simple frequency control and measurement, high resolution and sensitivity, and overall system reliability and simplicity.

The most important early development was the invention of the cross waveguide harmonic generator with an embedded nonlinear diode in the microwave structure [4]. Figure 7 shows a drawing of this device. Since it was designed for spectroscopy, broad spectral coverage was emphasized and only back shorts were included for tuning. Not only are these multipliers broadbanded across an entire waveguide input band, but many harmonics are generated simultaneously. Since linear molecules have series of absorptions in

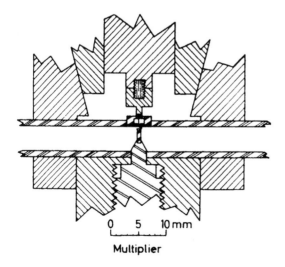

0   5   10 mm

Multiplier

Fig. 7. Cross waveguide harmonic generator with embedded nonlinear element

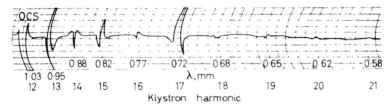

Fig. 8. "Harmonic series" of OCS absorptions. Shown in this figure are absorptions in the 12th (291 GHz) through 21st harmonics (517 GHz) of a klystron operating near 24.325 GHz

almost, but not exactly, harmonic relation. the multiple harmonic output can be separated and observed by scanning the input frequency of the multiplier through the closely spaced subharmonics of the different absorption frequencies. The resultant closely spaced (in source frequency) absorptions are shown in Fig. 8 [11,72]. Similar devices, some with matching optimized for narrow spectral intervals and operating in chains of low-order multiplication, are the sources of choice for many applications requiring local-oscillator power for receivers in remote-sensing instruments [73]. Examples of these instruments will be discussed in more detail in Sect. 4 below.

Because absorption spectroscopy is linear, increased detector sensitivity can be used to compensate for reduced source power. Thus the introduction [6,50] of sensitive cryogenic detectors [52,54,74] resulted in a significant advance for THz spectroscopy. The reliability, spectroscopic flexibility, and relative simplicity of systems similar to that shown in Fig. 9 led to their wide adoption by the community and a significant increase of spectroscopic activity in the THz region. With systems of this type, the spectra of most of the small, fundamental species that are ubiquitous in both man-made and natural systems have been studied. These include H_2O, O_2, O_3, CO, HOOH, NO, N_2O, NO, HCN, HNO_3, SO_2, NH_3, HCl, ClO, H_2CO, H_2S, CH_3OH, $HCOOCH_3$ and many more.

The resolution of most THz spectroscopy is limited by Doppler broadening to a Q of about 10^6. While this is satisfactory for most applications, higher resolution is often important. As in many Doppler-limited applications, molecular-beam approaches have been used in the THz both for molecular beam masers [75,76] and for molecular-beam electric-resonance machines. An especially nice application of the latter used Ramsey's method of separated oscillating fields to obtain a line Q of $\sim 10^9$ [77]. Figure 10 shows a spectrum obtained with this device.

Another important application of THz spectroscopy has been the study of molecular ions. Not only is their production of fundamental interest, their relative abundance in the interstellar medium has made them of significant astrophysical interest as well [78]. After the demonstration by Woods and his coworkers [25] that the positive column of a glow discharge contained observable quantities of molecular ions, a number of workers were attracted

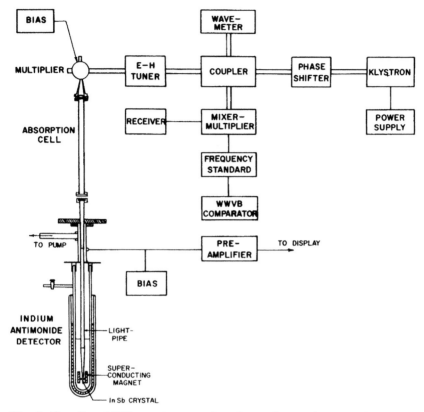

Fig. 9. Broadband THz spectrometer based on a harmonic-generation source and cryogenic detector

to the field. The development of the magnetically enhanced negative glow (MENG) ion source increased concentrations by two orders of magnitude [27], and this source has been used by a number of workers [79–83]. Figure 11 shows the cell and Fig. 12 shows the enhancement as a function of applied magnetic field.

The low density of the interstellar medium, coupled with energetic fluxes of photons and electrons, leads not only to a relative abundance of molecular ions but also of other reactive species. Among the most interesting of these have been species with long-carbon-chain backbones and rings studied by Thaddeus and his coworkers [84,85].

A second branch of laboratory astrophysics has involved studies of the collisional properties of astrophysically important species. In contrast to the spectroscopy just discussed (for which the line frequencies do not depend upon the laboratory environment), collision-induced spectroscopic properties (e.g. linewidths and inelastic transition rates) are fundamentally dependent upon the environment. Because interstellar temperatures are typically too

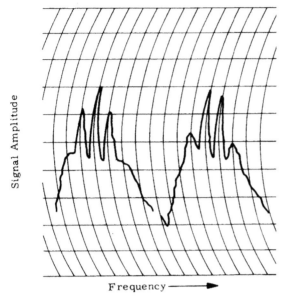

Signal Amplitude

Frequency ⟶

Fig. 10. Ramsey fringes in the spectrum of H$_2$S obtained with a molecular-beam electric-resonance system

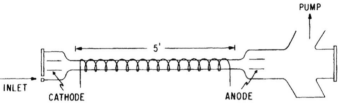

PUMP

5'

INLET

CATHODE   ANODE

Fig. 11. Cell for the study of molecular ions based on the magnetic enhancement of the negative-glow region of a discharge

low to allow adequate vapor pressure for experimental studies in the laboratory, much of the work has been theoretical work based on quantum-chemical scattering calculations [86]. However, a technique, namely collisional cooling, has been developed which allows direct measurements of collisions between gas phase species at very low temperatures [19]. This technique has been extended to allow the study of molecular ions as well [20] and direct observations of inelastic scattering rates have now been made [21].

Figure 13 shows such a system, including its vacuum enclosure, cryogen reservoirs, and sample gas injector. At the heart of this system is the collisional-cooling cell shown in Fig. 14. This cell is initially filled with either He or H$_2$ and cryogenically cooled to low temperature. Small amounts of the spectroscopically active gas (e. g. CO) are then injected into the cell at a temperature for which the sample has adequate vapor pressure. Upon collision with the He or H$_2$ the spectroscopically active gas rapidly ($\sim$ 10–100 collisions) cools to the temperature of the buffer gas. However, at typical operating pressures ($\sim$ 10 mTorr) the spectroscopic gas takes $\sim 10^4$ collisions

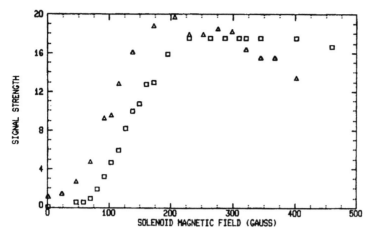

Fig. 12. Spectroscopic gain that results from the application of an axial magnetic field to an abnormal glow discharge for HCO$^+$ (*triangles*) and NH$_2^+$ (*squares*)

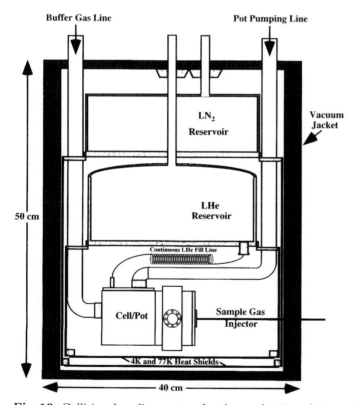

Fig. 13. Collisional-cooling system for the production of "space in a bottle"

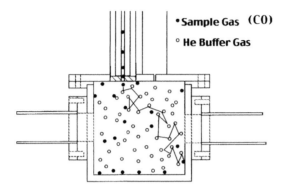

Fig. 14. Collisional-cooling cell

to reach the cell wall and condense. It is interesting that much the same quasi-equilibrium exists in the interstellar medium, with the cold dust grains there playing the role of the walls for the condensation of the spectroscopic gas. However, the interstellar timescale is of the order 10^5–10^6 years rather than a few milliseconds. The latter time scale has certain advantages for the completion of PhD theses! Similar systems have been developed with liquid nitrogen as a coolant, which are appropriate for the study of systems under conditions similar to those in the atmospheres of the outer planets [87].

Figure 15 shows a comparison between pressure broadening and rotationally inelastic cross sections for the 1_{10}–1_{01} transition of H_2S in collision with He. The scientifically interesting result shown here is the divergence between the cross sections at low temperature. This is a direct result of the transition between an essentially classical collision process at high temperature and a distinctly quantum mechanical one at low temperature.

Perhaps the most dramatic result in this field has been produced by Willey and his coworkers [88], who have shown that an ammonia maser can be produced in such a cell as a result of differential rotational relaxation. Figure 16 shows a comparison of the emission at 10 K with the absorption at 35 K. This result is particularly interesting, both in that ammonia is observed as a maser in the interstellar medium and in that the result represents an experimental realization of an early "thermal" maser concept.

Spectroscopy in the THz region can be a powerful probe of molecular systems. In addition to the spectroscopic study of energy levels, the THz region has been used to probe both discharge [31] and optically pumped lasers [89,90], as well as chemical reactions in plasmas and collision dynamics in systems far from thermal equilibrium [91]. Figure 17 shows a recent result which combines many of these experimental technique to establish an experimentally well-characterized environment for the study of inelastic collisions of molecular ions at very low temperatures.

More specifically, this figure shows an example of the results of a THz study of the chemistry and physics which result from a pulsed flux of electrons being focused through a mixture of H_2 and CO at $\sim 50\,K$ to form HCO^+.

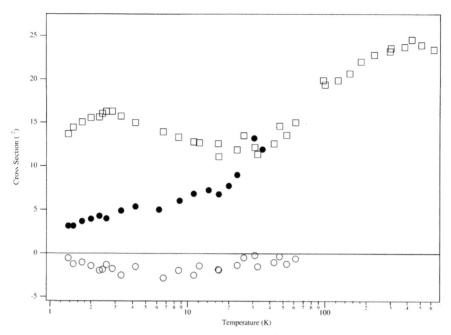

Fig. 15. A comparison between pressure broadening (*open squares*) and rotationally inelastic (*solid circles*) cross sections for the 1_{10}–1_{01} transition of H_2S in collision with He. The open circles are the pressure shift cross sections

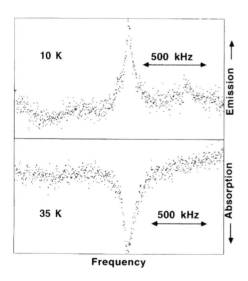

Fig. 16. The $(J = 4,\ K = 3)$ inversion transition of NH_3 at temperatures of 10 and 35 K

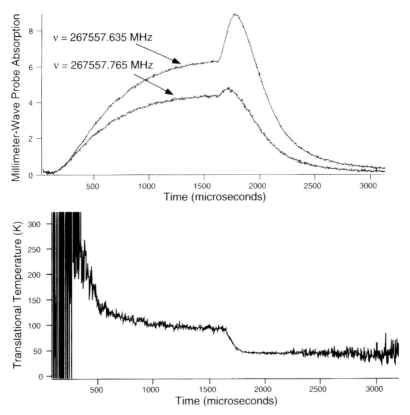

Fig. 17. THz probe of a pulsed-e-beam-induced ion molecule reaction in a low-temperature H_2–CO mixture, showing (*upper portion*) the initial formation of HCO^+ (the initial rise in probe absorption), the rotational and translational cooling of the average temperature after the end of the e-beam pulse (the sharp rise at 1.8 ms), and the eventual recombination of the HCO^+ and the slow electrons. The *lower portion* shows the translational temperature calculated from the Doppler lineshape information provided by the two probes shown in the upper portion, which probe different parts of the Gaussian lineshape

This experimental system is a combination of that shown in Figs. 11 and 13 except that the electron flux is produced by an externally gated e-beam source rather than an abnormal glow discharge. The lower trace, which shows the measured translational temperature, is particularly interesting. Because the ions are formed with high translational temperature, early in the pulse the average over the ions in the cell is dominated by the hotter ions that have just been formed. Later an equilibrium average is reached. When the electron flux ends, the average translational temperature approaches that of the CO and H_2. It is satisfying that this latter temperature is in good agreement with the results of standard thermometry applied to the cell walls.

3.2   Sources Based on Mixing of Optical Sources

It has also long been recognized that difference frequency mixing of "optical" sources could be used to produce THz radiation [92,93]. Although the concept is straightforward, practical implementations depend upon the frequency stability and calibration of the laser sources as well as upon the efficiency of the mixing process.

One of the earliest spectroscopically successful implementations is shown in Fig. 18. This system was based on the mixing of two Lamb-dip-stabilized CO_2 lasers, along with a microwave synthesizer to provide tunability [49,60]. Because the frequencies of the CO_2 lasers are about two orders of magnitude higher than that of the THz radiation produced, this system requires careful stabilization of each laser to provide measurement accuracy. However, because it is possible to choose CO_2 laser lines that are widely separated, it is possible to get wide spectral coverage, with good sensitivity. Figure 19 shows a nice example of the observation of the electronic quadrupole hyperfine structure of DI [94].

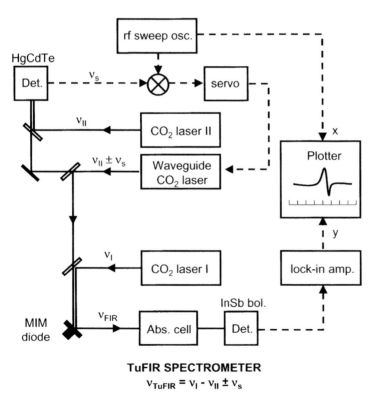

TuFIR SPECTROMETER

$$\nu_{TuFIR} = \nu_I - \nu_{II} \pm \nu_s$$

Fig. 18. THz spectrometer based on the difference frequency mixing of two stabilized CO_2 lasers

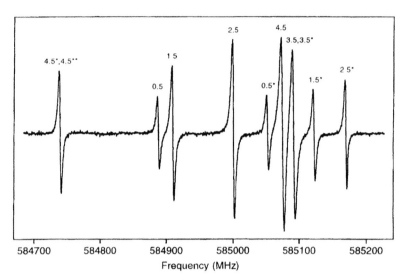

Fig. 19. The $J = 3$-2 transition of DI, which shows the nuclear hyperfine structure due to ^{127}I

Tunable laser sources in the visible and near infrared have also been used for difference-frequency mixing sources. These have typically used photoconductive mixers in which radiating THz antennas have been integrated with the semiconductor switch [61,63,64,95,96]. Since these sources operate at two to three orders of magnitude higher frequency than the THz radiation produced, a number of frequency control and measurement schemes have evolved.

Figure 20 shows an example of a spectrum taken with a system in which two dye lasers were mixed in a photoconductive switch [62]. Because this work was focused on measurements of pressure broadening, it was possible to calibrate the difference frequencies by the observation of the optical fringes produced by the scanning dye laser.

Figure 21 shows an example of a spectrum taken with a system driven by an all-solid-state source, whose schematic diagram is shown in Fig. 22 [64]. This system is particularly interesting in that it provides a novel scheme for absolute frequency measurement. In this system, two CW diode lasers at 850 nm are locked to different orders of a Fabry–Pérot (FP) cavity whose free spectral range is 3 GHz, while a third is offset locked to one of the cavity-locked lasers via a tunable 3–6 GHz microwave oscillator. By means of knowledge of the free spectral range of the cavity, the difference in mode order between the two fixed, locked lasers, and measurement of the microwave offset, an absolute frequency calibration of 10^{-7} has been achieved.

"Size" as measured in wavelengths is an important limiting factor in "electronic" approaches to THz technology. However, the "dimensionality" of the device also plays an important role. Most harmonic generators, solid-state sources, and frequency converters are "0-dimensional" devices, small in all

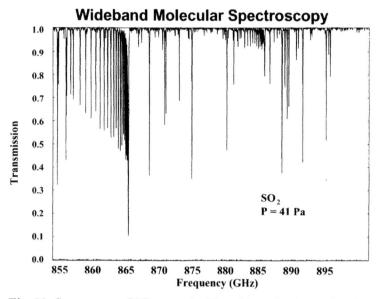

Fig. 20. Survey scan of SO_2 near the $^rQ_8$ subbranch, obtained with a photomixing source driven by two dye lasers

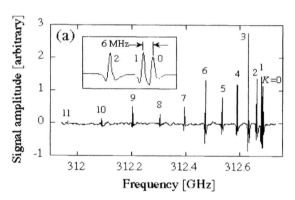

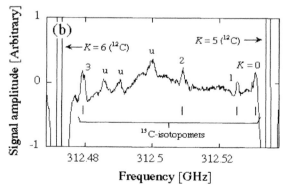

Fig. 21. Spectrum of CH_3CN $J_K = 16_K–17_K$ transition near 312 GHz (*upper portion*). The expanded trace (*lower portion*) shows ^{13}C isotopic features

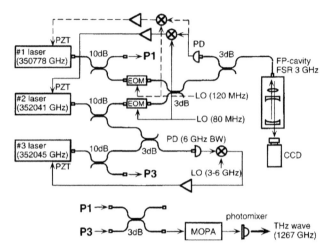

Fig. 22. An all-solid-state THz source based on photomixing. This system provides absolute frequency calibration by locking two CW diodes to different modes of a fixed Fabry–Pérot cavity, while locking a tunable diode laser to one of the fixed lasers via microwave techniques

three dimensions in comparison with the wavelength. This is also true for some electron beam tubes (e. g. klystrons), but not all. An important example of a "1-dimensional" tube is the backward-wave oscillator (BWO) discussed below. Much of its success can be traced to its macroscopic (in terms of wavelengths) length along the direction of the electron beam. Figure 23 shows an example of a solid-state mixer that takes advantage of this concept. Not only does its distributed interaction region allow the use of higher laser pump powers without burnout, but also, because the circuit elements are distributed, capacitance limitations on bandwidth are reduced [63].

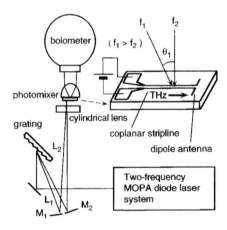

Fig. 23. A traveling-wave photomixer system

3.3   Tunable Sideband Sources

The invention of gas discharge [97] and optically pumped far-infrared (OPFIR) lasers [98] provided relatively powerful sources for the THz region. However, because these are essentially fixed-frequency sources, they are of limited spectroscopic use. Schemes based on mixing tunable microwave sources with these lasers have been developed and have proved to be useful in many applications [65–69,99].

Figure 24 shows an example of an early system [65]. In it, a crossed-waveguide harmonic generator was modified to accept drive from both the FIR laser and a sideband-generating klystron. A key element in this and similar systems is the passive filtering necessary to separate the relatively weak tunable sideband from the more powerful pump laser. Similar systems have

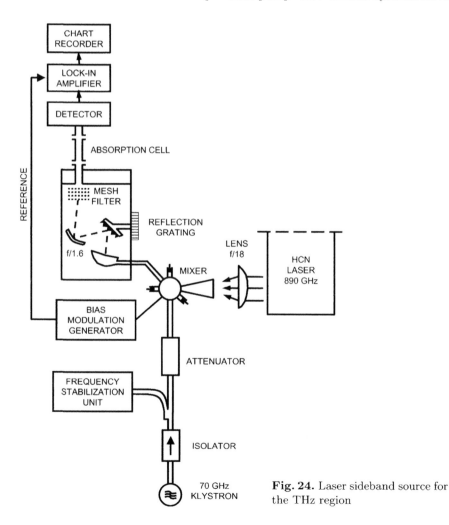

Fig. 24. Laser sideband source for the THz region

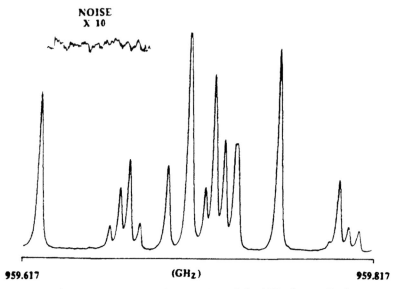

Fig. 25. The $1_{11}-0_{00}$ rotational transition of the NH_2 free radical

been used for a number of important spectroscopic observations including the study of molecular ions [26], the measurement of dipole moments of molecular ions [100], and the study of weakly bound complexes [28]. Figure 25 shows an example of the structure of the $1_{11}-0_{00}$ transition of the NH_2 free radical near 960 GHz [69].

Because the frequency of the FIR gas lasers can vary with operating conditions by 1–5 MHz, for good spectroscopic accuracy, provision to stabilize the laser frequency is ordinarily required. A recent example is shown in Fig. 26 [101]. In this system, the FIR laser is locked to a reference provided by harmonic mixing of a portion of its output with a phase-locked Gunn diode. Tunability is provided in a similar fashion as in the system shown in Fig. 24, except that broader tunability is obtained by the use of a BWO sideband generator. Figure 27 shows an example of the spectra obtained with this system.

There exists a powerful alternative to the generation of tunable sidebands from FIR lasers by the electronic methods just described: the tuning of the molecular resonances to the fixed frequencies of the FIR lasers by means of large electric (LER technique) or magnetic (LMR technique) fields [23]. Although the use of electric fields is more general in its applicability (essentially all of the species of interest have electric dipole moments, whereas few have significant magnetic moments), the use of magnetic field is much more common. This is largely because LMR is especially well suited to the study of free radicals and reactive open-shell ions and because the tunability of LER is often relatively small. Because of the fixed frequency and fundamentally

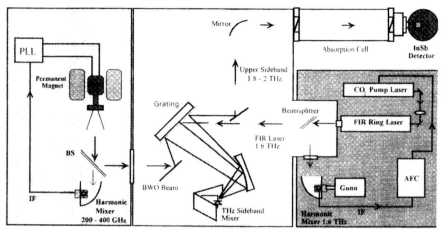

Fig. 26. A frequency-stabilized laser sideband source for the THz region

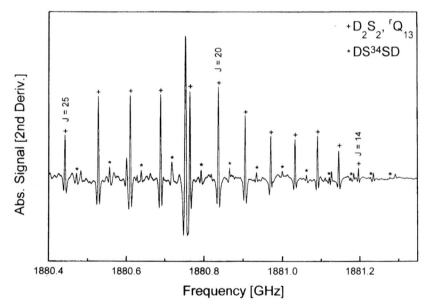

Fig. 27. A portion of the band head of the $^rQ_{13}$ branch of the slightly asymmetric internal rotor DSSD

more powerful radiation, exquisite sensitivity has been obtained from systems in which these reactive species were contained directly in the FIR laser cavity. Great complexity of the spectra is often characteristic of the method. Combined with complexity of the theoretical methods required to obtain the zero-field frequencies, this has caused the method to be focused on relatively simple species, but often reactive species, where its great sensitivity can be used to maximum advantage. A notable example of the application of LER has been the study of van der Waals bonds in the weakly bound Ar HCl complex [29.102].

3.4 Electron Beam Sources

Although the emphasis for the last several decades has been on solid-state source development, electron beam devices have played an important role in THz spectroscopy. In fact, as can be seen in Table 1 BWOs significantly exceed the broadband power-producing capabilities of solid-state sources shown in Fig. 6. Although this is beyond the scope of this chapter, it is probably worth noting that there is virtually no limit (other than economic) on the amount of power that can be produced in the THz region by e-beam sources. For example, a new free-electron laser (FEL) designed at University of California, Santa Barbara to be driven by a 2 MeV electron beam would produce 1 kW CW at 300 GHz.

Table 1. Characteristics of BWOs manufactured by ISTOK

	OB-24	OB-30	OB-32	OB-80	OB-81	OB-82	OB-83	OB-84	OB-85
Operating range (GHz)	179- 263	258 375	370 535	526- 714	667 857	789- 968	882- 1111	1034- 1250	1153- 1500
Power (mW)	1-10	1-10	1-5	1-5	1-5	1-3	1-3	0.5-2	0.5-2

As shown by Madey, FELs are formally equivalent to classical tubes. Arbitrarily high power at high frequency can be obtaining by raising the energy of the electron beam, especially into the relativistic regime, where the "size" limitations on THz sources are transcended by relativistic effects. While the limits on electron energy in solid-state sources preclude such an approach, it is likely that a functionally equivalent solution will ultimately be obtained by the integration of large arrays of individually small solid-state sources.

For the region between about 100 and 1000 GHz, BWOs [103] have been extremely useful sources. The most successful commercial devices have been produced by ISTOK (Russia) [57] and Thomson-CSF (France), although production of the latter devices ceased several years ago owing to economic/market considerations. The former devices have attracted a wide following and are particularly attractive spectroscopic sources, offering wide voltage tunability and milliwatt power levels. Table 1 shows typical characteristics. A comprehensive review of Soviet-era systems has been written [104], and a number of more recent articles have appeared [105,106].

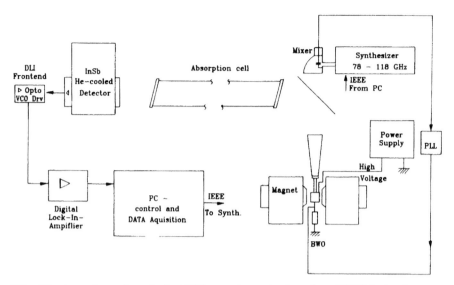

Fig. 28. A modern phase-locked THz spectrometer based on BWO technology

Figure 28 shows a modern implementation of the basic spectroscopic scheme [106]. In this system, the frequency reference is provided by a step-tunable millimeter wave synthesizer, whose output is harmonic-mixed with a portion of the BWO signal in a quasi-optical harmonic mixer. This reference is then used to phase-lock the BWO, which in turn tracks the frequency scan of the synthesizer. Figure 29 shows a comparison of a Doppler-limited spectrum obtained with this system and that from a large, high-resolution FTIR system.

Although under equilibrium conditions linewidths around 1 THz are ordinarily Doppler limited to ~ 1 MHz, the spectral purity and power of BWO sources make possible the use of saturation spectroscopy to significantly re-

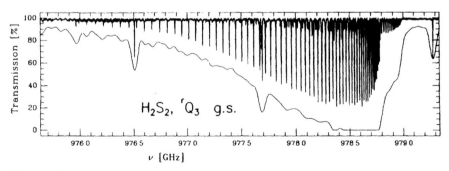

Fig. 29. Ground-state $^rQ_3$ branch of HSSH. Comparison between the instrument-limited Fourier transform and the Doppler-limited terahertz spectrum

duce this limit. In a particularly interesting demonstration. Belov and his coworkers have observed saturation features ∼ 0.05 MHz wide in the ammonia spectrum near 570 GHz [107]. Both the sensitivity and the linewidth of the system are dramatically illustrated in Fig. 30. which shows spectra for both $^{14}NH_3$ and $^{15}NH_3$ in natural abundance (0.366%).

Finally, we shall discuss in somewhat more detail a relatively new approach to THz spectroscopy. the fast-scan submillimeter spectroscopy technique (FASSST) [108]. Like the spectral region itself. it adopts methodologies

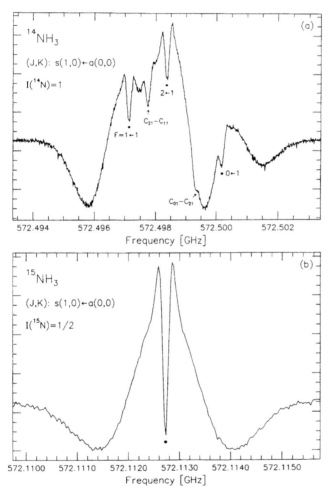

Fig. 30. Observed spectra of ammonia, showing Lamb dips and crossover resonances. The more complex structure in (**a**) is due to the nuclear hyperfine structure of the ^{14}N nucleus

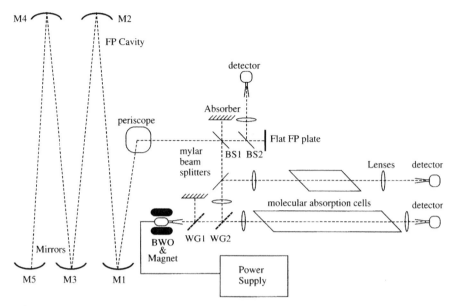

Fig. 31. The FASSST spectrometer

from both of the neighboring microwave and infrared regions. In Sect. 4 we shall discuss a particular application, analytical chemistry [7].

Figure 31 shows the FASSST system that was used to produce the series of spectral expansions shown in Fig. 3. In this example, an ISTOK OB-30 BWO is used to cover the 240–375 GHz region. Similar tubes are available from ISTOK for the ~ 100–1000 GHz region. The first wire grid polarizer (WG1) provides a well-defined polarization from the output of the overmoded BWO waveguide. The second polarizer (WG2) is used to split the output power of the BWO, with ~ 90% being directed quasi-optically through the molecular absorption cell and detected by an InSb hot-electron bolometer operating at 1.5 K. The remaining ~ 10% of the power is coupled into an FP cavity via a Mylar beam splitter (BS1), which provides fringes for frequency interpolation between reference spectral lines of known frequency. In order to provide a highly accurate basis for the analysis of the frequency–voltage characteristic of the BWO, a folded FP cavity of length ~ 38.89 m is used to provide modes every ~ 3.854 MHz. A second molecular-absorption cell that can be used for calibration purposes is also provided.

The philosophy behind this system is based on the fact that, as with most e-beam-based microwave oscillators, the source linewidth (short-term stability) of the BWO is much less than the Doppler width of a spectral line. Thus, the phase-lock approach discussed above is needed not to improve this spectral purity, but rather to stabilize against thermal and power supply-induced long-term (≥ 0.01 s) drift and ripple. It is straightforward to show that a combination of power supply development and a fast frequency scan can "freeze" these instabilities, so long as a frequency reference can be provided which will allow the BWO's frequency instability to be calibrated in software [108].

The size of the FP cavity is dictated by the need to interpolate between adjacent FP cavity modes and the details of the small-scale structure of the frequency–voltage characteristic of the BWO. Figure 32 shows a typical example. These fluctuations are caused by standing wave phenomena both within and external to the tube.

The key system elements include the following.

1. The most fundamental element of the FASSST system is the excellent short-term spectral purity of the BWO. From studies over many years, it has been observed that the short-term spectral purity of free.running ISTOK BWOs is < 20 kHz. Without this spectral purity, the FASSST system would not be possible.
2. Secondly, the BWOs can be voltage tuned continuously over an ~ 50% frequency range, which contains ~ 10^5 spectral resolution elements (Doppler limited).
3. The synthesized frequency reference system typical of high-resolution submillimeter spectrometers is replaced by a system more typical of opti-

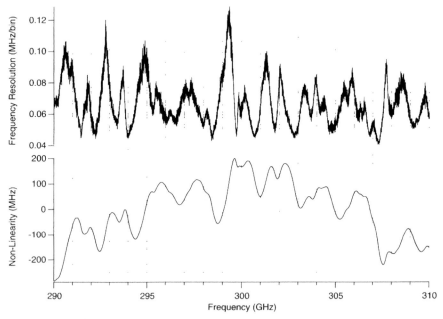

Fig. 32. Small-scale variation in frequency–voltage characteristic (*lower trace*) and frequency differences between adjacent digitized points (*upper trace*) as a function of frequency for a typical BWO

cal spectroscopy. However, the longer wavelength significantly relaxes the requirements for optical precision, and much greater frequency accuracy can be achieved.

4. A fast ($\sim 10^5$ spectral resolution elements/s, currently limited by detector bandwidth) sweep and data acquisition system "freeze" any drift in the source frequency over the time required to sweep from one reference fringe to the next. This eliminates the need for active frequency stabilization.

5. Fast data acquisition and calibration hardware and software. In a very general sense, the bandwidth of this system plays the same role as the bandwidth of the lock loops of more traditional systems.

The combination of these five elements makes it possible to measure thousands of spectral lines per second, with a frequency accuracy of a small fraction of a Doppler width ($\sim 0.1\,\mathrm{MHz}$ or $3 \times 10^{-6}\,\mathrm{cm}^{-1}$). Signal averaging is straightforward, and for equivalent integration times the sensitivity is the same as for slow-sweep, synthesized phase-locked systems. Finally, the system is very simple in both concept and execution and holds the promise of being used in a wide variety of applications.

3.5   Femtosecond Sources

All THz sources depend in one way or another on the production of current pulses whose timescale is of the order of the reciprocal of the frequency being generated. Since the reciprocal of 1 THz is 1 ps, it is natural to turn toward femtosecond lasers as a driving source for these current pulses. Two general schemes have evolved: one which uses the femtosecond pulse to generate a broad spectral pulse and Fourier transform techniques similar to those employed in FT-FIR systems to achieve spectral resolution, and another which uses a train of femtosecond pulses to produce a high-resolution comb of frequencies which is tuned by scanning the mode-lock frequency of the drive laser. Both use photoconductive switches to demodulate the optical pulse to provide the THz radiation [109].

Figure 33 shows an example of the former [110]. In this system the laser beam is split into two parts. The first drives the transmitter switch, and the second, which passes through a variable delay line, drives the detector switch. Since this detection process is coherent, the Fourier transform of the detector output as a function of delay time results in the spectrum of the gas sample. Figure 34 shows examples of the spectra which result [110,111]. Higher resolution than the value of $\sim 1 \times 10^{-2}$ to 1×10^{-3} shown can be achieved by use of a lower-pressure gas and a longer scan in the delay line, although at the expense of signal strength, as a smaller fraction of the THz pulse interacts with the sample.

Figure 35 shows an example of a system based on the demodulation of a train of femtosecond pulses [70,71]. The train of pulses on the photoconductive switch produces a comb of THz frequencies, separated by the mode-lock

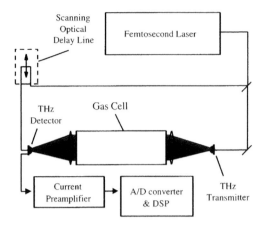

Fig. 33. Schematic of femtosecond THz time-domain spectrometer

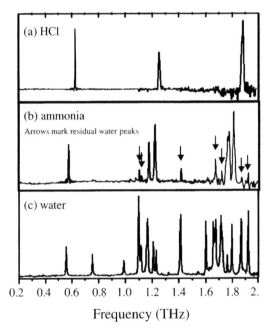

Fig. 34. Gas phase spectra observed with femtosecond THz time-domain spectrometer

frequency. Portions of this comb are selected by passive THz components such as gratings and filters, and the frequency of the source is continuously tuned by scanning the mode-lock frequency of the drive laser [112]. Figure 36 shows an example of a spectral line measured with this system. This system provides absolute frequency calibration via electronic counting of the mode-lock at a convenient microwave frequency and very high spectral purity, because it is based on a multiplication of a base microwave frequency rather than a difference between large optical frequencies. Measurements have shown a

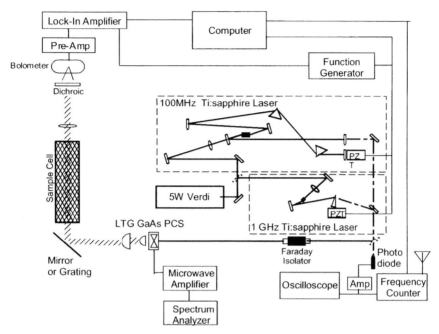

Fig. 35. Femtosecond demodulation system for high-resolution THz spectroscopy

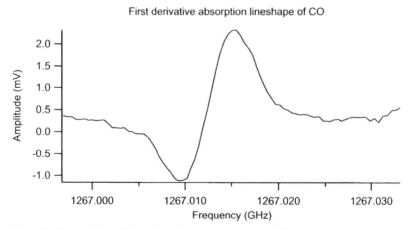

Fig. 36. Recording of the $J = 10\text{--}11$ transition of CO obtained with a femtosecond demodulation system

spectral purity of $\sim 3 \times 10^{-8}$ [113]. Measurement accuracy can be several orders of magnitude more precise. More recently, this comb of frequencies has also been used for absolute frequency measurement in the optical domain [114].

4   Applications

4.1   Atmospheric Spectroscopy

One of the principal applications of the THz spectral region has been the remote sensing of the atmosphere of the Earth. This has been an extremely successful application because the characteristics of THz spectroscopic interactions are especially well suited to this task. As a result, a number of sophisticated and powerful instruments have been developed [34,35,115–120].

Most of the focus of this work has been on the study of the complex processes which lead to the formation and destruction of ozone in the upper atmosphere [121–125]. Because of the complexity of the models designed to predict future trends and the effects of policy decisions on these trends, remote-sensing data is vital to monitor the concentration not only of the ozone itself, but also of the many contributors to the ozone cycle.

Figure 37 shows that major contributors to ozone destruction are catalytic cycles of the type: $X + O_3 \rightarrow XO + O_2$ and $XO + O \rightarrow X + O_2$, with the

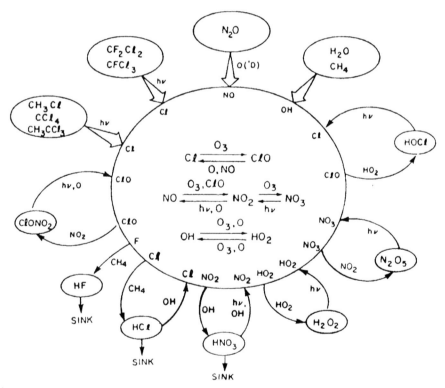

Fig. 37. Simplified diagram which shows the ozone production and destruction cycles in the upper atmosphere

net result $O + O_3 \rightarrow 2O_2$. The catalyst X is typically NO, OH, or Cl, which is often produced by either photochemistry or reactions with atomic oxygen from long-lived precursors.

Figure 38 illustrates the mixing ratios of a number of these species as a function of altitude. Because many of the processes are catalytic, even small concentrations of key species have a major impact on the overall cycle. As a result, it has been necessary to develop instruments capable of measurement of concentrations below the ppb (10^{-12}) level.

THz instruments have been able to meet this challenge because of the fundamental characteristics of molecular interactions discussed in Sect. 2. Of equal importance, it has been possible to develop instruments that in many cases approach fundamental theoretical limits. This has been, in many ways, a remarkable achievement in a spectral region which is often considered to be technology limited, and speaks of the progress that can be achieved by requirement-driven and tested technology development.

The most obvious foundation for these successes is the strong interactions between THz radiation and molecules. However, an important additional factor is that the THz region lies energetically below kT at atmospheric temperatures. As a result, it is possible to observe the thermal emissions from these molecular species in a limb-sounding geometry against the colder background of space. In these experiments, the beam of the instrument is scanned ver-

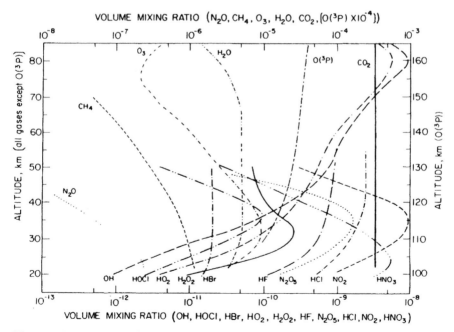

Fig. 38. Mixing ratios for a number of important contributors to the ozone cycle

tically through the atmosphere. and sophisticated deconvolution techniques make possible the recovery of vertical profiles similar to those shown in Figure 38 at near-diffraction-limited angular resolution. In contrast, most infrared instruments work in absorption. using the sun as a source, and thus limiting their observations to sunrise and sunset.

Another contributor to this success is that much of the ozone cycle occurs at relatively high altitude. This reduces the pressure broadening (~ 5 GHz at sea level) and gives rise to a spectroscopic richness which can approach that obtained in the laboratory studies discussed above. Not only does this provide unique spectroscopic signatures, but it also makes it possible to choose spectral regions for the study of trace species that are not contaminated by spectra from the much more abundant species such as O_3 and H_2O. In order to take advantage of this complexity, considerable effort has been devoted to the characterization and ultimately cataloging [13] of the spectra of atmospheric species. This effort has to be comprehensive because it is necessary to know not only the spectroscopy of the target species, but also that of any possible interlopers.

In this section, two complementary approaches to atmospheric spectroscopy in the THz region will be discussed: systems based on the extension of "microwave" technology to higher frequencies and systems based on the extension of "infrared" technology to longer wavelengths. There are a number of examples of both approaches [34,35,115 120], but we shall focus on the Microwave Limb Sounder (MLS) [119] as an example of the former and FIRS-2 [120] as an example of the latter. Both are important and interesting in their own right but, more generally, they serve as examples of the different approaches to spectroscopy in the THz region.

The similarities and contrasts of systems based on each approach are noteworthy. Because sensitivity, resolution. and measurement speed are major issues for both, optimal use of each photon emitted by the atmosphere is required. As a result both of the system mentioned above are multiplex instruments. but with very different implementations. Similarly, both approaches represent technological challenges as their core technologies are extended into the THz region, but again these challenges are very different.

4.1.1   Microwave-Like Instruments

One of the most successful instruments which makes use of the THz spectral region for atmospheric remote sensing is the Microwave Limb Sounder. The current version of MLS was launched as a part of the Upper Atmospheric Research Satellite (UARS) (http://umpgal.gsfc.nasa.gov/uars-science.html) in 1991 and has been operating successfully for almost a decade. It measures thermal emission from the limb of the Earth's atmosphere to retrieve vertical profiles of selected atmospheric gases, as well as temperature and pressure. A second generation of MLS, designed to observe a broader range of molecular species, is scheduled to be launched in 2002 as a part of the AURA

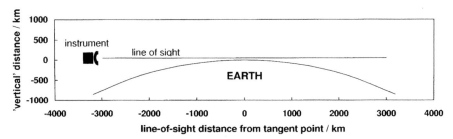

Fig. 39. Limb-viewing geometry of MLS

(http://eos-chem.gsfc.nasa.gov/) mission of NASA's Earth Observing System (EOS) (http://eospso.gsfc.nasa.gov/).

Figure 39 shows the limb-viewing geometry of MLS. A significant feature of MLS is that because it operates in the THz region, the natural thermal emission of the atmosphere provides the source of the observed radiation, whereas at infrared and shorter wavelengths molecular species are ordinarily observed as absorptions in the solar radiation. This latter viewing geometry restricts observations to sunrise and sunset.

Figure 40 shows a block diagram of the EOS MLS, which contains many features common to THz remote-sensing instruments, and illustrates the symbiosis between the development of technology and scientific applications in the THz region. In most THz applications, photons are precious and not to be wasted. In MLS, after a 1.6 m telescope collects the photons from the atmospheric limb and coupling to the thermal loads needed for accurate calibration of the instrument is provided for, the photons are efficiently split into separate channels by a passive optical multiplexer, consisting of dichroics and a polarizing grid. Details of the optical multiplexer and mixers are shown in Fig. 41. The mixers downconvert the spectral information to an intermediate frequency (IF) of $\sim$ 2–20 GHz, which is amplified and sent to multichannel filter banks. In Fig. 41, the downconverted frequency of each spectral feature to be observed is shown in parentheses below each IF amplifier. Figure 42 shows the signals downconverted to the IF in more detail, and Fig. 43 shows, a simulated spectrum for the 640 GHz channel. To increase the signal-to-noise ratio, a smaller 0.25 m antenna is used to feed the 2.5 THz radiometer separately.

Finally, in Fig. 44 the overall measurement capability of the instrument is shown. The low-altitude limits are set by the increasing pressure broadening that can be observed in Fig. 2 and by continuum contributions. At high altitudes, the decreasing number density of the atmosphere with altitude finally dominates the advantages that come with narrower linewidths, and signal levels drop below the noise.

Thus, the initial splitting of the photon stream in the optical multiplexer, the subsequent mixing followed by amplification at intermediate frequencies,

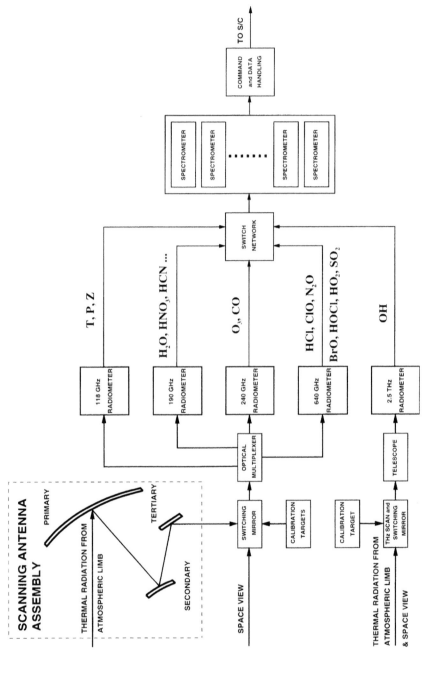

Fig. 40. Block diagram of the EOS MLS

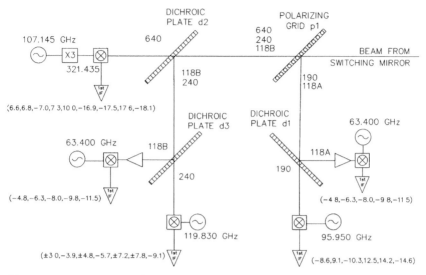

Fig. 41. Block diagram of the optical multiplexer

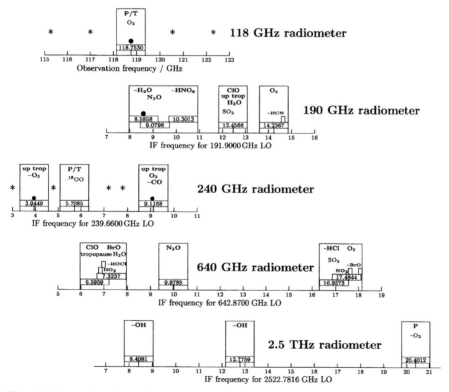

Fig. 42. Spectral regions downconverted to the IF

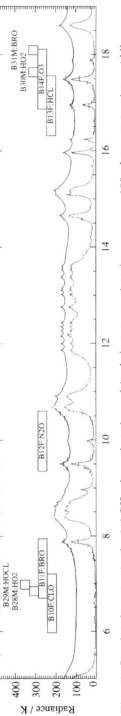

Fig. 43. Simulated spectra in the 640 GHz channel at three altitude/pressure points (*upper curve,* 100 mb tangent point; *middle curve,* 30 mb tangent point; *lower,* 10 mb tangent point)

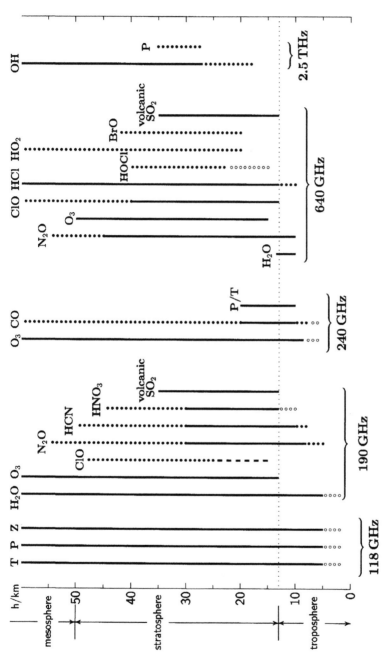

Fig. 44. Geophysical measurement capabilities of radiometer. The *solid lines* indicate profiles which can be obtained from single observations or daily averages, the *dotted lines* represent measurements which require zonal or other averages, and the *open circles* goals for difficult measurements. The *dashed line* for ClO indicates enhancements associated with polar winter vortices

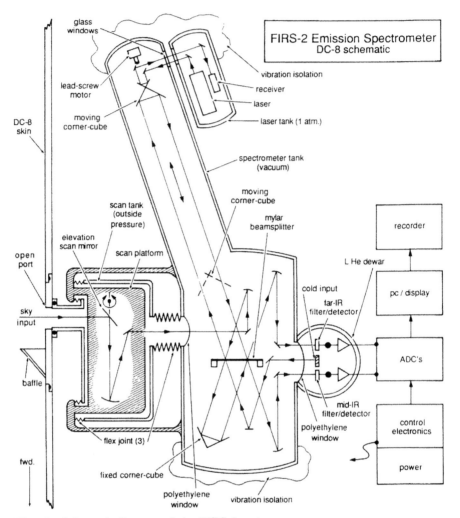

Fig. 45. Schematic diagram of the FIRS-2 system

and the availability of multichannel filter banks at the IF to simultaneously measure and record the atmospheric spectra result in an instrument of high resolution and large multiplex advantage.

Because broadly tunable instruments in this spectral region are not yet possible, an important design consideration for these instruments has been a careful analysis of the spectral characteristics of all known atmospheric molecules. This analysis makes possible the selection of center frequencies for each channel which maximize the number of species that can be observed within the available IF and filter bank width. While many of the strong lines used for retrievals have been known to spectroscopists for some time, the elimination of the possibility of spectral coincidences from unknown lines, especially the weaker lines (excited vibrational states, isotopic species, high-lying rotational states, etc.) of species abundant in the atmosphere (e. g. ozone or nitric acid) has been a significant spectroscopic challenge.

4.1.2 Infrared-Like Instruments

An excellent example of an "infrared-like" instrument which operates in the THz region is the FIRS-2 Fourier transform spectrometer operated by the Smithsonian Astrophysical Observatory Far-Infrared Spectroscopy Group (http://cfa-www.harvard.edu/firs) [126]. It is balloon- or aircraft-borne and covers the far-infrared ($80-340\,\mathrm{cm}^{-1}$) and mid-infrared ($330-1220\,\mathrm{cm}^{-1}$) with an apodized resolution of $0.008\,\mathrm{cm}^{-1}$ ($240\,\mathrm{MHz}$). The longer wavelengths of these two regions correspond to $2.67-11.33\,\mathrm{THz}$ and overlap nicely with the region typically thought of as the high-frequency end of the THz. In this region, thermal sources produce significant power even at atmospheric temperatures, and (as does MLS) FIRS-2 observes the stratosphere in thermal emission. This enables it to make abundance measurements during both day and night, in any azimuth heading. Additionally, although obtained in a very different way, FIRS-2, like MLS, is a multiplex instrument that simultaneously observes many species.

The heritage of FIRS-2 is that of an optical FTIR instrument, and its layout is shown in Fig. 45. The configuration shown here is that designed for use on the NASA DC-8 platform. FTIR instruments are multiplex in that all of the energy over a very wide band is simultaneously present on the detectors. The spectrum is then recovered via a Fourier transform of the signal which is recorded as a function of the optical path difference as one of the corner cube reflectors is translated.

More specifically, in FIRS-2, the light enters the system at the left, where vertical resolution in the atmosphere is obtained with an elevation scan mirror. The system is a dual-input, dual-output system, with one image being the sky image and the other a cold source in the detector dewar. The two beams are combined on the left half of the Ge-coated Mylar beam splitter. The reflected beam goes to the moving corner cube and the transmitted beam to the fixed corner cube. The beams are then recombined on the right of the beam splitter and focused onto the two detectors. Both are liquid-helium photoconductors, with one being for the region $\sim 80-320\,\mathrm{cm}^{-1}$ and the other for the region $\sim 320-1560\,\mathrm{cm}^{-1}$. The $\sim 80\,\mathrm{cm}^{-1}$ limit is set by the cutoff of the photoconductors. The separation into two bands increases the sensitivity of the long-wavelength portion by isolating it from the strong mid-infrared flux.

Figure 46 shows sample spectra on four different frequency scales. All spectra were recorded at a tangent height of $\sim 33\,\mathrm{km}$ and are from a 1997 Alaska flight. In its balloon mode, the interferometer scan takes 6 minutes to complete.

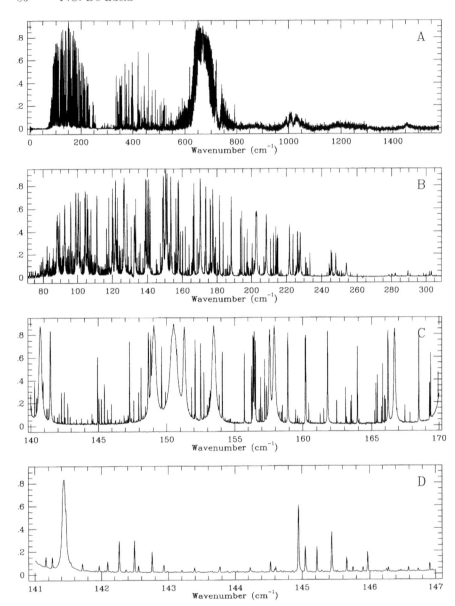

Fig. 46. Spectra from FIRS-2 in the region 80–$1400\,\mathrm{cm}^{-1}$ (A), with a series of graphical expansions (B–D) to show the resolution and sensitivity of the FTIR technique. The *bottom scan* shows less than 1% of the recovered spectrum

As can be seen in Fig. 46, instruments of this type record the complex spectra from many molecules simultaneously. FIRS-2 has been used to retrieve mixing-ratio profiles for an impressive and diverse set of species, including H_2O, CO_2, O_2, O_3, N_2O, HCN, CO, HF, HCl, HOCl, NO_2, HNO_3, OH, H_2O_2, HO_2, HBr, $ClNO_3$ N_2O_5, C_2H_6, C_3H_6O, HNO_4, CFC_{12}, CFC_{11}, $HCFC_{22}$, CCL_4, CH_3Cl, SF_6, OCS, and CH_4. These simultaneous measurements make it possible to place tight constraints on photochemical models. Table 2 shows details of the precision of atmospheric measurements from a balloon flight on April 29, 1997.

4.2 Astronomical Spectroscopy

Astronomical spectroscopy has attracted perhaps the largest body of scientists and engineers working in the THz (or, as it is more commonly referred to in that community, the millimeter and submillimeter or far-infrared) spectral region. That the line between scientists and the engineers is often blurred in this field is perhaps one of the chief reasons it has been so successful. The challenging technical problems of astronomy have not only driven and motivated the development of technology, but have also provided generally accepted and verifiable benchmarks.

The genesis of this astronomical activity can be traced back to the early days of microwave spectroscopy, when techniques developed for millimeter spectroscopy were adapted in 1955 to detect radiation from the sun in the atmospheric window around 94 GHz [127]. The first observation of a polyatomic species in the interstellar medium, ammonia (NH_3), was in 1968 by Townes and his coworkers Welsh, Rank, Thornton, and Cheung at the Hat Creek observatory [128]. This was followed closely by their discovery of water (H_2O) [129]. Interestingly, two young postdocs, Lew Snyder and David Buhl, at the National Radio Astronomical Observatory had previously applied for telescope time on their West Virginia facility, but their application to search for molecules in the interstellar medium had been turned down as a foolish waste of telescope time [45]. Shortly thereafter they were awarded telescope time and discovered formaldehyde (H_2CO) [130]. While it is fortuitous that these early searches in the centimeter spectral region were for molecules which turned out to be relatively abundant, none of these molecules are as widespread or as abundant as CO, whose 115 GHz $J = 1 \rightarrow 0$ transition was the first to be discovered in the millimeter spectral region [131]. We shall see below that the abundance of CO, along with its high correlation with the cosmically much more abundant but difficult to detect H_2 and He, have made it an important beacon for astrophysical study of the structure and dynamics of both our galaxy and active star-forming regions. An excellent history and overview of this field has been written by Winnewisser, Herbst, and Ungerechts [132].

At a basic level, astronomical spectroscopy and atmospheric spectroscopy are very similar. Both detect the emission from molecules and, to first order, both use similar technology to do so: heterodyne receivers being extended from lower frequencies and optical interferometric techniques being extended from shorter wavelengths.

Table 2. Measurement precision obtained in a FIRS-2 balloon flight

| Species | Precision (detection limit)[a] | | | Spectroscopic error[b] |
	Mixing ratio	Slant column	Other	
Atmospheric parameters				
T			1 K	
P		2%		
CO_2[c]	3%			2%
HO_r				
OH	1.5 ppt			4%
HO_2	5 ppt			2%
H_2O_2	10 ppt			4%
Cl_y				
HCl	70 ppt			3%
$ClNO_3$	10% (20 ppt)			10%
HOCl	7 ppt			2%
Bry				
HBr	(6 ppt)			3%
HOBr	(7 ppt)			2%
NO_y				
HNO_3	150 ppt			10%
HNO_4	50 ppt at 20 km	$(5 \times 10^{15}\,cm^{-2})$	10%	
N_2O_5	30% (50 ppt)			10%
NO_2	900 ppt			2%
Tracers				
N_2O	2%			3%
HF	30 ppt			4%
CCl_4	7 ppt at 15 km	$7\%\ (7 \times 10^{14}\,cm^{-2})$	10%	
CFC11	15 ppt at 15 km	$6\%\ (1 \times 10^{15}\,cm^{-2})$	10%	
CFC12	30 ppt at 15 km	$6\%\ (2 \times 10^{15}\,cm^{-2})$	10%	
CH_3Cl	300 ppt	$50\%\ (1 \times 10^{16}\,cm^{-2})$	10%	
$HCFC_{22}$	20 ppt at 15 km	$20\%\ (4 \times 10^{15}\,cm^{-2})$	10%	
OCS	70 ppt at 15 km	$15\%\ (1 \times 10^{16}\,cm^{-2})$	5%	
C_2H_6	250 ppt at 15 km	$(4 \times 10^{16}\,cm^{-2})$	5%	
C_3H_6O	150 ppt at 15 km	$(2 \times 10^{16}\,cm^{-2})$	50%	
HCN	80 ppt			2%
SF_6	1 ppt at 15 km	$16\%\ (2 \times 10^{14}\,cm^{-2})$	5%	
CH_4	500 ppb			2%
H_2O	160 ppb			7%
δD	20%			35%
$\delta^{18}O$	40%			70%
O_3	3% (50 ppb)			3%

[a] Precision obtained on a balloon flight on April 29. 1997, with an integration time of 12 minutes per spectrum. For this level of precision, 120 minutes are required to obtain a profile from 10 to 40 km with 4 km vertical resolution. Minimum integration time per spectrum is presently 6 minutes.

[b] Systematic errors are dominated by the uncertainty in spectroscopic line parameters.

[c] Used to retrieve T and P.

It is instructive to start by considering the differences in the science (as well as in the resulting technology) for atmospheric and astronomical applications. In general, the interstellar medium is colder, with temperatures typically not too many times that of the microwave background (2.7 K), but with hotter regions as protostellar cores are approached. Additionally, the astronomical collision times are much longer in all circumstances. This long collision time, combined with fluxes of energetic particles, produces molecular systems which are far from equilibrium in rotational-state populations, partial pressures of gases (which for almost all species would approach zero under conditions dictated by vapor pressure), and abundances of ions, free radicals, and other reactive species. A useful measure of this nonequilibrium is that the lifetime of gaseous species in the interstellar medium is $\sim 10^5$–10^6 years before they freeze out on dust grains.

From the point of view of laboratory astrophysics, this leads to a rather different set of problems from those motivated by atmospheric science. Table 3 shows a list of the molecular species that have been detected in the interstellar medium. A comparison of this list with the species found in Figs. 37 and 38 is instructive. An important first-order effect in this comparison is simply one of atomic abundance. The atmospheric species are derivatives of the major atmospheric components (N_2, O_2, and H_2O) and man-made injections into the atmosphere. The interstellar species are driven more by cosmic abundances

Table 3. Molecules detected in the interstellar medium

H_2	KCl	HNC	NH_3	C_3S	C_5	C_6H
			CH_3			HC_4CN
CH	AlCl	HCO	H_3O^+	CH_4	CH_3OH	C_7H, C_6H_2, C_8H
CH^+	AlF	HCO^+	H_2CO	SiH_4	CH_3SH	$HCOOCH_3$
						CH_3COOH
NH	PN	HOC^+	H_2CS	CH_2NH	C_2H_4	CH_3C_2CN
						$H_2C_6(lin)$
OH	SiN	HN_2^+	HCCH	$H_2C_3(lin)$	CH_3CN	$(CH_3)_2O$
						$H_2COHCHO$
C_2	SiO	HNO	$HCNH^+$	c-C_3H_2	CH_3NC	C_2H_5OH
CN	SiS	HCS^+	H_2CN	CH_2CN	HC_2CHO	C_2H_5CN
CO	CO^+	C_3	$C_3H(lin)$	NH_2CN	NH_2CHO	CH_3C_4H
CSi	SO^+	C_2O	c-C_3H	CH_2CO	HC_3NH^+	HC_6CN
		CO_2			C_4H_2	
CP	H_3^+	C_2S	HCCN	HCOOH	$H_2C_4(lin)$	$(CH_3)_2CO$
CS	CH_2	SiC_2	HNCO	C_4H	C_5H	$CH_3C_4CN?$
HF		SiCN	SiC_3		C_5N	
NO	NH_2	SO_2	$HOCO^+$	HC_2CN	CH_3NH_2	$NH_2CH_2COOH?$
NS	H_2O	OCS	HNCS	HCCNC	CH_3CCH	HC_8CN
SO	H_2S	MgNC	C_2CN	HNCCC	CH_3CHO	c-C_6H_6
HCl	C_2H	MgCN	C_3O	C_4Si	CH_2CHCN	$HC_{10}CN$
NaCl	HCN	N_2O	NaCN	H_2COH^+	CH_2OCH_2	+isotopomers

(H, C, O, N, ...). Additionally, the interstellar list has many prominent ions and free radicals whose lifetimes under terrestrial conditions are very short.

While spectral-line frequencies are independent of the molecular environment after correction to rest velocity, their shapes and widths are not. Moreover, the inelastic rotational-energy transfer rates, which are closely related to pressure broadening (which is absent in the interstellar medium but of great significance in the recovery of atmospheric parameters from remote-sensing data), are similarly necessary for the recovery of astronomical information from the nonequilibrium interstellar medium.

Thus much of the emphasis in laboratory astrophysics has been on the development of laboratory environments, both environments for the production of reactive species and the environments in which low-temperature collisional studies can be carried out. These have already been considered in more detail in Sect. 3.

A web site with excellent discussions of these molecules and the interstellar medium can be found at http://cfa-www.harvard.edu/ rplume/vhs/vhs.html.

4.2.1   Some Telescope Facilities

Table 4 lists a number of major millimeter, submillimeter, and far-infrared telescope facilities. These include ground-, aircraft-, and space-based telescopes. Especially noteworthy for the future of the field are the large new systems in the active development stage: the Atacama Large Millimeter Array (ALMA) (http://www.alma.nrao.edu/) to be placed at Llano de Chajnantor in Chile, the Far Infrared and Submillimeter Telescope (FIRST) satellite (http://sci.esa.int/first/), the Submillimeter Array (SMA) atop Mauna Kea (http://sma2.harvard.edu/index.html), and the recently launched Submillimeter-Wave Astronomy Satellite (SWAS) (http://cfa-www.harvard.edu/ cfa/oir/Research/swas.html). By the historical standards of the spectral region, these represent an enormous investment and are a testimony both to the importance of the science that drives these projects and to the technological advances that have made them possible.

4.2.2   Two Representative THz Telescopes

In this section we shall provide an overview of two recent satellite instruments, the Submillimeter-Wave Astronomy Satellite (SWAS) and the Far Infra-Red and Submillimetre Telescope (FIRST), which has recently been renamed the Herschel Space Observatory. These two telescopes differ by about a decade in their launch dates and technological development. For each, we shall focus on the relation between the THz technologies and the astronomical problems that motivated their design.

SWAS. The Submillimeter-Wave Astronomy Satellite was launched in December of 1998 as a part of NASA's Small Explorer Project (SMEX) and can

Table 4. Some representative THz astronomy facilities

Ground-based telescopes

Location	Size	Name	URL
Mauna Kea	15 m	James Clerk Maxwell (JCMT)	http://www.jach.hawaii.edu/JACpublic/JCMT/index.html
	10 m	Caltech Submillimeter Observatory (CSO)	http://www.submm.caltech.edu/cso/
	8 × 6 m	Submillimeter Array (SMA)	http://sma2.harvard.edu/index.html
Boston	1.2 m	The Cfa	http://cfa-www.harvard.edu/cfa/mmw/mini.html
Goernergrat (Switzerland)	3 m	KOSMA	http://www.ph1.uni-koeln.de/kosma.html
Llano de Chajnantor (Chile) (an NRAO telescope)	64 × 12 m	Atacama Large Millimeter Array (ALMA)	http://www.alma.nrao.edu/
Mt Graham	10 m	Max-Planck-Institut fr Radioastronomie University of Arizona	http://maisel.as.arizona.edu:8080/
South Pole	1.7 m	Center for Astrophysical Research in Antarctica (CARA)	http://cfa-www.harvard.edu/ adair/AST_RO/

Airborne telescopes

Stratospheric Observatory of Infrared Astronomy (SOFIA)	http://sofia.arc.nasa.gov/

Space-based telescopes

Submillimeter-Wave Astronomy Satellite (SWAS)	http://cfa-www.harvard.edu/cfa/oir/Research/swas.html
Far Infrared and Submillimeter Telescope (FIRST)	http://sci.esa.int/first/

serve as an example to illustrate the characteristics of astronomical systems in the THz spectral region [133]. SWAS was designed to study the Holy Grail of interstellar astrophysics, star formation from dense clouds. It does this by specific observation of water, molecular oxygen, atomic carbon, and isotopic carbon monoxide, species that are central to the cooling processes in these clouds that are necessary for gravitational collapse. To accomplish its goals, SWAS operates in the 487–556 GHz spectral region, a region in which observations cannot be made from the ground because of atmospheric attenuation, primarily due to the strong 1_{10}–1_{01} transition of water itself.

A block diagram is shown in Fig. 47. After the photons are collected by a 68×53 cm aluminum primary mirror and provision is made for comparison with a calibration load, the power is split with a wire grid polarizer into two channels and sent to separate heterodyne detectors. Because the background against which the molecular emissions are viewed is in general very cold (often that of the cosmic microwave background at 2.7 K), it is advantageous to use the gain in sensitivity that can be obtained with cooled mixers. However, in order to reduce size and cost and increase the mission lifetime, in SWAS they are only passively cooled to ~ 170 K rather than to cryogenic temperatures. Thus, the receiver consists of two (one for each polarization) second harmonic mixer diodes, each pumped by its own frequency tripled InP Gunn oscillator. The first of these systems is centered near 490 GHz to observe O_2 and atomic carbon and the second near 555 GHz for the study of ^{13}CO and H_2O. To further conserve precious photons, multiplex operation is achieved by the use of an acousto-optical spectrometer which allows the simultaneous observation of 700 1 MHz channels between 1.4 and 2.1 GHz.

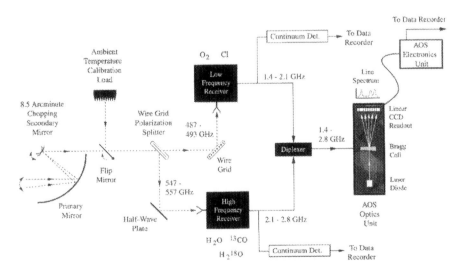

Fig. 47. Block diagram of the SWAS system

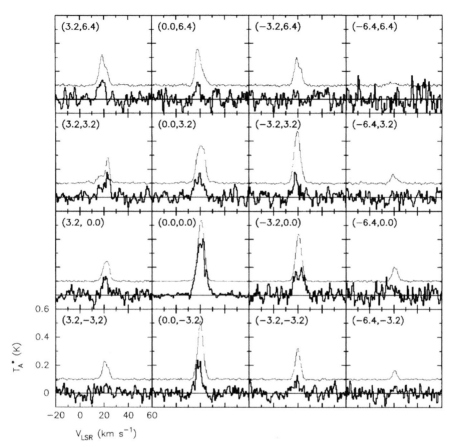

Fig. 48. Spectra of the 1_{10}–1_{01} transition of H_2O (*thick lines*) and of the $J = 5$–4 transition of ^{13}CO (*thin lines*) obtained for the astronomical source M17SW. The ^{13}CO spectra are divided by 10 in all cases except at positions (0.0, 0.0) and (−3.2, 0.0), where the spectra are divided by 20. The 16 spectra make up a 4×4 map on a regular grid with a spacing of 3.2′

Figure 48 shows the result of a SWAS study of the prototypical giant molecular cloud core M17SW [134]. From these spectra, the distribution in space, the abundance (calculated from integrated line intensity), and the velocity profiles (calculated from Doppler shifts) of both species can be obtained. Each location in the figure represents typically 3–7 hours of integration time, required to observe these weak signals that are only a few tenths of a degree above the 2.7 K microwave background. Another interesting feature of these results is that the local-oscillator and IF frequencies are selected so that H_2O and ^{13}CO can be observed simultaneously by use of the response in each sideband of the heterodyne receiver. Finally, it should be noted how much stronger the isotopic ^{13}CO line is than that of H_2O. In fact, CO is so abundant in the interstellar medium that its less abundant isotopic species are often used for astrophysical studies to avoid the problems associated with recovering remote-sensing data from optically thick data.

FIRST. The Far Infra-Red and Submillimetre Telescope is a cornerstone observatory mission of the European Space Agency (ESA) and is projected for launch in about 2007 [135]. It is designed to study the formation and evolution of galaxies outside of our own galaxy and the energy sources of particularly luminous galaxies in the early universe. Because of the absorption and reradiation of ultraviolet radiation by dust grains, galaxies emit large fractions ($\sim$ 30–100%) of their energy in the far infrared. Additionally, distant galaxies have large redshifts that further shift this radiation to long wavelength. Within our own galaxy, FIRST will search for protostars, study the formation of stars from the interstellar medium, and explore the evolution of planetary systems. This is a much wider range of scientific goals than that of SWAS, and as a result FIRST carries a suite of complementary instruments.

FIRST is a much larger (7 m high, 4.3 m wide, with a 3.5 m telescope primary and a weight of 3.25 tons) instrument than SWAS and will represent about an additional ten years of THz technology development. FIRST will have three instruments, employing photoconductor (PACS – photoconductor array camera and spectrometer) [136], bolometer (SPIRE – spectral and photometric imaging receiver) [137], and superconducting mixer (HIFI – heterodyne instrument for FIRST detectors) [138] technologies. In order to optimally detect the molecular emissions against the cold background, these detectors and parts of their optics and electronics will be cooled to cryogenic temperatures by a superfluid-liquid-helium cryostat with a three year minimum lifetime. As an important by-product of this cooling, much smaller local-oscillator powers can be used in the heterodyne instruments, and operation to much higher frequencies will result.

The three complementary instruments on FIRST typify systems in the THz region and comparison can illustrate the trade-offs associated with each. The heterodyne system will provide very high spectral resolution from 610 μm (490 GHz) down to at least 250 μm (1.2 THz), with a probable channel near 175 μm (1700 GHz). However, local-oscillator and single-mode matching requirements restrict both its high-frequency limit and its broadbandedness. In comparison, the two "optical" systems will provide medium-resolution spectroscopy and photometry between 60 μm (5 THz) and 600 μm (500 GHz). Because these systems use "optical" rather than heterodyne detection, they do not face increasing difficulties in the fabrication of single-mode mixers and in the provision of local-oscillator power with increasing frequency. Indeed, PACS employs a 16 × 25 stressed Ge : Ga array for imaging photometry and spectroscopy that has a long-wavelength cutoff near 210 μm.

HIFI. It is useful to compare the FIRST heterodyne instrument HIFI with that of SWAS. HIFI uses helium-temperature superconductor–insulator–superconductor (SIS) and hot-electron bolometer (HEB) mixers. The much lower local-oscillator power requirements of these mixers aid in the design of a relatively broadband system, with continuous frequency coverage of the 480–1250 GHz band, and also of the 1410–1910 and 2400–2700 GHz bands. Figure 49 shows a functional block diagram. Inspection of this figure shows that the instrument on FIRST is similar to that on SWAS and other heterodyne THz astronomical receivers. However, it is much more complex, with seven pairs (one for each polarization) of mixers spread over its much wider spectral range. Its IF section also uses acousto-optic modulators (AOMs) for multiplex operation, but supplements them with autocorrelator spectrometers to provide a variety of combinations of bandwidth and resolution.

In many respects, the greatest challenge for a heterodyne system operating at THz frequencies is the production of adequate local-oscillator power. HIFI plans to use 14 separate local-oscillator subbands. These synthesized local oscillators will consist of K- to W-band multipliers, high-power MMIC amplifiers operating in five bands between 71 and 113 GHz, and high-frequency planar diode multipliers [139].

The high spectral resolution of this heterodyne instrument, combined with the narrow spectral features of low-temperature astrophysical molecules, makes FIRST an ideal instrument to study a variety of astrophysical processes. Its wide spectral coverage can provide massive redundancy for the determination of cosmic abundances and interstellar chemistry. Its high resolution is essential to avoid spectral confusion and line blending in studies of dynamically evolving regions.

Photoconductor Array Camera and Spectrometer. The PACS instrument [136] is based on a photoconductive detector that uses two 25×16 Ge : Ga unstressed/stressed arrays [140] to cover two bands, 60–130 µm and 130–210 µm. A key element of this instrument is the very high sensitivity (NEP $\sim 5 \times 10^{-18}$ W/Hz$^{-1/2}$) obtainable with Ge : Ga photoconductive detectors at ~ 2 K [141,142]. However, these detectors have photoconductive thresholds at 130 µm (unstressed) and 210 µm (stressed), and for observations at longer wavelength, FIRST uses either heterodyne (HIFI) or bolometer (PACS) detectors. Figure 50 shows the relative response of these detectors as a function of wavelength.

In its photometry mode, PACS will perform photometry ($\lambda/\Delta\lambda \sim 2$) simultaneously in the two bands, with a mesh-filter-selected choice (60–90 µm/ 90–130 µm) available on the unstressed, shorter-wavelength array. This two-color system is designed for the study of broad emission features of external galaxies.

In its spectrometer mode, PACS has a resolution of ~ 1500. This is accomplished with a diffraction grating in a Littrow configuration and a dichroic

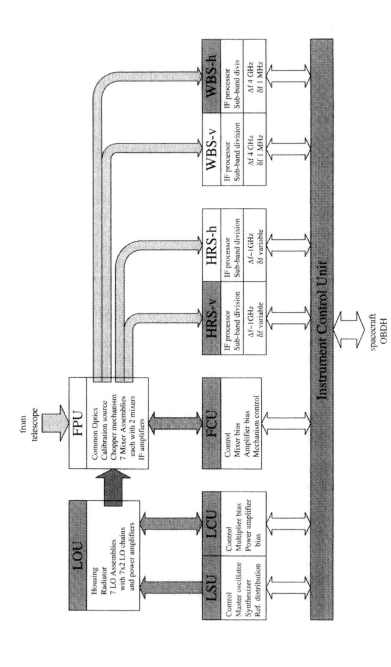

Fig. 49. Functional block diagram of the heterodyne receiver HIFI on FIRST

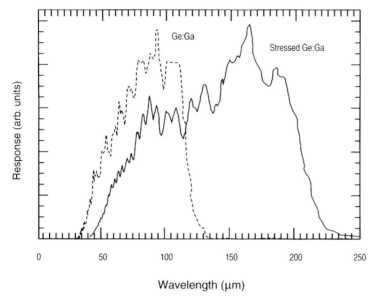

Fig. 50. Relative response of stressed (*solid line*) and unstressed (*dashed line*) Ge : Ga detectors

beam splitter to separate diffraction orders. In this configuration, multiplex spatial imaging is retained by the use of a 5×5 pixel detector array.

PACS complements HIFI in two important aspects and is a good illustration of the technology trade-offs that are still necessary in the THz region. First, because it does not require local-oscillator power for mixers, the difficulties of producing power at high frequency are eliminated. Secondly, it is much easier to build arrays of photodetectors than it is to build focal-plane arrays of heterodyne mixers. Thus, for astrophysical projects which require large-scale photometric surveys, the arrays of PACS provide *spatial* multiplexing and a large gain in the overall photometric efficiency for the system.

Spectral and Photometric Imaging Receiver. The SPIRE instrument is a bolometer detector array instrument with an imaging photometer and an imaging Fourier transform spectrometer (FTS). Rather than the Ge : Ga 2 K photoconductive arrays used in PACS, this instrument is based on arrays of "spider-web" Ge bolometers. This eliminates the long-wavelength cutoffs of the photoconductive detectors, but to obtain the required sensitivity ($\sim 3 \times 10^{-17}$ W/Hz$^{-1/2}$) the detectors are operated at 0.3 K via a closed-cycle ^{3}He sorption refrigerator.

SPIRE's photometer mode ($\lambda/\Delta\lambda \sim 3$) provides broadband response simultaneously in bands centered on $250\,\mu$m, $350\,\mu$m, and $500\,\mu$m, with arrays of 149, 88, and 43 elements, respectively. Since the bolometers are broadband detectors, the pass bands are determined by a series of filters and

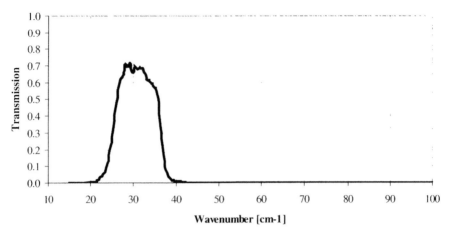

Fig. 51. Transmission response for a typical 350 μm photometer band

dichroic beam splitters. Figure 51 shows a typical response for the 350 μm band. Again, as with PACS, this low-resolution response is designed for the study of the broad emission features of distant galaxies.

In its spectrometer mode, SPIRE uses an FTS to obtain a resolution which can be adjusted between 0.04 and $2\,\mathrm{cm}^{-1}$ ($\lambda/\Delta\lambda$ from 20 to 1000 at 250 μm). It has two separate detector arrays of 37 and 19 elements to cover the 200–200 μm and 300–670 μm bands, respectively.

The main scientific goals of SPIRE are deep extragalactic and galactic imaging surveys. Because the bolometer arrays make possible imaging surveys to longer wavelengths than with PACS, SPIRE provides an important complement in the spectral region where distant galaxies, with large redshifts, are expected to radiate much of their energy.

FIRST Summary. The suite of instruments on FIRST is a particularly interesting example of the current technological state of the art in the THz region. While, taken as a whole, FIRST admirably addresses the scientific issues at hand, there is clearly still a gap between the extension of microwave-like technology up from long wavelength and optical-like technology down from short wavelength.

It is probably fair to say that the single technological advance which would most impact similar systems in the future would be the development of widely tunable THz sources with enough power to use as local oscillators for arrays of mixer elements throughout this spectral region. This would make possible the development of widely tunable heterodyne receivers (in contrast to receivers in carefully selected spectral regions) whose mixer arrays would have the spatial multiplex advantage of the "optical" systems in PACS and SPIRE. This "ideal" system of the future would then be able to use both spatial and frequency multiplex principles to fully utilize the precious and expensive photons collected by the telescope.

4.2.3   Examples of Other Results

In this section we shall briefly discuss a selection of astrophysical results based on THz observations to illustrate both the current technological state of the art and some of the science.

Mapping the Galaxy with CO. Figure 52 shows a comparison between an optical image of the galaxy (top) and an emission map of the $J = 1 \rightarrow 0$ transition of CO at 115 GHz (bottom). The astrophysical significance of this comparison is the close correlation between the CO emission and the optical obscuration, showing that regions where the gas and dust densities are high enough to allow molecules to form also produce optical "dark clouds" [143]. Figure 53 shows a more detailed map of the CO emission from giant molecular clouds (GMCs) located in the Perseus arm of our galaxy [144]. These GMCs contain a significant fraction of the mass of the interstellar matter in the Milky Way, each with a mass of the order 10^6 that of our sun.

These maps are based on data obtained with the 1.2 m telescope listed in Table 4 and currently located at the Center for Astrophysics. This telescope has been dedicated to CO surveys for about 25 years and has produced a significant body of results. Briefly, it is a multiplex heterodyne system, with an SIS mixer, a Gunn local oscillator, a 1.4 GHz IF frequency, and a 256 channel spectrometer.

These maps are also a good example of more general THz capabilities: the ability to penetrate particulate matter and to accurately measure small frequency shifts. Although the dust makes up only a relatively small fraction of the interstellar matter, it scatters light so as to make much of the most interesting galactic regions unobservable. These maps of CO are possible and astrophysically significant for three reasons. First, CO is relatively abundant and its presence is highly correlated with the numerically more abundant (but difficult to observe) H_2 and He. Second, its radiation at 115 GHz readily penetrates the dust content of the clouds. Finally, its sharp spectral resonance makes possible not only its unambiguous identification, but also the measurement of Doppler shifts, which result from both local galactic turbulence and the overall rotation of the galaxy.

Line Surveys. In addition to using a single molecule such as CO to study the astrophysical dynamics of galactic properties such as star formation, astronomers are also keenly interested in the chemistry of the interstellar clouds themselves. The molecular cloud in Orion is particularly well suited for study because of both its size and its proximity (~ 500 pc, or 2000 l.y.). There have been many line surveys of this source over the years, but a good example is that of Schilke, Groesbeck, Blake, and Phillips [145]. For this study, the

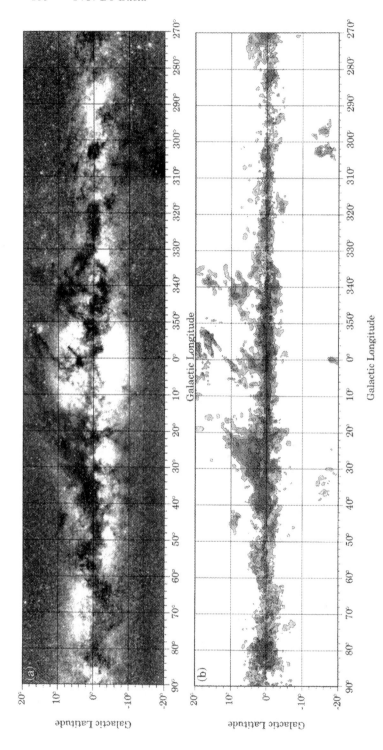

Fig. 52. A comparison between the galactic optical emission (*top*) and the $J = 1 \rightarrow 0$ emission of CO (*bottom*)

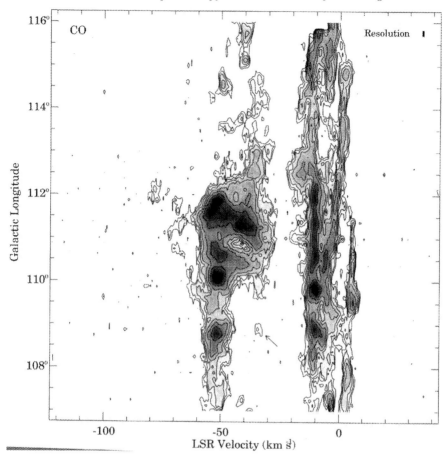

Fig. 53. Emission of CO integrated over the observed range in galactic latitude as a function of velocity and longitude. The lowest contour is at 0.30 K, and each contour is 1.58 times higher than the previous. for five contours per decade

CSO instrument atop Mauna Kea listed in Table 4 was used to survey the 325–360 GHz region. About 2200 lines were observed; an overview is shown in Fig. 54 and an expanded region around 338.5 GHz is shown in Fig. 55. This latter figure shows that the richness and denseness of the spectrum are such that spectral confusion rather than instrument sensitivity is becoming the limiting factor in these strong sources. More recently, a survey in the 607–725 GHz region has also been completed [146].

Figure 54. especially below 330 GHz, shows another important feature of the THz region: atmospheric absorption. Although Mauna Kea is one of the best submillimeter observing sites in the world. as telescopes push to ever higher frequencies. atmospheric absorption becomes an increasing limitation.

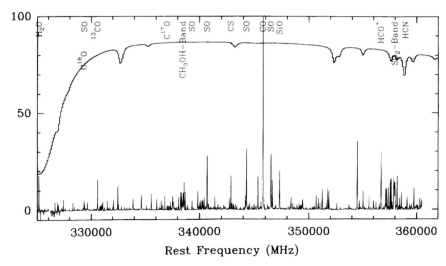

Fig. 54. Composite spectrum of the 325–360 GHz survey of the Orion molecular cloud taken by the CSO instrument on Mauna Kea (lower) and the atmospheric transmission (upper)

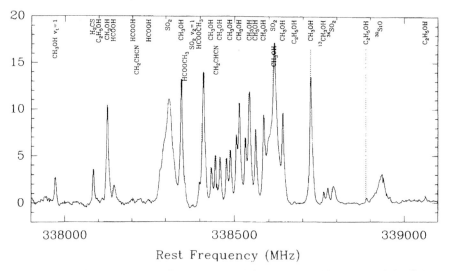

Fig. 55. Detail of the 338–339 GHz portion of the 325–360 GHz survey of the Orion molecular cloud taken by the CSO instrument on Mauna Kea

Figure 56 shows zenith observations of atmospheric transmission at Mauna Kea taken with a Fourier transform spectrometer with $\sim$ 200 MHz resolution [147]. The notable feature of this comparison is that the data of 1998 were taken under unusually dry conditions (ground-level relative humidity of

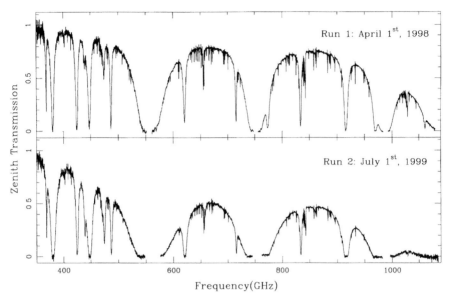

Fig. 56. Zenith atmospheric transmission from Mauna Kea

$\sim 2\%$) and that transmission windows as high as 1035 GHz provided as much as 35% transmission.

Analytical Spectroscopy. Almost from the beginning of microwave spectroscopy, its potential for the chemical analysis of gases has been recognized (e.g. Chap. 18 of the classic monograph by Townes and Schawlow, *The Use of Microwave Spectroscopy for Chemical Analysis*) [39]. However, while techniques based on the ultraviolet, optical, and infrared regions of the electromagnetic spectrum have become standard analytical tools, this early promise has not been realized at longer wavelength.

In the late 1950s and early 1960s a commercial instrument, operating in the general frequency region between 10 and 40 GHz, was developed and marketed by Hewlett-Packard. This instrument experienced some commercial success, being sold primarily to spectroscopists with an interest in molecular structure. In hindsight, its modest success with the analytical community was due to the instrument's size, cost, and complexity.

However, analytical systems based on the FASSST concept described in Sect. 3 overcome these limitations. In addition, they offer significantly greater generality, speed, and sensitivity. This has been achieved by the use of (1) the THz rather than the more classic microwave spectral region, (2) a fast scanning "optical" technique rather than the more traditional microwave phase-lock methodology, and (3) modern data acquisition, signal processing, and computing.

In the THz spectral region, fingerprints arise from the rotational energy levels of molecules. These fingerprints are ordinarily complex and unique because many rotational levels are thermally populated. Additionally, as discussed in Sect. 2, in the THz region the underlying physical interactions between radiation and matter are strong and the small Doppler broadening in the THz region in comparison with the IR allows resolution of the complex rotational signature.

As an example, consider Fig. 57. It shows FASSST spectra obtained with an ISTOK OB-80 BWO in an $\sim 30\,\mathrm{GHz}$ region centered near $500\,\mathrm{GHz}$, obtained by first adding $10\,\mathrm{mTorr}$ of pyrrole (C_4H_5N) to the sample cell (upper trace), then $20\,\mathrm{mTorr}$ of pyridine (C_5H_5N) (middle trace), and finally $20\,\mathrm{mTorr}$ of SO_2 (lower trace). Figure 58 shows an expansion of the $\sim 1\,\mathrm{GHz}$ shaded region near $512\,\mathrm{GHz}$ in Fig. 57. Finally, Fig. 59 shows an expansion of the $\sim 0.2\,\mathrm{GHz}$ shaded region near $511.8\,\mathrm{GHz}$ in Fig. 58. The spectral region in Fig. 59 is $\sim 0.1\%$ of the BWO bandwidth and represents $\sim 0.01\,\mathrm{s}$ of data acquisition. In each figure the sensitivity is such that no noise can be displayed on the graph. It is clear, even in the small expanded spectral interval displayed in the last figure, that each has a unique signature, as it

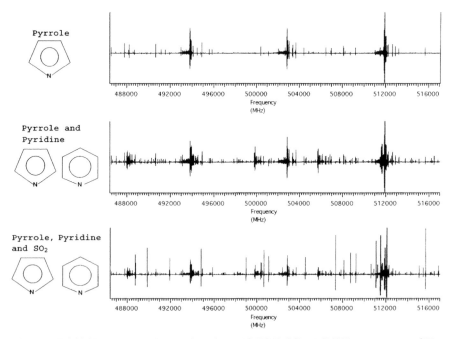

Fig. 57. FASSST spectra obtained with an ISTOK OB-80 BWO in an $\sim 30\,\mathrm{GHz}$ region centered near $500\,\mathrm{GHz}$, obtained by first adding $10\,\mathrm{mTorr}$ of pyrrole (C_4H_5N) to the sample cell (*upper trace*), then $20\,\mathrm{mTorr}$ of pyridine (C_5H_5N) (*middle trace*), and finally $20\,\mathrm{mTorr}$ of SO_2 (*lower trace*)

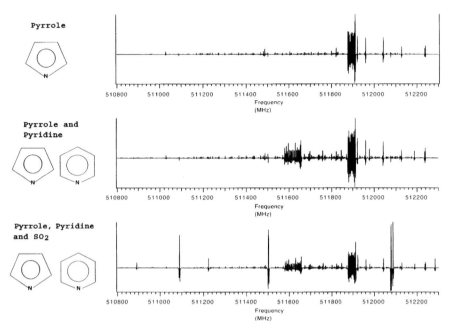

Fig. 58. An expansion of the ~ 1 GHz shaded region near 512 GHz in Fig. 57

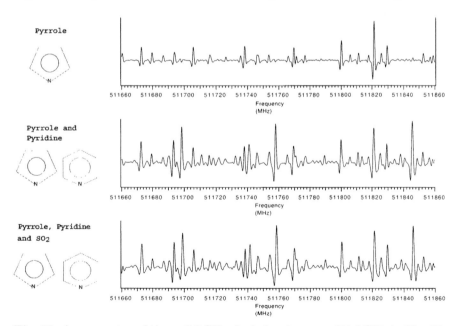

Fig. 59. An expansion of the ~ 0.2 GHz shaded region near 511.8 GHz in Fig. 58. The spectral region in this figure is $\sim 0.1\%$ of the BWO bandwidth and represents ~ 0.01 s of data acquisition

would in almost any other randomly chosen small interval throughout the submillimeter region.

One way of viewing the information content and power of FASSST as an analytical tool is to recognize that if Fig. 57 were expanded to show all of its resolution elements in the horizontal direction and 1 mm of noise in the vertical, the resulting graph would be approximately 10 m high by 100 m long for each 1 s of data acquisition.

The technique's sensitivity, specificity, and quantitative properties can be summarized as follows.

1. *Sensitivity.* THz frequencies are typically 10–100 times higher than the region historically referred to as the microwave. The distinction between systems operating in the microwave and the THz regions is very important. As discussed above, molecular absorption strengths increase with frequency with a functional dependence between ν^2 and ν^3, peaking somewhere in the THz region. As a result, spectroscopic systems in this region are between 10^2 and 10^6 times more sensitive than the much more common microwave systems. Additionally, the larger spectral region results in virtually universal coverage of molecular species, subject only to the requirement that the species have a dipole moment or a large-amplitude vibration in the THz region.

   Sensitivity comparisons with infrared systems are more complex because of the diversity of infrared technologies that have been used for analytical purposes, as well as the variety of analytical applications. However, because (1) the FASSST system is based on powerful electronic oscillators which are fundamentally very bright and very quiet, (2) at long wavelengths it is possible to build very quiet detectors, (3) at long wavelengths noise due to microphonics, etc. is much reduced, and (4) rotational transition moments are related to the total electric dipole moment, the FASSST system is inherently very sensitive.

2. *Specificity and rotational spectroscopy.* The large number of resolvable channels in the THz region, the large number of thermally excited rotational lines which populate these resolvable channels, the high signal-to-noise attainable even with very short integrating times, and the absolute measurement of absorption coefficients available in the submillimeter region make it possible in virtually all scenarios to set false-alarm rates that are so low as to be considered absolutely specific. Because of the larger Doppler-width in the infrared, even Doppler limited instruments are not capable of resolving the rotational fingerprints of molecules much larger than $ClONO_2$.

3. *Quantitative analysis.* Because the strengths of the molecular rotational transitions observed in the THz region depend upon the permanent molecular dipole moment (which can be measured to great accuracy via the Stark effect) and angular-momentum quantum mechanics (for which "exact" solutions are obtainable), measured fractional absorption can be

translated into absolute concentration by straightforward calculation. Most importantly, because of the high resolution of the FASSST spectra, any one of many isolated rotational lines can ordinarily be used for quantitative purposes, thus providing massive redundancy and eliminating the possibility of contributions from interfering and overlapping lines.

A more detailed discussion of a THz FASSST system used as an analytical instrument can be found in an A-pages article in *Analytical Chemistry* [7].

Acknowledgments

We would like to thank the many workers in the field who provided information, figures, and tables. Without their cooperation, this chapter would not have been possible. We would also like to thank the Army Research Office, the Air Force Office of Scientific Research, the Defense Advanced Projects Agency, the National Aeronautics and Space Administration, and the National Science Foundation for their support.

References

1. W. Gordy, "Microwave spectrocopy", *Rev. Mod. Phys.* **20**, 668 (1948).
2. W. Gordy, "Microwave spectroscopy above 60 kMC", *N.Y. Acad. Sci.* **55**, 744 (1952).
3. C. A. Burrus Jr., W. Gordy, "Submillimeter wave spectroscopy", *Phys. Rev.* **93**, 897 (1954).
4. W. C. King, W. Gordy, "One to two millimeter wave spectroscopy. I", *Phys. Rev.* **90**, 319 (1953).
5. R. R. Unterberger, W. V. Smith, "A microwave secondary frequency standard", *Rev. Sci. Inst.* **19**, 580 (1948).
6. P. Helminger, F. C. De Lucia, W. Gordy, "Extension of microwave absorption spectroscopy to 0.37-mm wavelength", *Phys. Rev. Lett.* **25**, 1397 (1970).
7. S. Albert, D. T. Petkie, R. P. A. Bettens, S. P. Belov, F. C. De Lucia, "FASSST: a new gas-phase analytical tool", *Anal. Chem.* **70**, 719A (1998).
8. H. J. Liebe, "Atmospheric water vapor: a nemesis for millimeter wave propagation", in *Atmospheric Water Vapor*, ed. by A. Deepak, Th. D. Wilkerson, L. H. Ruhnke (Academic Press, New York, 1980).
9. C. A. Burrus Jr., W. Gordy, "One-to-two millimeter wave spectroscopy. II. H_2S", *Phys. Rev.* **92**, 274 (1953).
10. W. C. King, W. Gordy, "One-to-two millimeter wave spectroscopy. IV. Experimental methods and results for OCS, CH_3F, and H_2O", *Phys. Rev.* **93**, 407 (1954).
11. W. Gordy, "Microwave spectroscopy in the region of 4–0.4 millimeters", *J. Pure Appl. Chem.* **2**, 403 (1965).
12. F. C. De Lucia, "Millimeter and submillimeter-wave spectroscopy", *Mol. Spectrosc., Mod. Res.* **2**, 73 (1976).

13. H. M. Pickett. R. L. Poynter, E. A. Cohen, M. L. Delitsky, J. C. Pearson, H. S. P. Muller, "Submillimeter, millimeter, and microwave spectral line catalog", *J. Quant. Spectrosc. Radiat. Transfer* **60**. 883 (1998).

14. L. S. Rothman, C. P. Rinsland, A. Goldman, S. T. Massie, D. P. Edwards, J.-M. Flaud, A. Perrin, C. Camy-Peyret, V. Dana, J.-Y. Mandin, J. Schroeder, A. Mccann, R. R. Gamache, R. B. Wattson, K. Yoshino, K. V. Chance, K. W. Jucks, L. R. Brown, V. Nemtchinov, P. Varanasi, "The HITRAN molecular spectroscopic database and HAWKS (HITRAN atmospheric workstation): 1996 edition", *J. Quant. Spectrosc. Radiat. Transfer* **60**, 665 (1998).

15. P. Helminger, F. C. De Lucia. "Pressure broadening of hydrogen sulfide". *J. Quant. Spectrosc. Radiat. Transfer* **17**. 751 (1977).

16. H. M. Pickett. "Determination of collisional linewidths and shifts by a convolution method", *Appl. Opt.* **19**, 2745 (1980).

17. A. Bauer, M. Godon, M. Kheddar, J. H. Hartmann, J. Bonamy, D. Robert, "Temperature and perturber dependences of the water-vapor 380 GHz-line broadening", *J. Quant. Spectrosc. Radiat. Transfer* **37**, 531 (1987).

18. T. M. Goyette, F. C. De Lucia, J. M. Dutta, C. R. Jones, "Variable temperature pressure broadening of the $4_{1,4}$-$3_{2,1}$ transition of H_2O by O_2 and N_2", *J. Quant. Spectrosc. and Rad. Transfer* **49**. 485 (1993).

19. J. K. Messer, F. C. De Lucia, "Measurement of pressure-broadening parameters for the CO-He system at 4 K". *Phys. Rev. Lett.* **53**. 2555 (1984).

20. J. C. Pearson, L. C. Oesterling, E. Herbst, F. C. De Lucia, "Pressure broadening of Gas Phase Molecular Ions at Very Low Temperature", *Phys. Rev. Lett.* **75**. 2940 (1995).

21. C. D. Ball, F. C. De Lucia, "Direct measurement of rotationally inelastic cross sections at astrophysical and quantum collisional temperatures", *Phys. Rev. Lett.* **81**. 305 (1998).

22. M. Winnewisser, K. V. L. N. Sastry, R. L. Cook, W. Gordy, "Millimeter wave spectroscopy of unstable molecular species II. Sulfur monoxide", *J. Chem. Phys.* **41**. 1687 (1964).

23. K. M. Evenson, R. J. Saykally, D. A. Jennings, R. F. Curl, J. M. Brown, *Far Infrared Laser Magnetic Resonance* (Academic Press, New York, 1980).

24. A. Charo, F. C. De Lucia, "The millimeter and submillimeter spectrum of HO_2: The effects of the unpaired electronic spin in a light asymmetric rotor", *J. Mol. Spectrosc.* **94**. 426 (1982).

25. R. C. Woods, T. A. Dixon, R. J. Saykally, P. G. Szanto, "Laboratory microwave spectrum of HCO^+". *Phys. Rev. Lett.* **35**, 1269 (1975).

26. F. C. van den Heuvel, A. Dynamus, "Observation of far-infrared transitions of HCO^+, CO^+. and NH_2^+". *Chem. Phys. Lett.* **92**. 219 (1982).

27. F. C. De Lucia, E. Herbst, G. M. Plummer, G. A. Blake, "The production of large concentrations of molecular ions in the lengthened negative glow region of a discharge". *J. Chem. Phys.* **78**. 2312 (1983).

28. K. L. Busarow, G. A. Blake, K. B. Laughlin, R. C. Cohen, Y. T. Lee, R. J. Saykally, "Tunable far-infrared laser spectroscopy in a planar supersonic jet: The Σ bending vibrations of $Ar-H^{35}Cl$". *Chem. Phys. Lett.* **141**. 289 (1987).

29. M. D. Marshall, A. Charo, H. O. Leung, W. Klemperer, "Characterization of the lowest-lying Π bending state of Ar HCl by far infrared laser-Stark spectroscopy and molecular beam electric resonance". *J. Chem. Phys.* **83**, 4924 (1985).

30. F. C. De Lucia, "The study of laser processes by millimeter and submillimeter microwave spectroscopy". *Appl. Phys. Lett.* **31**, 606 (1977).

31. D. D. Skatrud, F. C. De Lucia, "Dynamics of the HCN discharge laser", *Appl. Phys. Lett.* **46**, 631 (1985).
32. H. O. Everitt, D. D. Skatrud, F. C. De Lucia, "Dynamics and tunability of a small optically pumped CW far-infrared laser", *Appl. Phys. Lett.* **49**, 995 (1986).
33. K. V. Chance, D. G. Johnson, W. A. Traub, K. W. Jucks, "Measurement of the stratospheric hydrogen peroxide concentration profile using far infrared thermal emission spectroscopy", *Geophys. Res. Letts.* **18**, 1003 (1991).
34. J. W. Waters, *Atmospheric Remote Sensing by Microwave Radiometry* (Wiley, New York, 1993).
35. B. Carli, J. H. Park, "Simultaneous measurement of minor stratospheric constituents with emission far-infrared spectroscopy", *J. Geophys. Res.* **93**. 3851 (1988).
36. E. Herbst, "Chemistry in the interstellar medium", *Ann. Rev. Phys. Chem.* **46**, 27 (1995).
37. T. G. Phillips, in *The Physics and Chemistry of Interstellar Molecular Clouds,* ed. by G. Winnewisser, G. C. Pelz (Springer, New York. 1995).
38. W. Gordy, W. V. Smith, R. F. Trambarulo, *Microwave Spectroscopy* (Wiley, New York, 1953).
39. C. H. Townes, A. L. Schawlow. *Microwave Spectroscopy* (McGraw-Hill Dover, New York, 1955).
40. A. Carrington. *Microwave Spectroscopy of Free Radicals* (Academic Press, London, 1974).
41. H. W. Kroto, *Molecular Rotation Spectra* (Wiley, London, 1975).
42. W. Gordy, R. L. Cook. *Microwave Molecular Spectra*, 3rd ed. (Wiley, New York, 1984).
43. R. L. Cook, "Microwave spectroscopy", in *Macmillan Encyclopedia of Chemistry* (Macmillan, 1997) New York.
44. R. L. Cook, "Molecular microwave spectroscopy", in *Encyclopedia of Physical Science and Technology.* ed. by R. A. Meyers (Academic Press, 2000) San Diego.
45. C. H. Townes, *How the Laser Happened: Adventures of a Scientist* (Oxford University Press, New York, 1999).
46. E. F. Nichols, J. D. Tear, "Short electric waves", *Phys. Rev.* **21**, 587 (1923).
47. E. Brown, private communication (2001).
48. C. A. Burrus Jr., "Millimeter-wave point-contact and junction diodes", *Proc. IEEE* **54**, 575 (1966).
49. K. M. Evenson, D. A. Jennings, K. R. Leopold, L. R. Zink, in *Seventh International Conference on Laser Spectroscopy*, ed. by T. W. Hansch, Y. R. Shen (Springer, Berlin. 1985), p. 300.
50. P. Helminger, J. K. Messer, F. C. De Lucia, "Continuously tunable coherent spectroscopy for the 0.1- to 1.0-THz region", *Appl. Phys. Lett.* **42**. 309 (1983).
51. W. B. Lewis, "Fluctuations in streams of thermal radiation", *Proc. Phys. Soc.* **59**, 34 (1947).
52. E. H. Putley, "The ultimate sensitivity of Sub-mm detectors", *Infrared Phys.* **4**, 1 (1964).
53. E. H. Putley, "Far infra-red photoconductivity", *Phys. Status Solidi* **6**, 571 (1964).
54. E. H. Putley, "Indium antimonide submillimeter photoconductive detectors", *Appl. Opt.* **4**, 649 (1990).

55. F. Lewen, S. P. Belov, F. Maiwald, T. Klaus, G. Winnewisser, "A quasi-optical multiplier for terahertz spectroscopy", *Z. Naturforsch.* **50a**, 1182 (1995).
56. A. Karp, "Backward-wave oscillator experiments at 100 and 200 kilomegacycles", *Proc. IRE.* **45**, 496 (1957).
57. M. B. Golant, Z. T. Alekseenko, Z. S. Korotkova, L. A. Lunkina, A. A. Negirev, O. P. Petrova, T. B. Rebrova, V. S. Savel'ev, "Wide-range oscillators for the submillimeter wavelengths", *Prib. Tekh. Eksp.* **3**, 231 (1969).
58. M. Ino, T. Ishibashi, M. Ohmori, "Submillimeter wave Si p^+–p n^+ IMPATT diodes", *Jpn. J. Appl. Phys. Suppl.* **16-1**, 89 (1977).
59. E. R. Brown, J. R. Soderstrom, C. D. Parker, L. J. Mahoney, K. M. Molvar, T. C. McGill, "Oscillations up to 712 GHz in InAs/AlSb resonant-tunneling diodes", *Appl. Phys. Lett.* **58**, 2291 (1991).
60. K. M. Evenson, D. A. Jennings, F. R. Peterson, "Tunable far-infrared spectroscopy", *Appl. Phys. Lett.* **44**, 576 (1984).
61. E. R. Brown, K. A. McIntosh, K. B. Nichols, C. L. Dennis, "Photomixing up to 3.8 THz in low-temperature-grown GaAs", *Appl. Phys. Lett.* **66**, 285 (1995).
62. A. S. Pine, R. D. Suenram, E. R. Brown, K. A. McIntosh, "A terahertz photomixing spectrometer – applications to SO_2 self-broadening", *J. Mol. Spectrosc.* **175**, 37 (1996).
63. S. Matsuura, G. A. Blake, R. A. Wyss, J. C. Pearson, C. Kadow, A. W. Jackson, A. C. Gossard, "A traveling-wave THz photomixer based on angle-tuned phase matching", *Appl. Phys. Lett.* **74**, 2872 (1999).
64. S. Matsuura, P. Chen, G. A. Blake, J. C. Pearson, H. M. Pickett, "A tunable cavity-locked diode laser source for terahertz photomixing", *IEEE Trans. Microwave Theory Tech.* **48**, 380 (2000).
65. D. D. Bicanic, B. F. J. Zuidberg, A. Dymanus, "Generation of continuously tunable laser sidebands in the submillimeter region", *Appl. Phys. Lett.* **32**, 367 (1978).
66. W. A. M. Blumberg, H. R. Fetterman, D. D. Peck, P. F. Goldsmith, "Tunable submillimeter sources applied to the excited state rotational spectroscopy and kinetics of CH_3F", *Appl. Phys. Lett.* **35**, 582 (1979).
67. J. Farhoomand, G. A. Blake, M. A. Frerking, H. M. Pickett, "Generation of tunable laser sidebands in the far-infrared region", *J. Appl. Phys.* **57**, 1763 (1985).
68. P. Verhoeve, E. Zwart, M. Versluis, J. ter Meulen, W. L. Meerts, A. Dymanus, D. Mclay, "A far infrared laser sideband spectrometer in the frequency region 550–2700 GHz", *Rev. Sci. Instrum.* **61**, 1612 (1990).
69. G. A. Blake, K. B. Laughlin, R. C. Cohen, K. L. Busarow, D.-H. Gwo, C. A. Schmuttenmaer, D. W. Steyert, R. J. Saykally, "Tunable far infrared laser spectrometers", *Rev. Sci. Instrum.* **62**, 1693 (1991).
70. F. C. De Lucia, B. D. Guenther, T. Anderson, "Microwave generation from picosecond demodulation sources", *Appl. Phys. Lett.* **47**, 894 (1985).
71. T. M. Goyette, W. Guo, F. C. De Lucia, J. Swartz, H. O. Everitt, B. D. Guenther, E. R. Brown, "Femtosecond demodulation source for high resolution submillimeter spectroscopy", *Appl. Phys. Lett.* **67**, 3810 (1995).
72. M. Cowan, W. Gordy, "Precision measurements of millimeter and submillimeter wave spectra", *Phys. Rev.* **111**, 209 (1958).
73. H. Rothermel, T. G. Phillips, J. Keene, "A solid-state frequency source for radio astronomy in the 100 to 1000 GHz range", *Int. J. Infrared Millim. Waves* **10**, 83 (1989).

74. F. J. Low, "Low temperature germanium bolometer", *J. Opt. Soc. Am.* **51**, 1300 (1961).

75. F. C. De Lucia, W. Gordy, "Molecular beam maser for the shorter-millimeter-wave region: spectral constants of HCN and DCN", *Phys. Rev.* **187**, 58 (1969).

76. R. M. Garvey, F. C. De Lucia, "Extension of high-resolution beam maser spectroscopy into the submillimeter wave region", *Can J. Phys.* **55**, 1115 (1977).

77. R. G. Strauch, R. E. Cupp, V. E. Derr, J. J. Gallagher, "Millimeter electric resonance spectroscopy", *Proc. IEEE* **54**, 506 (1966).

78. G. Winnewisser, G. C. Pelz, *The Physics and Chemistry of Interstellar Molecular Clouds* (Springer, Berlin, Heidelberg, 1995).

79. C. Demuyuck, "Millimeter-wave spectroscopy in electric discharges. Rare molecules show themselves only if you look in the other direction", *J. Mol. Spectrosc.* **168**, 215 (1994).

80. H. E. Warner, W. T. Conner, R. H. Petrmichl, R. C. Woods, "Laboratory detection of the 1_{10}–1_{11} submillimeter wave transition of the H_2D^+ ion", *J. Chem. Phys.* **81**, 2514 (1984).

81. M. Bogey, C. Demuynck, M. Denis, J. L. Destombes, B. Lemoine, "Laboratory measurement of the 1_{10}–1_{11} submillimeter line of H_2D^+", *Astron. Astrophys.* **137**, L15 (1984).

82. G. A. Blake, K. B. Laughlin, R. C. Cohen, K. L. Busarow, R. J. Saykally, "Laboratory measurement of the pure rotational spectrum of vibrationally excited HCO^+ ($\nu_2 = 1$) by far-infrared laser sideband spectroscopy", *Astrophys. J.* **316**, L45 (1987).

83. T. Amano, A. Maeda, "Double-modulation submillimeter-wave spectroscopy of HOC^+ in the ν_2 excited vibrational state", *J. Mol. Spectrosc.* **203**, 140 (2000).

84. M. C. McCarthy, M. J. Travers, A. Kovacs, C. A. Gottlieb, P. Thaddeus, "Eight new carbon chain molecules", *Astrophys. J. Supp. Ser.* **113**, 105 (1997).

85. M. C. McCarthy, J.-U. Grabow, M. J. Travers, W. Chen, C. A. Gottlieb, P. Thaddeus, "Laboratory detection of the carbon chains $HC_{15}N$ and $HC_{17}N$", *Astrophys. J. Lett.* **494**, L231 (1998).

86. S. Green, P. Thaddeus, "Rotational excitation of CO by collisions with He, H, and H_2 under conditions in interstellar clouds", *Astrophys. J.* **205**, 766 (1976).

87. T. M. Goyette, F. C. De Lucia, "Collisional cooling as an environment for planetary research", *J. Geophys. Res.* **96**, 17455 (1991).

88. D. R. Willey, R. J. Southwick, K. Ramadas, B. K. Rapela, W. A. Neff, "Laboratory observation of maser action in NH_3 through collisional cooling", *Phys. Rev. Lett.* **74**, 5216 (1995).

89. R. I. McCormick, H. O. Everitt, F. C. De Lucia, D. D. Skatrud, "Collisional energy transfer in optically pumped far-infrared lasers", *IEEE J. Quantum Electron.* **23**, 2069 (1987).

90. H. O. Everitt, F. C. De Lucia, "Rotational energy transfer in small polyatomic molecules", *Adv. Atom. Mol. Phys.* **35**, 331 (1995).

91. L. C. Oesterling, E. Herbst, F. C. De Lucia, "Millimeter-wave time-resolved studies of HCO^+–H_2 inelastic collisions", *Spectrochim. Acta. A* **57**, 705 (2001).

92. S. S. Sussman, B. C. Johnson, J. M. Yarborough, H. E. Puthoff, R. H. Pantell, J. SooHoo, in *Submillimeter Waves*, ed. by J. Fox (Polytechnic, New York, 1971), p. 211.

93. R. L. Aggarwal, B. Lax, "CW generation of tunable narrow-band far-infrared radiation", *J. Appl. Phys.* **45**, 3972 (1974).

94. T. D. Varberg, J. C. Roberts, K. A. Tuominen, K. M. Evenson, "The far-infrared spectrum of deuterium iodide", *J. Mol. Spectrosc.* **191**, 384 (1998).

95. E. R. Brown, K. A. McIntosh, F. W. Smith, K. B. Nichols, M. J. Manfra, C. L. Dennis, "Milliwatt output and superquadratic bias dependence in a low-temperature-grown GaAs photomixer", *Appl. Phys. Lett.* **64**, 3311 (1994).

96. S. Verghese, K. A. McIntosh, E. R. Brown, "Highly tunable fiber-coupled photomixers with coherent terahertz output power", *IEEE Trans. Microwave Theory Tech.* **45**, 1301 (1997).

97. H. A. Gebbie, N. W. B. Stone, J. E. Chamberlain, "A stimulated emission source at 0.34 millimeter wave-length", *Nature* **202**, 685 (1964).

98. T. Y. Chang, T. J. Bridges, "Laser action at 452, 496, and 541 μm in optically pumped CH_3F", *Opt. Commun.* **1**, 423 (1970).

99. H. R. Fetterman, P. E. Tannenwald, B. J. Clifton, W. D. Fitzgerald, N. R. Erickson, "Far-ir heterodyne radiometric measurements with quasioptical Schottky diode mixers", *Appl. Phys. Lett.* **33**, 151 (1978).

100. K. B. Laughlin, G. A. Blake, R. C. Cohen, D. C. Hovde, R. J. Saykally, "Determination of the dipole moment of ArH^+ from the rotational Zeeman effect by tunable far-infrared laser spectroscopy", *Phys. Rev. Lett.* **58**, 996 (1987).

101. R. Gendriesch, F. Lewen, G. Winnewisser, J. Hahn, "Precision broadband spectroscopy near 2 THz: frequency-stabilized laser sideband spectrometer with backward-wave oscillators", *J. Mol. Spectrosc.* **203**, 205 (2000).

102. R. J. Saykally, "Far-infrared laser spectroscopy of van der Waals bonds: a powerful new probe of intermolecular forces", *Accts. Chem. Res.* **22**, 295 (1989).

103. R. Kompfner, N. T. Williams, "Backward-wave tubes", *Proc. IRE* **41**, 1602 (1953).

104. A. F. Krupnov, A. V. Burenin, "New methods in submillimeter microwave spectroscopy", in *Molecular Spectroscopy: Modern Research*, ed. by K. Narahari Rao (Academic Press, New York, 1976), Vol. 2, p. 93.

105. G. Winnewisser, A. F. Krupnov, M. Y. Tretyakov, M. Liedtke, F. Lewen, A. H. Saleck, R. Schieder, A. P. Shkaev, S. V. Volokhov, "Precision broadband spectroscopy in the terahertz region", *J. Mol. Spectrosc.* **165**, 294 (1994).

106. G. Winnewisser, "Spectroscopy in the terahertz region", *Vib. Spectrosc.* **8**, 241 (1995).

107. S. P. Belov, T. Klaus, G. M. Plummer, R. Schieder, G. Winnewisser, "Subdoppler spectroscopy of ammonia near 570 GHz", *Z. Naturforsch.* **50a**, 1187 (1995).

108. D. T. Petkie, T. M. Goyette, R. P. A. Bettens, S. P. Belov, S. Albert, P. Helminger, F. C. De Lucia, "A fast scan submillimeter spectroscopic technique", *Rev. Sci. Instrum.* **68**, 1675 (1997).

109. D. H. Auston, K. P. Cheung, P. R. Smith, "Picosecond photoconductive hertzian dipoles", *Appl. Phys. Lett.* **45**, 284 (1984).

110. D. M. Mittleman, R. H. Jacobsen, R. Neelamani, R. G. Baraniuk, M. C. Nuss, "Gas sensing with terahertz time-domain spectroscopy", *Appl. Phys. B* **67**, 379 (1998).

111. R. H. Jacobsen, D. M. Mittleman, M. C. Nuss, "Chemical recognition of gases and gas mixtures using terahertz waveforms", *Opt. Lett.* **21**, 2011 (1996).

112. J. R. Demers, F. C. De Lucia, "Modulating and scanning the mode-lock frequency of an 800 MHz femtosecond Ti:sapphire laser", *Opt. Lett.* **24**, 250 (1999).

113. J. R. Demers, T. M. Goyette, K. B. Ferrio, H. O. Everitt, B. D. Guenther, F. C. De Lucia, "Spectral purity and sources of noise in femtosecond-demodulation THz sources driven by Ti: sapphire mode-locked lasers", *IEEE J. Quant. Electron.* **37**, 595 (2001).

114. S. A. Diddams, D. J. Jones, J. Ye, S. T. Cundiff, J. L. Hall, J. K. Ranka, R. S. Windeler, R. Holzwarth, T. Udem, T. W. Hansch, "Direct link between microwave and optical frequencies with a 300 THz femtosecond laser comb", *Phys. Rev. Lett.* **84**, 5102 (2000).

115. R. M. Bevilacqua, J. J. Olivero, P. R. Schwartz, C. J. Gibbins, J. M. Bologna, D. J. Thacker, "An observational study of water vapor in the mid-latitude mesosphere using ground-based microwave techniques", *J. Geophys. Res.* **88**, 8523 (1983).

116. B. Carli, F. Mencaraglia, A. Bonetti, "Submillimeter high-resolution FT spectrometer for atmospheric studies", *Appl. Opt.* **23**, 2594 (1984).

117. J. W. Barrett, P. M. Solomon, R. L. de Zafra, M. Jaramillo, L. Emmons, A. Parrish, "Formation of the Antarctic ozone hole by the ClO dimer mechanism", *Nature* **336**, 455 (1988).

118. M. L. Santee, G. L. Manney, L. Froidevaux, W. G. Read, J. W. Waters, "Six years of UARS microwave limb sounder HNO_3 observations: seasonal, interhemispheric, and interannual variations in the lower stratosphere", *J. Geophys. Res.* **104**, 8225 (1999).

119. J. W. Waters, W. G. Read, L. Froidevaux, R. F. Jarnot, R. E. Cofield, D. A. Flower, G. K. Lau, H. M. Picket, M. L. Santee, D. L. Wu, M. A. Boyles, J. R. Burke, R. R. Lay, M. S. Loo, N. J. Livesey, T. A. Lungu, G. L. Manney, L. L. Nakamura, V. S. Perum, B. P. Ridenoure, Z. Shippony, P. H. Siegel, R. P. Thurstans, R. S. Harwood, H. C. Pumphrey, M. J. Filipiak, "The UARS and EOS microwave limb sounder experiments", *J. Atmos. Sci.* **56**, 194 (1999).

120. K. W. Jucks, D. G. Johnson, K. V. Chance, W. A. Traub, J. J. Margitan, G. B. Osterman, R. J. Salawitch, Y. Sasano, "Observation of OH, HO_2, H_2O, and O_3 in the upper stratosphere: implications for HO_x photochemistry", *Geophys. Res. Lett.* **21**, 3935 (1998).

121. G. Dobson, D. Harrison, J. Lawrence, "Measurements of the amount of ozone in the Earth's atmosphere and its relation to other geophysical conditions", *Proc. R. Soc. London* **A122**, 456 (1929).

122. S. Chapman, "On ozone and atomic oxygen in the upper atmosphere", *Philos. Mag.* **10**, 369 (1930).

123. D. Bates, M. Nicolet, "Atmospheric hydrogen", *Astron. Soc. Pac. Leafl.* **62**, 106 (1950).

124. P. Crutzen, "The influence of nitrogen oxide on the atmospheric ozone content", *Q. J. R. Meteorol. Soc.* **96**, 320 (1970).

125. M. Molina, F. Rowland, "Stratospheric sink for chlorofluoromethanes: chlorine atom catalysed destruction of ozone", *Nature* **249**, 810 (1974).

126. D. G. Johnson, K. W. Jucks, W. A. Traub, K. V. Chance, "Smithsonian stratospheric far-infrared spectrometer and data reduction system", *J. Geophys. Res.* **100**, 3091 (1995).

127. W. Gordy, S. J. Ditto, J. H. Wyman, R. S. Anderson, "Three-millimeter wave radiation from the sun", *Phys. Rev.* **99**, 1905 (1955)

128. A. C. Cheung, D. M. Rank, C. H. Townes, D. D. Thornton, W. J. Welch, "Detection of NH_3 molecules in the interstellar medium by their microwave emission", *Phys. Rev. Lett.* **21**, 1701 (1968).

129. A. C. Cheung, D. M. Rank, C. H. Townes, W. J. Welch, "Further microwave emission lines and clouds of ammonia in our galaxy", *Nature* **221**, 917 (1969).

130. L. E. Snyder, D. Buhl, B. Zuckerman, P. Palmer, "Microwave detection of interstellar formaldehyde", *Phys. Rev. Lett.* **22**, 679 (1969).

131. R. W. Wilson, K. B. Jefferts, A. A. Penzias, "Carbon monoxide in the orion nebula", *Astrophys. J.* **161**, L43 (1970).

132. G. Winnewisser, E. Herbst, H. Ungerechts, "Spectroscopy among the stars", in *Spectroscopy of the Earth's Atmosphere and Interstellar Medium*, ed. by K. Narahari Rao, A. Weber (Academic Press, Boston, 1992), p. 423.

133. G. J. Melnick, J. R. Stauffer, M. L. N. Ashby, E. A. Bergin, G. Chin, N. R. Erickson, P. F. Goldsmith, M. Harwit, J. E. Howe, S. C. Kleiner, D. G. Koch, D. A. Neufeld, B. M. Patten, R. Plume, R. Schieder, R. L. Snell, V. Tolls, Z. Wang, G. Winnewisser, Y. F. Zhang, "The submillimeter wave astronomy satellite: science objectives and instrument description", *Astrophys. J. Lett.* **539**, L77 (2000).

134. R. L. Snell, J. E. Howe, M. L. N. Ashby, E. A. Bergin, G. Chin, N. R. Erickson, P. F. Goldsmith, M. Harwit, S. C. Kleiner, D. G. Koch, D. A. Neufeld, B. M. Patten, R. Plume, R. Schieder, J. R. Stauffer, V. Tolls, Z. Wang, G. Winnewisser, Y. F. Zhang, G. J. Melnick, "The distribution of water emission in M17SW", *Astrophys. J. Lett.* **539**, L97 (2000).

135. G. L. Pilbratt, "The ESA FIRST cornerstone mission", *Proc. SPIE* **4013**, 142 (2000).

136. A. Poglitsch, C. Waelkens, N. Geis, "The photoconductive array camera and spectrometer (PACS) for FIRST", *Proc. SPIE* **4013**, 221 (2000).

137. M. Griffin, B. Swinyard, L. Vigroux, "The SPIRE instrument for FIRST", *Proc. SPIE* **4013**, 184 (2000).

138. N. Whyborn, "The HIFI heterodyne instrument for FIRST: capabilities and performance", in *The Far Infrared and Submillimeter Universe*, Eds. G. Pilbratt, S. Volonte, A. Wilson, ESA SP-401, p. 19 (1997).

139. J. C. Pearson, R. Guesten, T. Klein, N. D. Whyborn, "The local oscillator system for the heterodyne instrument for FIRST (HIFI)", *Proc. SPIE* **4013**, 264 (2000).

140. S. Kraft, O. Frenzl, O. Charlier, T. Cronje, R. Katterloher, D. Rosenthal, U. Grozinger, J. Beeman, "FIRST-PACS: design and performance of the sensor engineering models", *Proc. SPIE* **4013**, 233 (2000).

141. N. Hiromoto, T. Itabe, T. Aruga, H. Okuda, H. Matsuhara, H. Shibai, T. Nakagawa, T. Saito, "Stressed Ge Ga photoconductor with a compact and stable stressing assembly", *Infrared Phys.* **29**, 255 (1989).

142. G. J. Stacey, J. W. Beeman, E. E. Haller, N. Geis, A. Poglitsch, M. Rumitz, "Stressed and unstressed Ge Ga detector arrays for airborne astronomy", *Int. J. Infrared Millim. Waves* **13**, 1689 (1992).

143. T. M. Dame, D. Hartmann, P. Thaddeus, "The Milky Way in molecular clouds: a new complete CO survey", *Astrophys. J.* **547**, 792 (2001).

144. H. Ungerechts, P. Umbanhowar, P. Thaddeus, "A CO survey of giant molecular clouds near Cassiopeia A and NGC 7538", *Astrophys. J.* **537**, 221 (2000).

145. P. Schilke, T. Groesbeck, G. Blake, T. G. Phillips, "A line survey of the Orion KL from 325 to 360 GHz", *Astrophys. J. (Supp.)* **108**, 301 (1997).

146. P. Schilke, D. J. Benford, T. R. Hunter, D. C. Lis, T. G. Phillips, "A line survey of Orion-KL from 607–725 GHz", *Astrophys. J. (Supp.)* **132**, 281 (2001).

147. J. R. Pardo, E. Serabyn, J. Cernicharo, "Submillimeter atmospheric transmission measurements on Mauna Kea during extremely dry El Nino conditions: implications for broadband opacity contributions", *J. Quant. Spectrosc. Radiat. Transfer* **68**, 419 (2001).

Terahertz Imaging

Daniel Mittleman

1 Introduction

Recent years have seen a resurgence of interest in the science and technology of far-infrared radiation. A wide range of new techniques for the generation and detection of far-IR light have been proposed and demonstrated, many of which have relied on nonlinear optical techniques. Although the use of nonlinear optics for far-infrared generation was proposed as early as 1976 [1], widespread application has been hindered until recently by the limited availability of suitable laser sources. With the rapid advances in laser technology, the use of techniques such as optical rectification and difference frequency generation for the production of far-infrared radiation have proliferated. One of the earliest and most promising of these techniques is terahertz time-domain spectroscopy (THz-TDS), a method which relies on the generation of broadband electromagnetic transients using ultrafast laser pulses. This method, pioneered in the late 1980s by groups at AT&T Bell Laboratories [2] and IBM's T. J. Watson Research Center [3], has become increasingly popular with the widespread availability of femtosecond laser sources. Because of the challenges associated with more conventional methods of far-infrared research, this novel THz source has proven to be quite valuable for spectroscopy. Further, the ultrashort duration of the THz pulses, the associated broad bandwidth, and the fact that these pulses are inherently synchronized with a femtosecond pulse train have enabled several novel measurement techniques, which would otherwise be impractical. Examples include time-resolved techniques such as the optical pump, THz probe measurement, THz emission spectroscopy, and correlation spectroscopy. Nuss and Orenstein have recently reviewed the application of THz-TDS to the study of solid-state systems; this work contains numerous examples of both linear spectroscopy and these other techniques [4]. THz-TDS has also been used to study gases, most notably by Grischkowsky and coworkers [5–11].

In 1995, Hu and Nuss reported the first use of THz-TDS for imaging [12]. This work has sparked a flood of interest in the field and in the public media [13], as it immediately suggested a range of new and interesting applications for this spectrometer. The ability to form images in a reasonable time, combined with the unique properties of THz radiation, has inspired an amazing array of suggested applications areas, ranging from biomedical diagnosis to the monitoring of the water content of packaged food products and to fault detection in packaged integrated circuits. In this chapter, we review the rapid developments in THz "T-ray" imaging which have occurred since the initial report. We describe a number of different examples, which illustrate the capabilities and limitations of this novel imaging technique. This chapter does not seek to provide a tutorial on how to construct a THz imaging system: numerous different system configurations have been employed, and an exhaustive review of these is beyond the scope of this chapter. We provide a description of a typical system, highlighting those features which are important for the various imaging modes discussed later. The reader is referred to a number of reviews on this subject for more information [4,14 16]. We note that focal-plane imaging using free-space electro-optic sampling has also been demonstrated [17]. This technique is related to the methods described here. and is discussed in the chapter by Jiang and Zhang in this volume.

2 Key Components of a THz Imaging System

Figure 1 is a schematic diagram of a typical THz-TDS spectrometer used for T-ray transmission imaging. The spectrometer consists of a femtosecond laser, a computer-controlled optical delay line, an optically gated THz transmitter, optics for collimating and focusing the THz beam, the sample to be imaged, an optically gated THz receiver, a current preamplifier, and a digital signal processor (DSP) controlled by a personal computer. The gross features of this setup have not altered substantially since the initial report of the first THz-TDS system in 1989 [3]. However, there have been a number of significant advances, which have enabled the imaging applications. Several of these, including the use of a rapid scanning optical delay line and DSP data acquisition, were first noted by Hu and Nuss [12]. Others, such as the

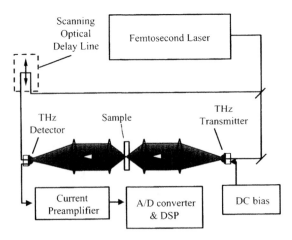

Fig. 1. Schematic of a terahertz time-domain spectrometer used for transmission imaging. The sample is located at an intermediate focal plane, and can be scanned in the two transverse dimensions for image formation

use of structured electrodes on the THz emitter and detector antennas, were described later [18,19]. Here, a brief comment on the state of the art of each component is presented.

2.1   The Femtosecond Laser Source

The vast majority of the work in this field has been carried out using mode-locked Ti:sapphire lasers, operating near a wavelength of 800 nm. These are the most widespread femtosecond laser sources, and the laser parameters (pulse duration, wavelength, and output power) are all ideally suited for driving GaAs-based THz emitters and detectors, as well as radiation-damaged silicon-on-sapphire (rd-SOS). These lasers typically provide excellent pulse-to-pulse and long-term stability, and are relatively easy to operate. Indeed, the first commercialized version of the THz-TDS system, announced in January 2000, is based on Ti:sapphire technology [20].

Despite this impressive level of success, the possibility of employing different femtosecond laser sources is of interest, for several reasons. First, although the noise performance of typical Ti:sapphire laser systems is quite good compared with most other femtosecond sources, it can be the limiting factor in the noise in THz waveform measurements. Second, these lasers are usually quite sensitive to small changes in optical alignment, and are thus not stable against mechanical vibrations. In order to construct a portable THz imaging system, a more robust femtosecond laser is desired. An excellent alternative is the mode-locked fiber laser, in which the light pulses propagate entirely within an optical fiber. Such lasers are commercially available, and a considerable amount of research is being directed towards the development of THz emitter and receiver antennas which can operate when driven at ~ 1550 nm, the wavelength of these fiber sources. One can no longer use GaAs, since this wavelength is below the band gap of the material. Other candidate materials include GaAs-based ternary and quaternary alloys [21,22], InAs [23], and ion-implanted germanium [24].

2.2   Optical Delay Line

The THz-TDS system requires a means for varying the delay of one optical beam relative to a second, in order to move the sampling gate across the waveform to be sampled. In nearly all cases, this is accomplished by varying the optical path length traversed by one of the beams. Often, the alignment of the optical beam onto the detector is more sensitive than the alignment of the generating beam, although this depends on the details of the antenna structures employed. In these cases, it makes more sense to place the variable delay in the generating arm. This optical delay is typically generated using a retroreflecting mirror arrangement mounted on a mechanical scanner. This can either be a slow stepper motor or a rapidly oscillating device such as a galvanometer. The speed of the scanner dictates the arrangement for the data acquisition, as described below.

In many imaging applications, it is desirable to scan the optical delay as rapidly as possible, in order to increase the waveform acquisition rate. There are a number of difficulties associated with this; one problem is that there are not many mechanical delay lines which can provide a sufficient delay window at higher than a few tens of hertz. To obtain a window of 100 ps, the retroreflecting mirror arrangement must oscillate with an amplitude of ~ 1.5 cm, ideally with a displacement which varies linearly with time over as much of this window as possible. This is difficult to achieve using mechanical shakers; one device which has been commonly employed, manufactured by Clark MXR, Inc., has a mechanical resonance near 20 Hz, and cannot be driven safely beyond ~ 100 Hz. For this reason, the majority of the point-by-point THz images taken to date have been acquired at a rate of 20 pixels per second, which translates to a few minutes per image. For certain applications, a much smaller delay window (e. g. a few picoseconds) is sufficient, in which case much higher waveform acquisition rates can be obtained by mounting a mirror on a piezoelectric transducer. Numerous other methods for varying the timing of the two pulse trains have been proposed, although none are yet in widespread use in THz systems.

2.3   Terahertz Optoelectronic Switches

While the THz transmitter and receiver designs were identical in early THz-TDS work, a wide range of different structures have been investigated, and optimized either for maximum signal or maximum bandwidth. Grischkowsky and coworkers pioneered the coplanar strip line for use as a THz emitter [25]. This structure is appealing both for the simplicity of the design and for the extremely broadband emission which can be produced. Using 60 fs pulses, spectral coverage to 6 THz can be achieved [26].

For many of the imaging applications explored to date, more complex structures have been used to produce a larger signal at the expense of the enhanced bandwidth. Figure 2a shows one commonly used design, in which the emitted signal is optimized through the use of a structured electrode [18]. This antenna consists of two 10 μm wide metal lines deposited on semi-insulating GaAs, with a separation of 100 μm. Two metal tabs extend out from these lines towards each other; these tabs may terminate in either flat or triangular ends. The gap between the two tabs can be as small as 5 μm, or as large as 100 μm. A voltage is applied across the two lines, creating a strong depletion field near the anode. Focusing a few tens of milliwatts of the femtosecond beam near the corners or points of the anode structure gives rise to a greatly enhanced THz emission, as a result of this enhanced depletion field. Figure 2b shows the enhancement of the emitted THz field, as the laser focus is moved relative to the position of the emitter antenna. Numerous other antenna designs have been employed [19,27].

(a)

(b)

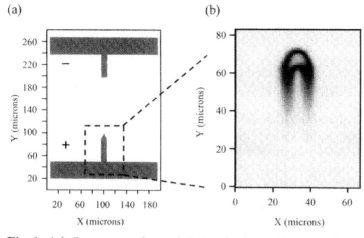

Fig. 2. (a) One commonly used design for high-efficiency THz emitters, with a pointed electrode to enhance the electric field strength. (b) Strength of the peak THz emission as a function of the location of the optical beam spot relative to the pointed electrode. The THz field is most intense when the optical beam is focused at the point [18]. Data supplied courtesy of I. Brener

The most commonly used THz receiver design has been a simple dipole antenna, roughly $50\,\mu m$ in length [2]. A number of variations on this design have been employed, such as the interdigitated structure shown in Fig. 3. For a Hertzian dipole antenna, with a length much less than the wavelength of the radiation, the collection sensitivity and radiation efficiency both vary inversely with the wavelength. Thus, smaller dipole antennas provide a broader-bandwidth response; dipoles as small as $30\,\mu m$ have been used. Of course, this $1/\lambda$ dependence no longer applies when the radiation wavelength in the substrate becomes comparable to the dipole length, so the details of the high-

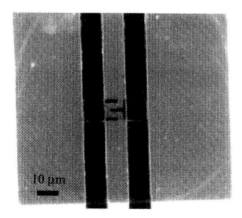

Fig. 3. Photograph of an interdigitated electrode design lithographically deposited on radiation-damaged silicon-on-sapphire, for use as a THz receiver

frequency response are more complex. In addition, the low-frequency roll-off in these quasi-optical systems is typically limited by diffraction effects due to the finite size of optical elements such as the substrate lenses mentioned below. As a consequence, the $1/\lambda$ dependence may apply over only a rather limited spectral range. A detailed analysis of a THz-TDS system can accurately predict the measured spectral response, as long as all of these effects are taken into account [15].

If very small dipole antennas are used for both transmission and detection, the high-frequency limit is generally determined by the temporal response of the system. One obvious manifestation of this limit is the duration of the laser pulses used to generate the THz radiation. The majority of THz-TDS experiments have employed pulses of $\sim 100\,\mathrm{fs}$ duration, since these are the easiest to obtain from commercial systems. Such pulses can be used to generate radiation via a number of different mechanisms, including the "current surge" mode, which produces the largest signals, and optical rectification, which provides the broadest bandwidth. Either way, the radiation cannot contain frequency components varying more rapidly than the derivative of the pulse envelope, which limits the bandwidth to $\sim 4.5\,\mathrm{THz}$ for 100 fs pulses. Consequently, shorter pulses are often used to generate bandwidth extending into the mid-infrared [28–30].

A second important limiting factor is the response time of the photoconductive material used as a substrate for the THz antennas. This limit is most pronounced in the receiver antenna, since one does not expect the highest measured frequency to exceed the inverse of the temporal width of the photogenerated sampling gate. In most cases, this duration is limited by the carrier-trapping time, although a number of schemes have been proposed to avoid this limitation [31,32]. As a consequence, the most commonly used materials are those which are characterized by subpicosecond free-carrier lifetimes, as well as strong optical absorption at $\sim 800\,\mathrm{nm}$, high carrier mobilities, and low dark currents. The two most common choices are radiation-damaged silicon-on-sapphire [14] and low-temperature-grown GaAs [33]. The former material has been demonstrated to provide an excellent high-frequency response, producing THz pulses shorter than 400 fs [25,34]. The latter provides performance comparable to that of rd-SOS, although with somewhat lower bandwidth [35].

It should be noted that it is possible to detect broadband THz radiation even using a photoconductive antenna with a slow carrier lifetime. In this case, one relies not on the width of the sampling gate, but merely on its fast-rising edge. If the photocurrent can be modeled as a step function, then the detector operates as a fast sampling gate in an integrating mode, and the measured signal is proportional to the integral of the incident THz field [31]. In this configuration, the bandwidth is limited by the speed of the rising edge of the current, which is determined by the duration of the optical pulse, and by the RC time constant of the antenna. Bandwidths up to 20 THz have been reported recently, using substrates with carrier lifetimes of more than 1 picosecond [36].

2.4 Terahertz Beam Optics

A key ingredient to a high-performance T-ray imaging system is an optical system that (1) allows one to focus the THz waves to a diffraction-limited focal point at the object, and (2) has the highest possible throughput. Unlike optical systems designed for visible light, the wavelength of the THz electromagnetic signals is not negligible compared with the size of the optical elements used, and diffraction effects can dominate ray propagation. This can be a significant complication in the design of optical systems for the THz frequency region. Additionally, because of the large spectral bandwidth, the optical system needs to be achromatic and exhibit a flat phase response over the frequency range of the pulse.

A critical component that enables these features of the THz beam system is the substrate lenses, which are attached to the backs of the transmitter and receiver chips [3]. The substrate lenses improve the coupling of the light into and out of the photoconducting antennas, and suppress the excitation of slab modes. These lenses are generally composed of high-resistivity (greater than $10^4\,\Omega\,\mathrm{cm}$) silicon, which is dispersionless over the whole THz range [37]: as a result, these lenses exhibit no chromatic aberration.

Two different lens designs are commonly used. The first of these places the transceivers at the focal point of the substrate lens [14,15,38]. With this design, the distance from the transceiver to the tip of the lens is given by $R[1 + 1/(n-1)]$, where R is the radius of the lens and n is the refractive index of the lens material, assumed to be the same as that of the substrate. With this collimating hyperhemispherical design, rays emitted near the optical axis emerge as a collimated beam. However, light emitted at larger angles emerges at substantial angles, so the lens is astigmatic. At still larger angles, radiation is internally reflected at the lens–air interface, and thus lost. A ray-tracing diagram illustrating these effects is shown in Fig. 4a. Further, the threshold angle of total internal reflection defines an effective aperture size for the lens, which in turn can lead to additional diffraction of the emerging beam.

A second commonly used substrate lens design is the aplanatic hyperhemisphere, for which the transceiver-to-lens-tip distance is $R(1 + 1/n)$ [4,16]. This arrangement ensures that the critical angle for total internal reflection corresponds to rays emitted at $90°$; that is, parallel to the plane of the substrate. Thus, the light lost to internal reflection is minimized, and simultaneously the effective lens aperture is maximized. It is relatively straightforward to show that, for silicon substrate lenses ($n = 3.42$), the effective aperture is larger using the aplanatic design, by roughly 13%. More significantly, the emerging beam does not exhibit the astigmatism of the collimating design but, rather, diverges as a Gaussian beam with a half-angle of $\sin^{-1}(1/n)$. Figure 4b shows a ray diagram for an 8 mm diameter aplanatic silicon hyperhemisphere.

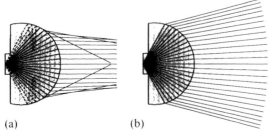

(a)   (b)

Fig. 4. Ray-tracing diagrams of the two lens designs described in the text. (a) Collimating lens design. The aberration of the wavefront, arising from rays propagating close to the critical angle inside the lens, is evident. The rays represented by dashed lines are those that are trapped inside the lens by total internal reflection. (b) Hyperhemispherical lens design, in which no rays are internally reflected, and the emitted radiation emerges with a divergence half-angle of about 15° and no wavefront aberration. These diagrams are shown to scale for a substrate lens of 4 mm radius, with a refractive index equal to that of silicon, $n_{Si} = 3.418$. In both cases, the antenna is fabricated on a 2 mm square GaAs substrate, also shown. In the hyperhemispherical design, a substrate of this size can interfere with the propagation of radiation at large angles, and may decrease the emission efficiency as a result. The diagrams neglect the small index difference between GaAs and Si

The implications of the substrate lens design are most important when one is considering the optical system used to collect and manipulate the THz beam [3,15,16]. For example, a common design would be to place a free-space collection lens or off-axis parabola in front of the emitter, to produce a collimated beam. In the case of the collimating substrate lens, the need for this collection lens can only be understood if diffraction is included, since in the ray-optics description this beam is already collimated. Since the beam that reaches the collection lens is strongly perturbed by diffractive effects, the beam after the collection lens has a frequency-dependent transverse spatial profile. Radiation with higher frequencies propagates closer to the optical axis. If this beam is re-imaged using a second free-space lens, the resulting focal spot is frequency-independent, which can be advantageous for such applications as coupling into waveguides, for example [39]. On the other hand, in the case of the aplanatic substrate lens, one might imagine that the beam is less affected by diffraction, since it emerges as a diverging source (see Fig. 4(b)). In this case, the beam after the collection lens has a transverse profile that is independent of frequency [16]. Re-imaging the beam produces a focal spot with a strongly frequency-dependent diameter. This can be exploited to improve resolution in imaging applications [40]. Also, in the ray-optics picture the beam after the collection lens is collimated in one case, but comes to a beam waist quite rapidly in the other. This could have important implications for the design of a THz system, since it would dictate the required separation between the emitter and receiver antennas [14]. However, when diffraction effects are included, it is no longer clear that this distinction is important.

Rudd and Mittleman have recently undertaken a thorough experimental and computational comparison of these two substrate lens designs [41]. Although both lens designs have been used extensively for many years, it is only relatively recently that direct comparisons have been performed. Several groups have described simulated and experimental results for the parameters of the beam emerging from the lens, as a function of the (cylindrical) extension of the lens beyond hemispherical [42–44]. These results cover cases similar, though not identical, to the two lens designs mentioned above. This work has usually been concerned with the generation or detection of narrowband emission. The situation in which the emitted radiation is broadband, of particular relevance to THz time-domain spectroscopy, has not previously been addressed. As a result, the dramatic influence of the substrate lens design on the achievable bandwidth a in THz-TDS system has not been generally appreciated.

Figure 5 illustrates measured amplitude spectra, as a function of frequency and emission angle, for the two lens designs described above. These measurements represent the p-polarized E-plane emission from the emitter antenna, a 90° bow-tie fabricated on low-temperature-grown GaAs. In these data, the persistence of higher-frequency components at low angles is more noticeable for the collimating lens than for the hyperhemisphere. This can be understood directly from the ray diagram (Fig. 4), which is increasingly valid at higher frequencies. Also, these data illustrate the pronounced interference fringes at larger angles in both cases.

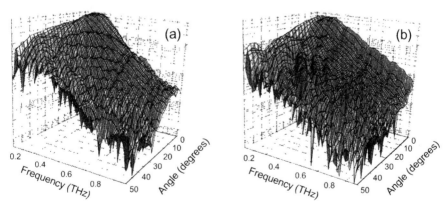

Fig. 5. Amplitude spectra of measured THz waveforms as a function of both frequency and emission angle. (**a**) E-plane emission from an antenna coupled to a hyperhemispherical lens. (**b**) E-plane emission from an antenna coupled to a collimating lens. These two plots are shown on common vertical scales, to facilitate comparisons of relative amplitudes. The vertical axes are logarithmic

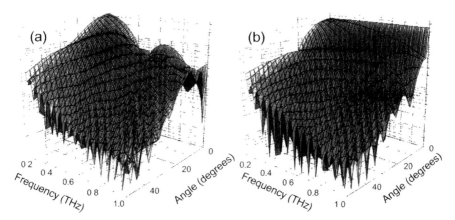

Fig. 6. Simulation of the E-plane emission pattern as a function of both angle and frequency, for (**a**) the hyperhemispherical lens design and (**b**) the collimating lens design. These show the amplitude spectra on a logarithmic scale, as in Fig. 5. The hyperhemispherical lens introduces strong interference fringes even at zero degrees, limiting the measurable emission bandwidth of the lens-coupled antenna. In contrast, the collimating design places no such limits on the spectrum, at least within the approximations of the calculation

Figure 6 shows a calculation which simulates these results. This computation is based on a numerical solution of the Fresnel–Kirchhoff diffraction integral, following the method of Jepsen and Keiding [38]. It is important to note that these calculations do not provide an accurate description of the relative amplitudes of different frequency components within the measured THz wave. These relative amplitudes are determined not only by the interference of the diffracting beam, but also by such factors as the duration of the optical pulses used to gate the antennas, the carrier lifetime in the detector antenna, and the size of the emitter and receiver dipoles [15]. None of these factors are included in these simulations. Nonetheless, it is instructive to see how much of the frequency response is determined solely by the diffraction effects modeled here. It is clear from Fig. 6 that, even in the absence of any other mechanism that could limit the measured THz bandwidth, the geometry of the substrate lens can have a dramatic effect. In the case of the hyperhemispherical lens, the spectral content is limited to ~ 0.6 THz even in the forward direction ($\theta = 0°$), in approximate agreement with the experiment (Fig. 5a). The bandwidth is severely constrained by the geometry of the substrate lens, independent of other limiting factors that are usually considered to be more significant. In contrast, with a collimating lens design, the geometry does not place any limits on the bandwidth along the optic axis, at least up to 1 THz. This highlights an important and usually neglected factor in the optimization of the THz bandwidth for spectroscopic measurements.

Once the radiation has emerged from the substrate lens, a bulk optical system is required to collimate and focus the THz beam in the region where it interacts with the sample to be imaged. This has most often been accomplished using off-axis paraboloidal reflectors, which are broadband and lossless. However, these optics are difficult to align properly, and as a result the THz beam can often exhibit severe spherical aberration. This not only degrades the spatial resolution in an image, but also reduces the measured signal at the detector, since the THz beam spot is distorted. For this reason, transmissive optics have become increasingly popular (see Fig. 1). For signals with less than 2 THz of bandwidth, high-density polyethylene (HDPE) is an excellent choice as an optical material, with very low absorption (less than $0.04\,\mathrm{cm}^{-1}$ at 300 GHz, rising to $\sim 1\,\mathrm{cm}^{-1}$ at 2 THz), little dispersion, and small Fresnel reflection losses (a refractive index of 1.52 leads to an insertion loss of order 9% per lens) [45,46]. HDPE exhibits a fairly narrow absorption resonance near 2.2 THz, which can be observed as oscillations following the single-cycle waveform (Fig. 7). For applications in which broad bandwidth is required, high-resistivity silicon lenses may be more appropriate. The disadvantages of silicon optics are the higher Fresnel losses ($\sim 50\%$ per lens) due to the larger index ($n = 3.418$ [37]) and the substantially higher cost. In order to alleviate the former difficulty, antireflection coatings may be employed. Several proposals have been made for such coatings [47,48], although no coatings have been reported which are sufficiently broadband to be useful for THz-TDS. A third alternative material is the polyolefine TPX, which has

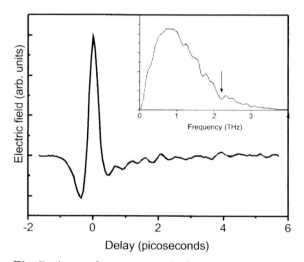

Fig. 7. A waveform measured after propagating through a piece of high-density polyethylene $\sim 1\,\mathrm{cm}$ thick. The oscillations which follow the initial transient result from the relatively narrow absorption feature in this material at $\sim 2.2\,\mathrm{THz}$. This can be observed as a dip in the amplitude spectrum (*inset*)

the advantage that it is transparent in the visible as well as in the THz range; further, the THz refractive index ($n = 1.46$) is quite similar to the index in the visible region ($n = 1.43$) [45]. This can greatly simplify the alignment of a complex THz optical system, since visible light can be used as a guide. TPX exhibits slightly more absorption than does HDPE at low frequencies, but the spectrum is devoid of features and the power absorption coefficient is still less than $1\,\text{cm}^{-1}$ even at $3\,\text{THz}$ [49].

The propagation of single-cycle THz pulses has been a very active area of research in recent years, in part because of the ease with which these pulses can now be generated and detected. The propagation of radiation pulses with large fractional bandwidths poses an interesting problem in diffractive optics. This problem was first modeled by Ziolkowski and Judkins, shortly after the development of THz-TDS [50], and subsequently explored by a number of groups [15,51 55]. Others have investigated the effects of apertures or waveguides [39,40,56,57]. Substantial reshaping of both the temporal and the spectral profile of the pulse can occur. In most cases, despite the large fractional bandwidths, Gaussian beam models have been sufficient to describe these effects.

2.5   Polarization of the THz Beam

As mentioned above, in most cases the emission from THz antennas has been described using the approximation of an ideal dipole [15], producing linearly polarized radiation. Of course, the dipoles used in real THz systems are not ideal, and so the polarization state of the radiation is not, in general, purely linear. The conventional wisdom in the field is that a typical emitter generates a cross-polarized component which is on the order of a few percent as large as the component polarized along the dipole. Cai and coworkers [19] found that the cross-polarized radiation had an amplitude roughly 7% as large as the dominant polarization component, although they provided no explanation for the origin of this small component. Garet et al. [58] reported a frequency-dependent variation in the linear polarization axis, attributed to substrate lens misalignment.

Rudd et al. [59,60] provided the first thorough characterization of the cross-polarized component of the field radiated from a lens-coupled THz antenna, similar to the antenna shown in Fig. 2. For the s-polarized E-plane emission, the largest peak-to-peak electric field amplitude is approximately 7% of the p-polarized emission, but this maximum emission occurs at an angle of approximately 6° away from the optical axis. The results are shown in Fig. 8. Figure 8a shows typical s-polarized waveforms measured at various emission angles, relative to the terahertz-beam optic axis $\theta = 0$. Figure 8b summarizes these results for all frequency components, demonstrating a pronounced minimum in the cross-polarized field on the optic axis.

These results have been interpreted as quadrupole, rather than dipole, emission, with the following mechanism [59]. The current flowing into the

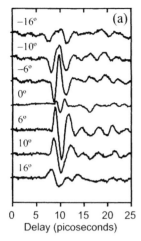
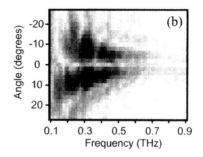

Fig. 8. (a) s-polarized E-plane waveforms emitted from a structured dipole emitter of the type illustrated in Fig. 2. These cross-polarized THz waveforms are more than an order of magnitude smaller than the component polarized along the dipole axis. Further, the waveforms are largest at angles of $\pm 6°$ relative to the terahertz-beam optic axis, $\theta = 0$. (b) Spectral amplitude of the measured cross-polarized THz radiation as a function of frequency and angle, illustrating the pronounced minimum along the optic axis

dipole is drawn from both ends of the strip-line on one side, and exits the dipole in both directions into the other strip-line. This current distribution is characterized by the length L of the dipole, and by the distance d from the dipole to the point at which the current vanishes. This static current distribution possesses zero dipole moment, but a quadrupole tensor with two nonzero elements $Q_{xy} = Q_{yx} \equiv Q_0$, where Q_0 is proportional to the product of d and L. This quadrupole tensor gives rise to an electric field, which is given by

$$\boldsymbol{E}_{\mathrm{Q}} \propto \sin\theta \, \cos\theta \, \sin 2\phi \, \hat{\theta} + \sin\theta \, \cos 2\phi \, \hat{\varphi} \,. \tag{1}$$

where ϕ is measured from the axis of the dipole. For E-plane emission ($\phi = 0$), $\boldsymbol{E}_{\mathrm{Q}}$ is an s-polarized wave with amplitude $\sin\theta$. It should therefore exhibit a null at $\theta = 0$, as well as a polarity reversal on either side of $\theta = 0$, just as observed in Fig. 8. The angular widths of the emission lobes can be understood in terms of Fresnel diffraction from the exit aperture of the substrate lens [59,60].

2.6   Signal Acquisition

The traditional method for acquiring THz waveforms relies on a photoconductive sampling technique. Here, the delay of the generating pulse is swept relative to that of the detecting pulse, and the average photocurrent generated in the receiver is measured as a function of the delay. The resulting signal is the convolution of the THz waveform with the temporal shape of the photoconductive sampling gate. In order to eliminate the majority of the external noise in these measurements, the signal was usually acquired with a lock-in amplifier, and one of the gating beams (or the THz beam) was modulated with a chopper wheel. Since this required a lock-in time constant in the range of tens to hundreds of milliseconds, the sweep time of the delay was quite slow, on the order of hundreds of milliseconds or more per data point. At this rate, it took several minutes to acquire a single 1024-point THz waveform.

In order to obtain images, where the entire waveform is measured and analyzed at each pixel, this acquisition time must be reduced dramatically. To accomplish this, Hu and Nuss replaced the slow stepper motor with a scanning optical delay line (ODL) [12]. This consists of a small retroreflecting corner cube mounted on a galvanometric shaker. It typically provides up to 100 ps of optical path delay on each sweep, and can sweep at up to 100 Hz. With this device, one directly detects the photocurrent as a function of time, using a synchronized signal from the ODL as a periodic trigger. In most cases, the antenna is connected directly to a current-to-voltage preamplifier, so that a voltage can be observed on an oscilloscope or digitized. Dispensing with the noise filtering provided by the lock-in detection does degrade the signal-to-noise ratio (SNR) somewhat, but allows much faster acquistion times. The amount by which the SNR is diminished depends on the details of the noise sources, but a factor of 100 is not unusual. Nonetheless, a waveform measured with a single sweep of the ODL can typically have an SNR of 1000 or more, which is sufficient for many imaging applications. Figure 9 shows a representative comparison between waveforms acquired using an ODL and lock-in data acquisition.

2.7   Data Processing

One of the most important aspects of THz imaging involves the processing of the waveforms acquired at each pixel of the image. Each waveform contains a large amount of information, and the reduction of this to a single, color-encoded pixel can be a challenging task. Because of the diversity of samples to be studied and properties to be investigated, the development of a single algorithm for signal processing is impractical. For similar reasons, it makes sense to discuss the processing issues which arise in each particular context individually. The unique set of signal-processing challenges associated with each imaging mode is deferred to the following section, where we provide a number of illustrative examples of images obtained with a THz-TDS system.

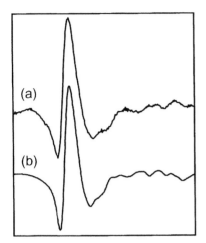

Fig. 9. A comparison of lock-in and ODL data which does not yet exist. (a) THz waveform acquired by averaging multiple sweeps of an optical delay line. (b) The same waveform, acquired using a conventional lock-in amplifier. The two waveforms were acquired with roughly equivalent levels of signal averaging

3 Imaging with THz-TDS

Numerous examples of submillimeter- and millimeter-wave imaging can be found in the literature which predate the first demonstration of T-ray imaging [61–63]. In at least one case, an imaging system using an optically pumped methanol laser was installed in a factory for on-line inspection of polyethylene high-voltage cable insulation [64]. However, the practicality of most of these examples for application in "real-world" situations has been limited at best. Even the aforementioned example, although installed in a manufacturing environment, employed an 80 watt CO_2 pump laser, high-speed scanning optics, and liquid-helium-cooled photodetectors, and occupied an entire room [65].

The THz "T-ray" imaging system offers a promising alternative to such cumbersome inspection systems, with a number of advantages. The THz-TDS system can be designed to be compact, efficient, and relatively inexpensive. Unlike the case for many far-infrared systems, no cryogenics are required. In addition, the sub-picosecond duration of the THz pulses and the phase-sensitive detection combine to provide a number of unique imaging modes. This range of advantages has prompted the development of a commercial version of the T-ray system. This device, announced in early 2000 [20], is the first commercially available far-infrared imaging system. In the following paragraphs, we provide several examples of T-ray images, in order to illustrate the unique sensing capabilities of such a system.

3.1   Amplitude and Phase Imaging

Once a THz-TDS system with an intermediate focus is constructed (see Fig. 1), one can place an object at this focus and measure the waveform of the THz pulse which has traversed the object. By translating the object and measuring the transmitted THz waveform for each position of the object, one can build an image pixel by pixel. The image formed in this fashion may represent any desired aspect of the measured waveforms, including amplitude [12,16], phase [16,66], or even combinations of these two [67]. This versatility provides a number of different methods for imaging any given object, each of which can reveal different properties of the sample.

The list of possible applications of such a system is quite extensive. Perhaps the most promising applications lie in the area of quality control of packaged goods. Figure 10 shows a THz image of a $\sim 2\,\mathrm{cm}$ square portion of a small (1-3/8 oz.) box of cereal. The cardboard box in which the cereal is packaged is nearly transparent to the THz radiation. The dark areas in the image are raisins, which exhibit a high contrast relative to the surrounding material due to their high water content. In this image, the thickness of the sample ($\sim 5\,\mathrm{cm}$) is somewhat larger than the confocal parameter of the THz beam ($\sim 1\,\mathrm{cm}$); as a result, the raisins (which were not situated at the THz beam focus) appear larger in the image than their actual size. Since one can in principle choose any set of THz beam optics for an imaging system, this effect should not substantially limit the utility of the technique. This imaging tool is well suited for inspection of sealed packages if the packaging is composed of transparent materials such as cardboard, most plastics, thin pieces of dry wood, etc.

Figure 11 illustrates another use of THz imaging as a quality control monitor. This figure shows a THz image of a portion of an automobile dashboard, representative of innumerable mold-fabricated plastic parts. This part consists of two parallel black plastic sheets, with a rubber foam padding between

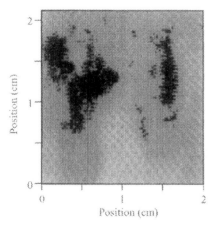

Fig. 10. THz transmission image of a portion $\sim 2\,\mathrm{cm}$ square of a small (1-3/8 oz.) box of cereal. The cardboard box in which the cereal is packaged is nearly transparent to the THz radiation. The dark areas in the image are raisins, which exhibit a high contrast relative to the surrounding material due to their high water content

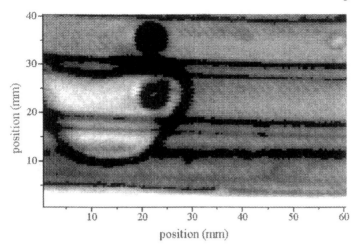

Fig. 11. THz transmission image of a portion of an automobile dashboard, showing a 1.5 cm diameter void in the foam padding between the two plastic surfaces. This type of manufacturing defect is very difficult to detect by any other means

them. In the manufacturing process, the foam is sprayed between the plastic sheets, holding them together when it dries. However, occasionally bubbles or voids develop in the foam, sometimes as large as an inch in diameter or more. Detection of these voids is a significant quality control issue, and no simple method currently exists. X-ray transmission does not provide a high contrast between the plastic–rubber foam and the air; in addition, health and safety issues preclude the use of x-ray diagnostics here. Ultrasound analysis is effective only with the use of an index-matching fluid, while other probing techniques such as magnetic resonance imaging, are too expensive and cumbersome. Probing the parts with microwaves would work quite nicely, except that it would be difficult to detect voids significantly smaller than the wavelength of the radiation used. Since, in this case, it is desirable to detect voids smaller than one centimeter in diameter, this effectively eliminates conventional microwave analysis from consideration.

The THz imaging technique exploits the fact that the rubber foam filling has many tiny air pockets, which act as good scattering sites for radiation in the 0.5 mm wavelength range. The solid plastic surfaces on the front and rear of the sample are fairly transparent to the THz radiation. As shown in Fig. 11, the THz image clearly locates the void, identified as an increase in the transmission through the sample wherever a smaller length of foam is traversed. THz imaging is extremely sensitive to small voids or other morphological variations which may occur inside solid plastic or nonconducting composite parts. It may seem somewhat fortuitous that the foam rubber in this particular example consists of scattering sites ideally suited for the wavelength range spanned by the THz pulses. However, one should recall that the THz radiation used in this imaging technique is extremely broad-band, spanning more than one order of magnitude in wavelength. Further, it should be emphasized that one can detect voids of this sort without relying on such a coincidence by observing the *transit time* of the THz pulse through the material, rather than the transmitted *amplitude*. The combination of the amplitude and phase information measured in THz-TDS makes for a very powerful tool for quality control measurements.

Figure 12 illustrates this, showing two THz images of a chocolate bar. In the upper image, the grayscale is determined by the peak-to-peak amplitude of the THz pulse at each pixel, as in Figs. 10 and 11. The chocolate does not absorb much THz radiation, but several other features are visible. First, the sample has a plano-convex cross-sectional profile, and is therefore thinner at the top and bottom than in the middle. Second, the embossed letters are visible only because of scattering effects at their stepped edges, and as a result are rather difficult to read. Finally, because almonds absorb more THz radiation than chocolate, they can be easily detected using this technique: four almonds are visible in this image.

The lower image shows the same sample, but this image is formed using the *phase* of the transmitted THz pulse, rather than the amplitude, to encode the false-color scale at each pixel. As a crude measure of phase changes, one may simply monitor the change in the arrival time of the peak (or first zero

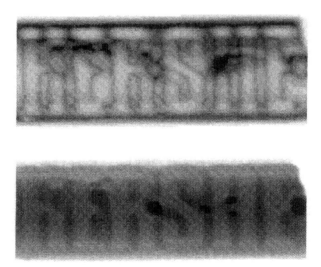

Fig. 12. THz images of a chocolate bar. In the *upper image*, the grayscale is determined by the peak-to-peak amplitude of the THz pulse at each pixel. The chocolate does not absorb much THz radiation, but several almonds are discernable. The *lower image* is encoded according to the travel time of the THz pulse through the image, which highlights variations in thickness more than changes in composition. Data courtesy of J.V. Rudd, Picometrix Inc.

crossing) of the THz waveform. Changes in this arrival time as a function of position on the sample indicate changes in the optical path length of the THz beam relative to the femtosecond beam which gates the receiver. This may result from either changes in the thickness of the sample as it is scanned transversely across the THz beam, or changes in the refractive index, or both. The change in arrival time Δt is given by $\Delta t = (1/c) \int n(z) \, dz$, where $n(z)$ is the refractive index sampled by the THz beam along its optical path, and where the integral is taken along the path. In the lower part of Fig. 12, the sample is fairly homogeneous (except for the almonds), and so the phase delay image primarily contains information about the thickness of the sample at each point. Thus, the embossed letters and the overall thickness variation are much more prominent. The almonds are nearly invisible, except for the very dark regions where the transmitted pulse was too small for an accurate determination of the arrival time.

Figure 12 demonstrates one way in which the phase information contained in the THz waveforms can be used to form images which contain different information from those encoded on the basis of amplitude alone. Another example, which illustrates the impressive sensitivity of the technique to small changes in delay, is shown in Fig. 13, where a small gas flame has been imaged in transmission. In this experiment, the ionized molecules which make up the flame may have a significant absorption at certain frequencies within the spectrum of the pulse [9], but the gas density is so low that this effect cannot be observed. However, the flame heats the air locally, which changes the refractive index $n(z)$ along the THz optical path. As a rough estimate of the magnitude of this effect, one may assume that the refractive index of

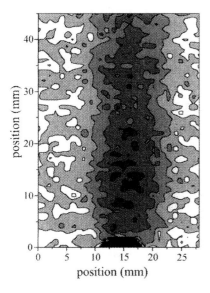

Fig. 13. THz image of a small gas flame, in which the grayscale is encoded according to the transit time of the THz pulse through the heated air. Adjacent contours correspond to shifts in transit time of only 5 fs

air varies inversely with temperature. If the flame has a thickness of roughly 5 mm, this would lead to a shift in the transit time of the THz pulse by $\sim 3\,\mathrm{fs}$ for a flame at 1000°C [68]. Each contour in Fig. 13 represents a shift of 5 fs in transit time. It is also worth noting that this and other THz-TDS measurements of flames [9] are unique for another reason. No other far-infrared detection system would be capable of performing these measurements, because the thermal emission from the sample would swamp the detector, and the much weaker probe beam would be nearly impossible to detect. Here, the gated detection and the averaging of incoherent fields combine to provide an impressive level of rejection for thermal radiation [14]. Not only does this permit spectroscopic measurements of flames, plasmas, or other thermally active samples, it also enables a host of interesting applications, such as in situ monitoring of sintering processes and engine exhaust testing.

These types of applications are not limited to quality control of consumer products. Figure 14 illustrates the use of THz imaging for the inspection of artwork. Figure 14a is a visible image of a piece of parchment, upon which an inscription has been written. The inscription has been painted over with a thick layer of black paint. The parchment was then imaged with T-rays, in an attempt to reveal the hidden writing. Because the black paint overlayer is not a lead-based paint, it is transparent to THz radiation. Depending on the writing implement used to form the original inscription, the THz image may or may not show the underlying writing. Figure 14b shows a THz image of one sample, in which the original inscription ("OK") was written with a pencil. The thin graphite layer of the writing on the parchment provides sufficient amplitude contrast to be visible in the THz image. Not surprisingly, if the inscription is written with conventional ink, one cannot detect it in the THz image, because the dye molecules in the ink are nonpolar and thus do not absorb strongly in the THz range. This test suggests the possibility of using THz imaging for the investigation of underdrawings beneath paintings. This could be an excellent complementary technology to the mid-infrared and x-ray imaging systems currently used for such studies. More broadly,

 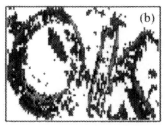

Fig. 14. (a) Visible and (b) THz images of a piece of paper on which some letters ("OK") have been written with a charcoal pencil, and then painted over with black latex paint. The paint layer is fairly transparent, but the graphite from the pencil provides sufficient contrast to form an image

one can imagine that this new technique could be useful in a wide range of fields, including art history, archaeology, forensics, etc., in which one requires noninvasive methods for imaging. In this context, it is worth emphasizing that the power generated in the T-ray beam is extremely low, roughly 20–100 times less than the same power emitted from a room-temperature black body in the frequency range 0–3 THz. Thus, the risks to delicate or sensitive samples are minimal.

3.2   Terahertz Imaging of Liquid Water

The simplest type of image one can generate in a transmission imaging setup is one in which the transmitted power determines the nature of the image. A good example of this is shown in Fig. 15, which shows a THz transmission image of a leaf from a common houseplant. The color scale is determined by the amplitude of the transmitted THz power, which in turn is related to the amount of moisture present at each point on the leaf. (The dashed line across the figure refers to the measurements of Fig. 16, discussed below.) This image illustrates the extreme sensitivity of this technique to water content. In this frequency range, water absorbs quite strongly. Recently, a number of groups have used THz-TDS techniques to measure the properties of liquid water, in both transmission [69,70] and reflection [71] geometries. Typical values of $\alpha \sim 230\,\mathrm{cm}^{-1}$ at $\nu \sim 1\,\mathrm{THz}$ have been measured. This value has direct implications for the ultimate sensitivity of THz imaging to changes in water content. With a signal-to-noise ratio of 100:1 in the measured electric field, the minimum detectable water concentration is given by $nx \sim 10^{16}\mathrm{cm}^{-2}$, where n is the density of water molecules and x is the length of the path traversed by the THz beam in the material. In a material with a thickness of 1 mm, this implies a detection limit of less than 10^{-5} of liquid density.

This extreme sensitivity to water content can be exploited in measurements such as those illustrated by Fig. 15. Indeed, this is of great interest as a method of measuring the water content of leaves on *living* plants. Currently, there is no accepted, nondestructive procedure for measuring the leaf

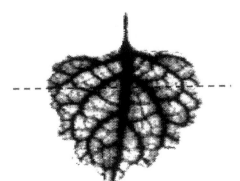

Fig. 15. THz transmission image of a living leaf from a common houseplant. This plant was water-starved for several days before this image, so the stems, containing most of the water, exhibit large contrast. The dashed line across the figure represents the location of the line scans shown in Fig. 16

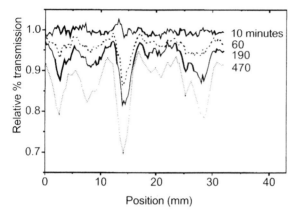

Fig. 16. Relative percentage transmission of the THz pulse through a living leaf, as a function of position along the dashed line in Fig. 15. The plant was water-starved for several days before these measurements, and watered at $t = 0$. Water uptake into the leaf is clearly observed, as the transmission drops steadily in the hours following the watering

water status of a transpiring plant [72]. Research in the field of plant water relations has been limited to point-in-time observations that provide average values across the tissue being studied [73]. Multiple spatial and repeated temporal observations have been required to account for the dynamic movement of water through the plant and for spatial variation in water status in individual leaves. Using THz imaging, it is possible to account for inhomogeneities in the sample by performing a spatially resolved measurement. Because these measurements are inherently nondestructive, repeated measurements may be made on the same tissue, thereby providing for the study of water flow dynamics. A demonstration of this application is shown in Fig. 16, in which the percentage change in amplitude of the transmitted THz radiation (relative to the transmission immediately before watering) is shown as a function of horizontal position across a leaf, at various delay times after the plant was watered. The approximate location on the leaf of these line scans is shown by the dashed line in Fig. 15. In Fig. 16, the variations observed as a function of position on the leaf, which are repeated at each delay time, reflect the different transmission coefficients through different parts of the sample. These result primarily from the stem structure; as seen in Fig. 15, the region of the line scans intersects the primary or central stem, as well as a number of subsidiary stems on either side. As a function of time after watering, the transmission decreases by several percent, indicating that water is absorbed into the leaf on a timescale of hours. Also, it is clear that water is absorbed more rapidly in the stems than in other parts of the leaf. Experiments such as this can be used to understand the early warning signs of plant water stress, and may be valuable aids in irrigation management for a wide range of crops. Recently, other far-infrared techniques have been used to validate the measurement of leaf water content using THz radiation [74], although in these experiments the spatial resolution of the images was limited by the shortest wavelength generated, approximately 0.6 mm.

3.3 Processing for Amplitude and Phase Imaging

All of the images shown in the preceding two subsections used fairly simple methods for extracting information from the measured waveforms. For phase or time delay imaging, it is a simple matter to extract the arrival time of the waveform at each pixel. For amplitude imaging, several methods are possible. One can simply use the peak-to-peak amplitude of the time-domain waveform, for a measure of bandwidth-averaged transmission. Alternatively, one may perform a fast Fourier transform on each waveform, and then integrate over any desired frequency band in order to retain some measure of spectral information. This method is of particular interest if high spatial resolution is desired. By integrating only over the high-frequency portion of the waveform, one obtains only the information corresponding to those frequency components. Since the focal-spot size is inversely proportional to the frequency, one can essentially tune the spatial resolution in software, without manipulating any optics [40].

3.4 Reflection Imaging

As noted, the extreme sensitivity to liquid water can be of substantial benefit in certain applications. Yet, it seems to precludes an entire area of imaging research, namely that of biomedical diagnostics. Because of the strong absorption of liquid water, the penetration depth into living tissue is only a few hundred microns, far too small for imaging of internal systems. However, a THz imaging system which operates in a reflection geometry could be useful for studies of surface or near-surface properties of a wide range of materials, including opaque objects such as humans. Further, because of the short coherence length of the THz radiation, the reflection geometry presents a number of new imaging possibilities, including that of THz tomography. For this reason, as well as for reasons of instrumental simplicity, it is expected that many of the aforementioned applications will eventually be implemented in a reflection geometry.

Figure 17 shows one possible implementation of a reflection geometry, in which the THz beam is focused at the sample, and is reflected at near-normal incidence [66,75]. One can imagine performing time-of-flight measurements in a reflection geometry similar to those described in the transmission experiment above. This is similar in principle to ultrasonic imaging. A commonly encountered problem in ultrasonic imaging is the large difference in acoustic impedance between air and liquid or solid objects, which necessitates some form of index matching. For terahertz waves, the dielectric constants of many

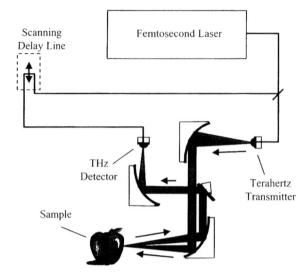

Scanning
Delay Line

Femtosecond Laser

THz
Detector

Terahertz
Transmitter

Sample

Fig. 17. Schematic of a reflection-mode THz imaging system

materials are not too different from that of air, and index matching is therefore not usually required.

THz time-of-flight imaging is also similar in certain respects to optical coherence tomography (OCT), in the sense that it is a reflection imaging technique which utilizes a broadband radiation source to achieve depth resolution. In OCT, a short-coherence-length optical source is split into two beams: one is reflected off the sample and the other off a reference. The two are then correlated to provide information about the reflectivity of the sample [76,77]. Distance measurements are possible because a correlation signal is only measured when the path length of the reference arm is equal to that of the sample arm, within the coherence length of the light source. As a result, the depth resolution is limited by the coherence length of the source.

As illustrated above, the THz measurement is not limited in the same way. The coherence length of a THz pulse is related to the pulse duration, which is on the order of several hundred femtoseconds. However, the measurement of the flame presented in Fig. 13 demonstrates a sensitivity to temporal shifts as small as 5 fs. This sensitivity stems from the phase-sensitive nature of the measurement, and from the fact that each THz pulse is a single cycle of the electromagnetic field. It is possible to measure the zero crossing of an electric field much more accurately than the peak of an intensity autocorrelation. Using this simple idea, it is possible to measure changes in either refractive index or thickness with extremely high accuracy. For example, one can detect, in transmission, a single sheet of paper added to a stack; in reflection, it is possible to achieve a depth resolution of a few microns. The limitation in this measurement is simply the accuracy with which the optical scanning delay line repeats its position on each scan. This is in contrast to the situation where two reflecting surfaces are close to one another; in that case, the ability to distinguish these two surfaces is limited by the coherence length of the radiation in the intervening medium. For typical THz imaging systems, this length is on the order of a few hundred microns. It is possible to overcome this limit, through the use of interferometric techniques. This method, which can provide a dramatic enhancement in the detectability of thin or subtle features in an object, is described below.

3.5 Burn Diagnostics

As noted above, a possible use of THz reflection imaging is in the area of biomedical diagnostics, where the samples may consist of only one reflecting surface, but with a complex morphology. An example of an application in this area is the study of surface or near-surface skin properties, such as in the diagnosis of burn depth and severity. A reliable noninvasive probe of burn depth would be of great value to clinicians, who currently have no such technology. The application of optical probes to this problem has been an active research area recently [78]. More examples of biological and biomedical applications of THz imaging can be found in the chapter by Koch in this volume.

Figures 18–20 show the results of a simple experiment to illustrate the idea of burn diagnosis using THz pulses. This demonstration experiment used chicken breast as a model tissue system. The tissue was burned using an argon ion laser, producing a series of circular burns of increasing severity. Figure 18 shows a terahertz reflection image of one such burn; here, the reflected THz energy is displayed in grayscale, with the center of the burn reflecting the least. Figure 19 shows a series of waveforms obtained by translating the sample relative to the THz spot, in 250 μm steps. The central waveform $E_{\mathrm{cent}}(t)$ (thick line) was reflected off of the center of the burn, while the waveforms at either end of the series (E_{end1} and E_{end2}) originated from reflections off undamaged tissue on either side of the burn. The distortions imposed upon E_{cent}, relative to E_{end1} and E_{end2}, reflect the modifications in the THz optical properties of the tissue as a result of the burn.

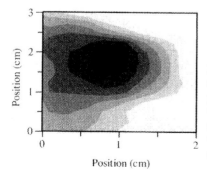

Fig. 18. THz reflection image of a burned region on a piece of chicken, used as a tissue phantom for these experiments. The burn was formed using the (circular) beam from an argon ion laser. The darker grayscale at the center of the burn indicates a decreased reflection of THz pulse energy

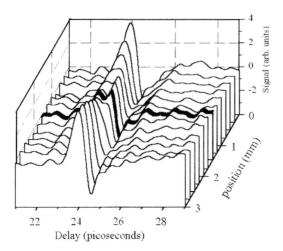

Fig. 19. A series of THz waveforms reflected from the burned region shown in Fig. 18. obtained by translating the sample across the THz beam spot in 0.25 mm steps. The waveform reflected from the center of the burn is the most distorted. and is highlighted

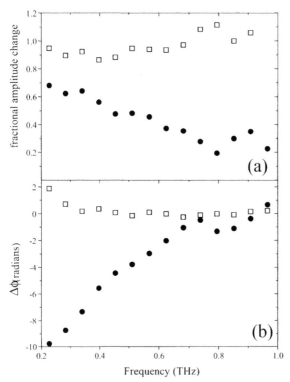

Fig. 20. (a) The *open squares* show the ratio of the spectral amplitudes of the two endmost waveforms in Fig. 19. This ratio is approximately unity. indicating that the unburned regions on either side of the burned tissue are nearly indistinguishable. The *solid circles* show the ratio of one of the endmost waveforms to the central waveform. showing the decreasing reflected amplitude at higher frequencies. (b) A comparison of the spectral phases of the two endmost waveforms (*open squares*). and of the endmost with the central waveform (*solid circles*)

We may analyze the waveform distortions by comparing E_{cent} with E_{end1} in the frequency domain. Figure 20 shows this comparison (solid circles) in both amplitude and phase. For reference. the two end waveforms are compared with each other as well (open squares). In Fig. 20a. it is clear that

$|E_{\text{cent}}/E_{\text{end1}}|$ is substantially less than one. and decreases with increasing frequency. This explains the relatively restricted bandwidth of this measurement: above 1 THz, the waveform reflected from the burn is attenuated to below the noise limit. Figure 20b shows that the relative phase of the two waveforms is also modified by the properties of the burned tissue, at least below 0.7 THz. The reasons for these modifications of $n(\omega)$ and $\alpha(\omega)$ remain unclear, although they cannot be attributed solely to changes in surface water content [71]. A number of chemical and morphological modifications occur when tissue is burned. and any one or a combination of several factors could be responsible. The fact that distinct, frequency-dependent effects are observable is an encouraging indication of the potential value of this technology.

3.6   Terahertz Tomography: The Third Dimension

For an object with multiple reflecting internal surfaces, the reflected waveform consists of a series of replicas of the input pulse of varying magnitude, polarity, and temporal distortion. This is illustrated here using the example of a 3.5 inch floppy disk. Typical waveforms are shown in Fig. 21. The upper waveform (Fig. 21a) was obtained by replacing the object with a mirror. and thus represents the pulse incident on the sample. The small oscillations which follow the main pulse in this waveform are a result of residual water vapor in the beam path [79]. and do not affect the measurement significantly. The second curve (Fig. 21b). a representative reflected waveform, consists of a series of replicas of the input waveform. These correspond to reflections from the dielectric interfaces of the floppy disk, either from air to plastic. from plastic to air. or from the surfaces of the magnetic recording material. The polar-

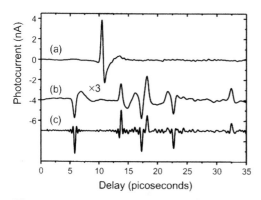

Fig. 21. THz waveforms (a) incident on. and (b) reflected from a conventional 3.5 inch floppy disk. Each buried dielectric interface within the disk generates a distinct reflected pulse. which can be discerned in (b). The waveform in (c) represents the Fourier deconvolution of the incident pulse shape (a) from the reflected waveform (b)

ity and magnitude of each reflection are given by the reflection coefficient at each interface, and are related to the size and sign of the corresponding index step. The four reflections resulting from the front and back plastic covers are clearly resolved. However, the thickness of the magnetic recording material is so small that the waveforms returned from its front and back surfaces cannot be distinguished, and appear as a single distorted waveform. In this example, the temporal waveforms hardly change shape while traversing the object because the plastic material has little absorption and dispersion. In a more general situation, reflected waveforms may be significantly altered in shape.

Figure 21c shows the waveform of Fig. 21b, after numerical Fourier deconvolution (i.e. division of the Fourier spectra of the incident and reflected waveforms, with a low-pass filter to remove noise above $\sim 2.5\,\mathrm{THz}$). Subsequently, a low-frequency background was removed by wavelet filtering [80]. This procedure produces a sharp spike at a time delay corresponding to the position of every reflecting interface. Thus, the procedure helps to determine more accurately the positions of the various interfaces. In contrast to Fig. 21b, the front and back surfaces of the thin ($\sim 120\,\mu\mathrm{m}$) magnetic recording material are clearly resolved in the deconvolved data. This is consistent with the expected resolution of $L_c/2$, where $L_c = 200\,\mu\mathrm{m}$ is the coherence length of the THz pulse in the intervening material in this particular example.

Figure 22 shows a conventional T-ray image of a section of the floppy disk obtained in reflection. This image was obtained by computing the total reflected power, as described above. The plastic cover with its various features, the circular recording disk, and the metallic hub in the center of the disk can be distinguished. Figure 23 illustrates a tomographic T-ray "slice" of the floppy disk at a particular vertical position ($y = 15\,\mathrm{mm}$) in Fig. 22. For each horizontal (x) position, a reflected waveform was acquired, and displayed as a function of delay in this tomographic image. The upper image shows the resulting image using the raw waveforms, with no post-processing. The amplitude of the processed waveforms has been translated into a grayscale, so that each reflecting surface appears as a pair of stripes, one for the positive

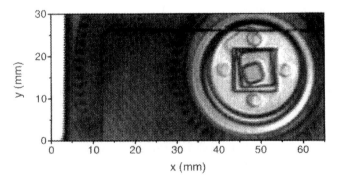

Fig. 22. A conventional THz reflection image of a portion of the floppy disk

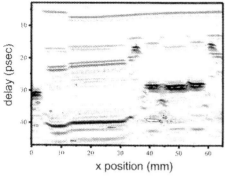

x position (mm)

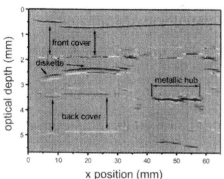

x position (mm)

Fig. 23. A THz tomographic slice through the floppy disk, showing the depths of various buried interfaces. The *upper image* is formed using raw waveforms, with no post-processing. Since the THz pulses are approximately single cycles, each reflecting surface generates a double stripe in this image, one for the positive lobe and one for the negative lobe. The *lower image* shows the result after the waveforms are deconvolved as in Fig. 21c. The internal structures of the floppy disk are easily resolved

lobe of the THz waveform and one for the negative lobe. The lower image illustrates the improvements which result when each waveform is processed as described above (Fig. 21c) prior to forming the image. The positions of the various parts of the floppy disk along the propagation direction of the THz beam can be observed clearly in this image, such as the front and back cover, the magnetic recording disk, and the metal hub. The image also shows some artifacts of the technique resulting from multiple reflections between the various interfaces, such as the features observed behind the (opaque) metal hub, and the apparent discontinuity in the magnetic recording medium caused by a change in the thickness of the front plastic cover at $x = 12\,\mathrm{mm}$.

3.7   Interferometric Tomography

In the image of the floppy disk, it is possible to resolve the front and rear surfaces of the thin recording material because they are separated by slightly more than one coherence length. In most tomographic imaging techniques, this is a strict bound on the achievable depth resolution. This limit can be understood most intuitively in the time domain: if the pulses used to distinguish between two closely spaced reflecting surfaces are broad compared with the temporal separation between them, then it becomes difficult to distinguish them as two distinct pulses. For transform-limited pulses (those whose temporal duration is as short as can be achieved with the given bandwidth), this translates directly into a criterion in the frequency domain, which is precisely analogous to the well-known Rayleigh criterion for resolving two closely spaced light sources. Because the THz pulses used in these studies are quite broadband, a coherence length of a few hundred microns is easily achieved.

It is possible, however, to obtain images which far exceed the Rayleigh criterion in depth resolution with coherent radiation, through the use of interferometry [81]. Figure 24 shows a schematic of an interferometric imaging arrangement. This has a simple Michelson geometry, in which the reference arm mirror is mounted on a manual translation stage for varying the relative delay between the two arms of the interferometer. A focusing lens is inserted in the sample arm, to provide spatial resolution at the sample surface. This lens has a second crucial function, which enables the enhanced imaging capability. Because the pulse which is reflected off the sample surface has passed through a focus, it acquires a geometrical phase known as the Gouy phase shift [55], which is approximately equal to π. In contrast, the pulse which traverses the reference arm of the interferometer is not focused, and thus does not acquire any additional phase shift. As a result, if the sample is replaced by a metal mirror, these two pulses, otherwise identical, are nearly 180° out of phase when they interfere at the detector, and almost no signal is detected. This cancellation is disrupted if the pulse is distorted in any way during its interaction with a sample at the focus of the lens. Small distortions in either

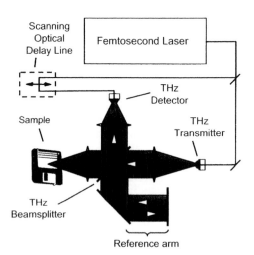

Fig. 24. A schematic of the apparatus used for interferometric tomography using THz pulses. The THz pulses are injected into a Michelson interferometer, with a lens in one arm. In addition to providing the lateral resolution for imaging, the lens also generates a phase shift of approximately π so that the pulses in the two arms of the interferometer can destructively interfere at the detector

amplitude or phase are thus converted into large changes in the measured signal amplitude.

Figure 25 shows typical waveforms which illustrate these effects. Curve (a) shows the waveform measured from the sample arm when the sample is simply a retroreflecting mirror, while (b) shows the waveform in the reference arm. Note that these two are nearly inverses of each other, denoting a relative phase shift of approximately π. Curve (c) shows the very small waveform which results when these two are allowed to interfere at the detector. This waveform is not precisely zero, because the Gouy phase shift is not precisely equal to π over the entire bandwidth of the THz pulse [55]. When a perturbation is introduced into the sample arm, such as by putting a thin ($\sim 75\,\mu m$) piece of tape across the retroreflecting mirror, the resulting waveform is dramatically altered, as shown in curve (d). The fractional change in the measured waveform upon introduction of this perturbation can be many times as large as the equivalent change observed without interferometry.

Figure 26 illustrates this enhancement in an imaging context [81]. In order to simulate thin air gaps between a metal and a dielectric material, a series of grooves of controlled depth were milled into a Teflon block. This block was pressed against a metal mirror, so that the focused THz beam

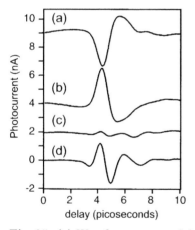

Fig. 25. (a) Waveform measured from the sample arm of the interferometer, with the reference arm blocked. (b) Waveform measured from the reference arm with the sample arm blocked. This is approximately π out of phase from the sample arm waveform shown in (a), owing to the Gouy phase shift. (c) The destructive interference of the waveforms in (a) and (b). The small residual signal results from the fact that the Gouy phase is not precisely equal to π at all frequencies. (d) When a thin piece of dielectric tape is placed on the mirror in the sample arm, the additional delay it introduces disrupts the destructive interference, leading to a large waveform which resembles the derivative of the waveform in (a). In this case, the thickness of the tape is roughly $1/4$ of the coherence length of the THz pulses used to make the measurements

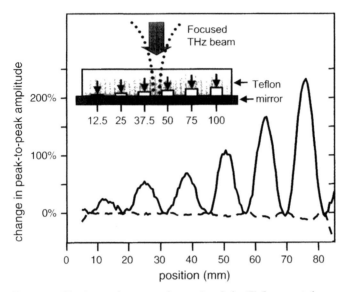

Fig. 26. The inset shows a schematic of the Teflon–metal composite sample, with air gaps of calibrated widths, fabricated for these measurements. The curves show the percentage change in peak-to-peak amplitude of a THz waveform reflected from this sample, relative to that of a waveform reflected from a position on the sample with no air gap. The dashed curve shows the result obtained with no interferometry, while the solid curve shows the enhanced contrast and sensitivity provided with the use of interferometry

was incident through the back of the dielectric block, as shown schematically in the inset of Fig. 26. The numbers 12.5 through 100 indicate the depths of the grooves, in microns. This sample was scanned transverse to the THz beam direction, and a reflected waveform was collected every 500 µm. The peak-to-peak amplitudes of these waveforms were then compared with the peak-to-peak amplitude of a waveform collected using a Teflon block with no grooves. The dashed curve shows this fractional change in contrast as a function of position along the line scan, without interferometry. Here, the 50 µm and 100 µm air gaps are discernable, but only barely. The solid line shows the equivalent measurement with interferometry. This demonstrates an enormous increase in contrast for the detection of thin features, as well as a reasonable sensitivity to even the smallest (12.5 µm) gap. This gap is about 100 times smaller than the coherence length of the radiation used in this measurement, which was a relatively modest 1.1 mm.

4   Future Prospects

One of the most exciting aspects of this field is the tremendous prospects for the future. It should be clear from the examples shown here that there are numerous potential "real-world" applications of THz imaging. In some cases, THz radiation may provide the only feasible option for certain tasks, whereas in others it might be only one of several competing technologies. It is clear that, in these latter cases, crucial factors in the successful implementation of systems of the kind described here include their cost and ease of use. These issues have historically been beyond the purview of research scientists, but modern research can no longer afford to ignore such practical concerns. In the case of THz imaging, much concerted effort has led to the development of a commercially available system based on photoconductive generation and detection techniques [20]. A photograph of the recently announced "T-Ray 2000" spectrometer is shown in Fig. 27. This system is reliable and easy to operate, and can be reconfigured for either transmission or reflection imaging. As of this writing, this system is already operating in one factory, as an on-line quality control monitor, with a number of other exciting prospects on the horizon.

Fig. 27. Photograph of the "T-Ray 2000", the first commercial THz time-domain spectrometer, manufactured by Picometrix, Inc

References

1. Y. R. Shen, "Far-infrared generation by optical mixing", *Prog. Quantum Electron.* **4**, 207 (1976).
2. P. R. Smith, D. H. Auston, M. C. Nuss, "Subpicosecond photoconducting dipole antennas", *IEEE J. Quantum Electron.* **24**, 255 (1988).
3. Ch. Fattinger, D. Grischkowsky, "Terahertz beams". *Appl. Phys. Lett.* **54**, 490 (1989).

4. M. C. Nuss, J. Orenstein, "Terahertz time-domain spectroscopy", in *Millimeter and Submillimeter Wave Spectroscopy of Solids*, ed. by G. Grüner (Springer. Berlin, Heidelberg 1998).

5. H. Harde. S. Keiding, D. Grischkowsky, "THz commensurate echoes: periodic rephasing of molecular transitions in free-induction decay", *Phys. Rev. Lett.* **66**, 1834 (1991).

6. H. Harde. D. Grischkowsky, "Coherent transients excited by subpicosecond pulses of terahertz radiation". *J. Opt. Soc. Am. B* **8**, 1642 (1991).

7. H. Harde, N. Katzenellenbogen, D. Grischkowsky, "Terahertz coherent transients from methyl chloride vapor". *J. Opt. Soc. Am. B* **11**, 1018 (1994).

8. H. Harde. N. Katzenellenbogen, D. Grischkowsky, "Line-shape transition of collision broadened lines". *Phys. Rev. Lett.* **74**, 1307 (1995).

9. R. A. Cheville, D. Grischkowsky, "Far-infrared terahertz time-domain spectroscopy of flames", *Opt. Lett.* **20** (1995).

10. H. Harde. R. A. Cheville. D. Grischkowsky. "Collision-induced tunneling in methyl halides", *J. Opt. Soc. Am. B* **14**, 3282 (1997).

11. R. A. Cheville, D. Grischkowsky, "Far-infrared foreign and self-broadened rotational linewidths of high-temperature water vapor". *J. Opt. Soc. Am. B* **16**. 317 (1999).

12. B. B. Hu, M. C. Nuss. "Imaging with terahertz waves", *Opt. Lett.* **20**, 1716 (1995).

13. T. Clancy. S. Pieczenik, *Games of State*, (Penguin Putnam, New York, 1996).

14. M. van Exter, D. Grischkowsky, "Characterization of an optoelectronic terahertz beam system", *IEEE Trans. Microwave Theory Tech.* **38**, 1684 (1990).

15. P. Jepsen, R. H. Jacobsen, S. R. Keiding, "Generation and detection of terahertz pulses from biased semiconductor antennas", *J. Opt. Soc. Am. B* **13**. 2424 (1996).

16. D. M. Mittleman, R. H. Jacobsen, M. C. Nuss, "T-ray imaging", *IEEE J. Sel. Top. Quantrum Electron.* **2**, 679 (1996).

17. Q. Wu, T. D. Hewitt, X.-C. Zhang, "2-dimensional electro-optic imaging of THz beams". *Appl. Phys. Lett.* **69** (1996).

18. I. Brener, D. Dykaar, A. Frommer, L. N. Pfeiffer, J. Lopata, J. Wynn, K. West, M. C. Nuss, "Terahertz emission from electric field singularities in biased semiconductors", *Opt. Lett.* **21**, 1924 (1996).

19. Y. Cai, I. Brener, J. Lopata, J. Wynn, L. N. Pfeiffer, J. Federici, "Design and performance of singular electric field terahertz photoconducting antennas". *Appl. Phys. Lett.* **71**, 2076 (1997).

20. J. V. Rudd. D. Zimdars, M. Warmuth, "Compact, fiber-pigtailed terahertz imaging system", *Proc. SPIE* **3934**, 27 (2000).

21. R. Takahashi, Y. Kawamura, H. Iwamura, "Ultrafast 1.55 μm all-optical switching using low-temperature-grown multiple quantum wells", *Appl. Phys. Lett.* **68**, 153 (1996).

22. L. Qian, S. D. Benjamin, P. W. E. Smith, B. J. Robinson, D. A. Thompson, "Picosecond carrier lifetime and large optical nonlinearities in InGaAsP grown by He-plasma-assisted molecular beam epitaxy", *Opt. Lett.* **22**, 108 (1997).

23. T. Kondo, M. Sakamoto, M. Tonouchi, M. Hangyo, "Terahertz radiation from (111) InAs surface using 1.55 μm femtosecond laser pulses", *Jpn. J. Appl. Phys. Pt. 2 (Lett.)* **38**, 1035 (1999).

24. N. Sekine, K. Hirakawa, F. Sogawa, Y. Arakawa, N. Usami, Y. Shiraki, T. Katoda, "Ultrashort lifetime photocarriers in Ge thin films", *Appl. Phys. Lett.* **68**, 3419 (1996).

25. N. Katzenellenbogen, D. Grischkowsky, "Efficient generation of 380 fs pulses of THz radiation by ultrafast laser pulse excitation of a biased metal–semiconductor interface", *Appl. Phys. Lett.* **58**, 222 (1991).

26. S. E. Ralph, D. Grischkowsky, "THz spectroscopy and source characterization by optoelectronic interferometry", *Appl. Phys. Lett.* **60**, 1070 (1992).

27. M. Tani, S. Matsuura, K. Sakai, S. Nakashima, "Emission characteristics of photoconductive antennas based on low-temperature-grown GaAs and semi-insulating GaAs", *Appl. Opt.* **36**, 7853 (1997).

28. A. Bonvalet, M. Joffre, J.-L. Martin, A. Migus, "Generation of ultrabroadband femtosecond pulses in the mid-infrared by optical rectification of 15 fs light pulses at 100 MHz repetition rate", *Appl. Phys. Lett.* **67**, 2907 (1995).

29. Q. Wu, X.-C. Zhang, "7 terahertz broadband GaP electro-optic sensor", *Appl. Phys. Lett.* **70**, 1784 (1997).

30. Q. Wu, X.-C. Zhang, "Free-space electro-optic sampling of mid-infrared pulses", *Appl. Phys. Lett.* **71**, 1285 (1997).

31. F. G. Sun, G. A. Wagoner, X.-C. Zhang, "Measurement of free-space terahertz pulses via long-lifetime photoconductors", *Appl. Phys. Lett.* **67**, 1656 (1995).

32. J. Bromage, I. A. Walmsley, C. R. Stroud, "Dithered-edge sampling of terahertz pulses", *Appl. Phys. Lett.* **75**, 2181 (1999).

33. S. Gupta, M. Y. Frankel, J. A. Valdmanis, J. F. Whitaker, G. A. Mourou, F. W. Smith, A. R. Calawa, "Subpicosecond carrier lifetime in GaAs grown by molecular beam epitaxy at low temperatures", *Appl. Phys. Lett.* **59**, 3276 (1991).

34. F. E. Doany, D. Grischkowsky, C.-C. Chi, "Carrier lifetime versus ion-implantation dose in silicon on sapphire", *Appl. Phys. Lett.* **50**, 460 (1987).

35. Y. Cai, I. Brener, J. Lopata, J. Wynn, L. Pfeiffer, J. B. Stark, Q. Wu, X.-C. Zhang, J. F. Federici, "Coherent terahertz radiation: direct comparison between free-space electro-optic sampling and antenna detection", *Appl. Phys. Lett.* **73**, 444 (1998).

36. S. Kono, M. Tani, G. Ping, K. Sakai, "Detection of up to 20 THz with a low-temperature-grown GaAs photoconductive antenna gated with 15 fs light pulses", *Appl. Phys. Lett.* **77**, 4104 (2000).

37. D. Grischkowsky, S. Keiding, M. van Exter, C. Fattinger, "Far-infrared time-domain spectroscopy with terahertz beams of dielectrics and semiconductors", *J. Opt. Soc. Am. B* **7**, 2006 (1990).

38. P. Uhd Jepsen, S. R. Keiding, "Radiation patterns from lens-coupled terahertz antennas", *Opt. Lett.* **20**, 807 (1995).

39. R. W. McGowan, G. Gallot, D. Grischkowsky, "Propagation of ultrawideband short pulses of terahertz radiation through submillimeter-diameter circular waveguides", *Opt. Lett.* **24**, 1431 (1999).

40. S. Hunsche, M. Koch, I. Brener, M. C. Nuss, "THz near-field imaging", *Opt. Commun.* **150**, 22 (1998).

41. J. V. Rudd, D. M. Mittleman, "The influence of substrate lens design in terahertz time-domain spectroscopy," *J. Opt. Soc. Am. B* **19**, 319 (2002).

42. D. F. Filipovic, S. S. Gearhart, G. M. Rebeiz, "Double-slot antennas on extended hemispherical and elliptical silicon dielectric lenses", *IEEE Trans. Microwave Theory Tech.* **41**, 1738 (1993).

43. W. B. Dou, G. Zeng, Z. L. Sun, "Pattern prediction of extended hemispherical lens/objective lens antenna system at millimeter wavelengths", *IEE Proc. Microwave Antennas Propag.* **145**, 295 (1998).

44. J. R. Bray, L. Roy, "Performance trade-offs of substrate lens antennas," in Proceedings of the Symposium on Antenna Technology and Applied Electromagnetics (ANTEM), Ottawa, Canada, 1998, pp. 321-324.

45. M. N. Afsar, "Precision millimeter-wave measurements of complex refractive index, complex dielectric permittivity, and loss tangent of common polymers", *IEEE Trans. Instrum. Meas.* **36**, 530 (1987).

46. J. R. Birch, "The far infrared optical constants of polyethylene", *Infrared. Phys.* **30**, 195 (1990).

47. K. Kawase, N. Hiromoto, "Terahertz-wave antireflection coating on Ge and GaAs with fused quartz", *Appl. Opt.* **37**, 1862 (1998).

48. C. R. Englert, M. Birk, H. Maurer, "Antireflection coated, wedged, single-crystal silicon aircraft window for the far-infrared", *IEEE Trans. Geosci. Remote Sens.* **37**, 1997 (1999).

49. J. R. Birch, E. A. Nicol, "The FIR optical constants of the polymer TPX", *Infrared Phys.* **24**, 573 (1984).

50. R. W. Ziolkowski, J. B. Judkins, "Propagation characteristics of ultrawidebandwidth pulsed Gaussian beams", *J. Opt. Soc. Am. A* **9**, 2021 (1992).

51. D. You, P. Bucksbaum, "Propagation of half-cycle far infrared pulses", *J. Opt. Soc. Am. B* **14**, 1651 (1997).

52. A. E. Kaplan, "Diffraction-induced transformation of near-cycle and subcycle pulses", *J. Opt. Soc. Am. B* **15**, 951 (1998).

53. P. Kuzel, M. A. Khazan, J. Kroupa, "Spatiotemporal transformations of ultrashort terahertz pulses", *J. Opt. Soc. Am. B* **16**, 1795 (1999).

54. S. Hunsche, S. Feng, H. G. Winful, A. Leitenstorfer, M. C. Nuss, E. P. Ippen, "Spatiotemporal focusing of single-cycle light pulses", *J. Opt. Soc. Am. A* **16**, 2025 (1999).

55. A. B. Ruffin, J. V. Rudd, J. F. Whitaker, S. Feng, H. G. Winful, "Direct observation of the Gouy phase shift with single-cycle terahertz pulses", *Phys. Rev. Lett.* **83**, 3410 (1999).

56. J. Bromage, S. Radic, G. P. Agrawal, C. R. Stroud, P. M. Fauchet, R. Sobolewski, "Spatiotemporal shaping of terahertz pulses", *Opt. Lett.* **22**, 627 (1997).

57. C. Winnewisser, F. Lewen, J. Weinzierl, H. Helm, "Transmission features of frequency-selective components in the far infrared determined by terahertz time-domain spectroscopy", *Appl. Opt.* **38**, 3961 (1999).

58. F. Garet, L. Duvillaret, J.-L. Coutaz, "Evidence of frequency dependent THz beam polarization in time-domain spectroscopy", *Proc. SPIE* **3617**, 30 (1999).

59. J. V. Rudd, J. L. Johnson, D. M. Mittleman, "Quadrupole radiation from terahertz dipole antennas", *Opt. Lett.* **25**, 1556 (2000).

60. J. V. Rudd, J. L. Johnson, D. M. Mittleman, "Cross-polarized angular emission patterns from lens-coupled terahertz antennas", *J. Opt. Soc. Am. B* **18**, 1524 (2001).

61. A. J. Bahr, "Nondestructive evaluation of ceramics", *IEEE Trans. Microwave Theory Tech.* **26**, 676 (1978).

62. T. S. Hartwick, D. T. Hodges, D. H. Barker, F. B. Foote, "Far infrared imagery", *Appl. Opt.* **15**, 1919 (1976).

63. D. B. Rutledge, M. S. Muha, "Imaging antenna arrays", *IEEE Trans. Antennas Propag.* **30**, 535 (1982).
64. A. J. Cantor, P. K. Cheo, M. C. Foster, L. A. Newman, "Application of submillimeter wave lasers to high voltage cable inspection", *IEEE J. Quantum Electron.* **17**, 477 (1981).
65. P. K. Cheo, "Far-infrared lasers for power cable manufacturing", *IEEE Circuits Devices.* **2**, 49 (1986).
66. M. Brucherseifer, P. Haring Bolivar, H. Klingenberg, H. Kurz, "Angle-dependent THz tomography characterization of thin ceramic oxide films for fuel cell applications", *Appl. Phys. B* **72**, 361 (2001).
67. D. M. Mittleman, J. Cunningham, M. C. Nuss, M. Geva, "Noncontact semiconductor wafer characterization with the terahertz Hall effect", *Appl. Phys. Lett.* **71**, 16 (1997).
68. J. C. Owens, "Optical refractive index of air: dependence on pressure, temperature and composition", *Appl. Opt.* **6**, 51 (1967).
69. J. T. Kindt, C. A. Schmuttenmaer, "Far-infrared dielectric properties of polar liquids probed by femtosecond terahertz pulse spectroscopy", *J. Phys. Chem.* **100**, 10373 (1996).
70. C. Rønne, L. Thrane, P.-O. Åstrand, A. Wallqvist, K. V. Mikkelsen, S. R. Keiding, "Investigation of the temperature dependence of dielectric relaxation in liquid water by THz reflection spectroscopy and molecular dynamics simulation", *J. Chem. Phys.* **107**, 5319 (1997).
71. L. Thrane, R. H. Jacobsen, P. Uhd Jepsen, S. R. Keiding, "THz reflection spectroscopy of liquid water", *Chem. Phys. Lett.* **240**, 330 (1995).
72. P. F. Scholander, H. T. Hammel, E. D. Bradstreet, E. A. Hemmingsen, "Sap pressure in vascular plants", *Science* **148**, 339 (1965).
73. N. C. Turner, R. A. Spurway, E. D. Schulze, "Comparison of water potentials measured by in situ psychrometry and pressure chamber in morphologically different species", *Plant Physiol.* **74**, 316 (1984).
74. S. Hadjiloucas, L. S. Karatzas, J. W. Bowen, "Measurements of leaf water content using terahertz radiation", *IEEE Trans. Microwave Theory Tech.* **47**, 142 (1999).
75. D. M. Mittleman, S. Hunsche, L. Boivin, M. C. Nuss, "T-ray tomography", *Opt. Lett.* **22**, 904 (1997).
76. D. Huang, E. A. Swanson, C. P. Lin, J. S. Schuman, W. G. Stinson, W. Chang, M. R. Hee, T. Flotte, K. Gregory, C. A. Puliafito, J. G. Fujimoto, "Optical coherence tomography", *Science* **254**, 1178 (1991).
77. G. J. Tearney, M. E. Brezinski, J. F. Southern, B. E. Bouma, M. R. Hee, J. G. Fujimoto, "Determination of the refractive index of highly scattering human tissue by optical coherence tomography", *Opt. Lett.* **20**, 2258 (1995).
78. Z. B. Niazi, T. J. Essex, R. Papini, D. Scott, N. R. McLean, M. J. Black, "New laser doppler scanner, a valuable adjunct in burn depth assessment", *Burns* **19**, 485 (1993).
79. M. van Exter, C. Fattinger, D. Grischkowsky, "Terahertz time-domain spectroscopy of water vapor", *Opt. Lett.* **14**, 1128 (1989).
80. J. Buckheit, S. Chen, D. Donoho, I. Johnstone, J. Scargle, "Wavelab" software package, http://playfair.stanford.edu/ wavelab/ (1997).
81. J. L. Johnson, T. D. Dorney, D. M. Mittleman, "Enhanced depth resolution in terahertz imaging using phase-shift interferometry", *Appl. Phys. Lett.* **78**, 835 (2001).

Free-Space Electro-Optic Techniques

Zhiping Jiang and Xi-Cheng Zhang

Abstract. Terahertz radiation occupies a large portion of the electromagnetic spectrum between the infrared and microwave bands from 0.1 to 10 THz. This frequency range presents the next frontier in imaging science and technology. Compared to the relatively well-developed imaging techniques at microwave and optical frequencies, however, basic research, new initiatives, and advanced technology developments in the terahertz band are very limited. The "THz gap" is a scientifically rich but technologically limited frequency band–largely because efficient terahertz emitters and receivers are a relatively recent invention. This chapter provides the fundamentals of free-space electro-optic technology for generation and detection of terahertz pulses. The free-space THz optoelectronic detection system, which uses photoconductive antennas or electro-optic crystals, provides diffraction-limited spatial resolution, femtosecond temporal resolution, DC-THz spectral bandwidth and mV/cm field sensitivity.

1 Introduction

In recent years, many free-space techniques of generating and measuring terahertz pulses have been developed. In this chapter, we shall discuss free-space THz techniques based on electro-optic (EO) effects, including generation by optical rectification in optical nonlinear crystals, detection by EO sampling via the Pockels effect. and applications.

2 Generation

Although many different physical principles, such as surge currents [1], Bloch oscillations [2.3], coherent phonon oscillations [4], and plasma oscillations [5], can be used to generate freely propagating electromagnetic waves in the terahertz region, THz generation by optical rectification has the unique advantage of extremely broad spectral bandwidth. Although the conversion efficiency obtained with optical rectification is not as high as that obtained with biased photoconductive switches. optical rectification still provides a signal strong enough for many applications, such as time-domain spectroscopy and imaging.

Optical rectification is a second-order nonlinear optical effect. It was first found in 1962 by Bass and coworkers [6], and also independently proposed by

A`skaryan [7], who considered the effect as a form of Cherenkov radiation [8] produced by electromagnetic beams (photons) rather than by moving particles. In the 1970s, Shen and coworkers successfully generated far-infrared radiation by illuminating LiNbO$_3$ with phase-matched picosecond pulsed optical beams [9 11]. The theory of optical mixing and self-modulation in EO crystals was generally in good agreement with the experimental results. Details of early experiments on radiation via optical rectification in materials can also be found in [12 16]. The radiation can also be coupled out into free space [17 21].

Optical rectification is essentially a difference-frequency generation process, in which the frequency difference is close to zero. Recently, femtosecond laser pulses have been used to generate THz radiation from EO crystals via the optical rectification effect. Because a femtosecond pulse contains many frequency components, any two frequency components can contribute to the difference-frequency generation, and the overall result is a weighted sum of all these contributions. The fact that one femtosecond laser pulse is enough to stimulate optical-rectification radiation makes the experiment very simple to implement.

The polarization $\boldsymbol{P}$ can be expanded into a power series of the electric field $\boldsymbol{E}$:

$$
\begin{aligned}
\boldsymbol{P}(\boldsymbol{r},t) = {}& \chi^{(1)}(\boldsymbol{r},t)\boldsymbol{E}(\boldsymbol{r},t) + \chi^{(2)}(\boldsymbol{r},t):\boldsymbol{E}(\boldsymbol{r},t)\boldsymbol{E}(\boldsymbol{r},t) \\
& + \chi^{(3)}(\boldsymbol{r},t):\boldsymbol{E}(\boldsymbol{r},t)\boldsymbol{E}(\boldsymbol{r},t)\boldsymbol{E}(\boldsymbol{r},t) + \dots .
\end{aligned}
\tag{1}
$$

where $\chi^{(n)}(\boldsymbol{r},t)$ is the nth-order nonlinear susceptibility tensor. Optical rectification comes from the second term of (1). If the incident light is approximated as a plane wave, then $\boldsymbol{E}$ can be expressed as

$$
\boldsymbol{E}(t) = \int_0^{+\infty} E(\omega) \exp(-\mathrm{i}\omega t)\,\mathrm{d}\omega + \mathrm{c.c.}
\tag{2}
$$

By putting (2) into (1), we obtain the polarization associated with optical rectification as follows:

$$
\begin{aligned}
\boldsymbol{P}_{\mathrm{OR}}^{(2)} &= 2\chi^{(2)}: \int_0^\infty \int_0^\infty \boldsymbol{E}(\omega_1)\boldsymbol{E}^*(\omega_2) \exp[-\mathrm{i}(\omega_1-\omega_2)t]\,\mathrm{d}\omega_1\,\mathrm{d}\omega_2 \\
&= 2\chi^{(2)}: \int_0^\infty \int_0^\infty \boldsymbol{E}(\omega+\Omega)\boldsymbol{E}^*(\omega)\,exp(-\mathrm{i}\Omega t)\,\mathrm{d}\Omega\,\mathrm{d}\omega .
\end{aligned}
\tag{3}
$$

where Ω is the frequency difference of two frequency components. In the far field, the radiated electric field $\boldsymbol{E}_\mathrm{r}(t)$ is proportional to the second derivative of $\boldsymbol{P}_{\mathrm{OR}}^{(2)}(t)$ with respect to the time t:

$$
\boldsymbol{E}_\mathrm{r}(t) \propto \frac{\partial^2}{\partial^2 t}\boldsymbol{P}_{\mathrm{OR}}^{(2)}(t) .
\tag{4}
$$

The susceptibility tensor $\chi^{(2)}$ is identically zero in a centrosymmetric medium. However, in certain crystalline dielectrics and near surfaces, it is nonzero.

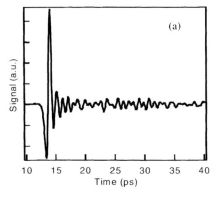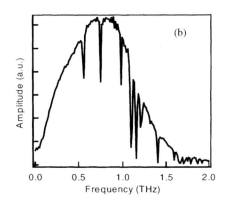

Fig. 1. Typical waveform (**a**) and spectral distribution (**b**) obtained using a 250 fs, 3 μJ laser and a thick (2 mm) ZnTe crystal as both the emitter and the sensor

Given the crystal structure and sufficient information about the incident light, (4) can in principle be used to calculate the far-field waveform of the radiation. In practice, many factors, such as the crystal orientation. its thickness, its absorption and dispersion [22], diffraction effects, and saturation effects [23–26], affect the radiation efficiency. the shape of the temporal waveform, and the frequency distribution of the emitted THz radiation.

Zinc telluride (ZnTe) is one of the most commonly used EO crystals for optical rectification in the THz range. It has only a moderately large EO coefficient. but a more compelling attribute is the velocity matching between the femtosecond optical beam (near 800 nm) and the THz beam. The refractive indices for the pump light and the THz dielectric constant match very well for frequencies from the sub-THz range up to several THz. Owing to this feature. very thick (up to several millimeters) ZnTe crystals can be used without significant walk-off effects. leading to a relatively large conversion efficiency. Figure 1 shows a typical waveform and the corresponding spectral distribution. The dips are absorption lines of water vapor. Electric fields as large as 10 V/cm have been obtained at a focal point.

It can also been seen that the central frequency of the radiation generated with a millimeter-thick ZnTe emitter is around 1 THz. Two main factors limit this bandwidth: the pulse duration of the excitation laser pulse and the phase matching. Roughly speaking, a laser pulse can generate a THz pulse with a bandwidth twice as broad as the laser pulse itself. Therefore shorter laser pulses are expected to extend the bandwidth of the radiation. With the development of ultrafast lasers in the sub-10-femtosecond range. bandwidths larger than 100 THz could be generated. In this case, the limiting factor is the phase matching. Because the frequency extent of the THz pulses is so broad. it is impossible to select an EO material that fulfills the condition of group velocity matching for all frequency components. In order to minimize walk-off between the THz and optical pulses. thinner crystals must be used.

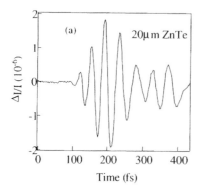
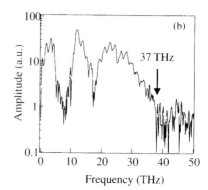

Fig. 2. Typical waveform and spectral distribution obtained using a 12 fs laser and a thin (20 μm) ZnTe crystal as both the emitter and the sensor. (**a**) Temporal waveform, (**b**) amplitude spectrum

Suppose Ω is the THz frequency, and ω is the optical frequency. Then, the frequency response function for the phase-matched process is

$$G(\Omega) = \frac{\exp(\mathrm{i}2\pi\Omega\delta) - 1}{\mathrm{i}2\pi\Omega\delta} . \tag{5}$$

where δ is the phase difference between the pump light and the THz light of frequency Ω, given by

$$\delta = \frac{d}{c}\left(\sqrt{\varepsilon(\Omega)} - n_{\mathrm{g}}\right) = \frac{d}{c}\Delta n . \tag{6}$$

Here, d is the thickness of the crystal, c is the light speed in vacuum, n_{g} is the group refractive index for the laser pulse, and $\varepsilon(\Omega)$ is the dielectric constant of the EO crystal at the THz frequency Ω. We note that the phase-matching equations (5) and (6) are the same for EO detection, described below. From these results, it is clear that, in order to obtain broad-bandwidth emission, it is essential to decrease the thickness d. Figure 2 shows a typical waveform and spectral distribution, obtained using a 12 fs laser and a very thin (20 μm) ZnTe crystal as both the emitter and the sensor. The useful frequency range extends from the sub-THz range to 37 THz [27–29].

Another factor that limits the useful spectrum is absorption in the EO crystals. As is evident in Fig. 2, a broad hole appears in the spectrum between ~ 5 and ~ 10 THz. This is a result of phonon absorption and the so-called reststrahlen band. Table 1 gives the phonon frequencies of some commonly used crystals.

It has been found that at low pump fluence, the amplitude of the radiated electric field generated via optical rectification is linearly proportional to the pump power. However, for organic crystals such as DAST and for ZnTe crystals, saturation has been observed [24,25]. No saturation has been reported for crystals such as $LiTaO_3$ and $LiNbO_3$. It was also found that the

Table 1. Phonon frequencies of commonly used crystals

	ZnTe	GaAs	InP	GaP	ZnS
ν_{TO} ($k = [\bar{1}, 0, 0]$) (THz)	5.3	7.6	10.0	10.8	9.8

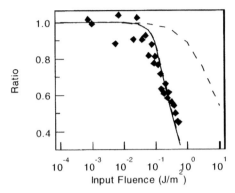

Fig. 3. The dependence of the peak THz signal on the optical excitation power, showing a saturation effect in ZnTe. The data points (*diamonds*) show the peak THz electric field divided by the pump laser fluence, as a function of the pump power. The *dashed curve* is calculated from a two-photon absorption depletion model, and the *solid curve* shows the prediction of a carrier absorption model [26]

optimal location for the crystal, for maximum conversion efficiency, is not at the focus of the pump laser beam, but a few millimeters away from the focal point. The saturation phenomenon can be explained by resonant two-photon absorption of the pump, and the subsequent free-carrier screening. While the band gaps of ZnTe and DAST are too large for linear absorption of 800 nm light, two-photon absorption leads to the depletion of the pump beam and the generation of free carriers, which in turn absorb THz radiation quite efficiently [26]. In Fig. 3, the diamond markers are the experimental data, showing the saturation effect. At low pump input fluence, the ratio of the THz peak signal to the input fluence remains constant. This ratio drops with increased input fluence, above a certain threshold. The two-photon absorption depletion model alone cannot explain the experimental results (dashed curve). However, when carrier screening is included, the theory is in good agreement with the experiment [26].

3 Detection

Historically, freely propagating THz pulses were measured by means of either photoconductive antennas [30–35] or far-infrared interferometric techniques using incoherent detectors such as bolometers [33,34]. Although the photoconductive antennas have excellent sensitivity, their frequency response is limited by the resonant behavior of the Hertzian dipole structure. For the interfero-metric techniques. the sensitivity is far worse than that of the photoconductive antennas, because the measurements is incoherent and its sensitivity is ultimately limited by the thermal background. Besides, the bolometers usually require liquid-helium cooling. On the other hand, since its first demonstration in 1982 [36.37], ultrafast electro-optic sampling has been widely used in the measurement of local transient electric fields in materials [38 47]. There existed a need to extend the local-field measurement to free space. In 1995, three groups reported their first results using free-space electro-optic sampling (FS-EOS) independently, nearly at the same time [48 50]. Although the preliminary results were very poor, rapid progress has been made in the intervening years. It turns out that FS-EOS is a powerful tool for THz pulse measurement. providing many advantages. such as high sensitivity, ultra-broad frequency response. ease of use, and parallel measurement capability [51 66].

Figure 4 shows a typical setup for THz measurement by free-space electro-optic sampling. The ultrafast laser pulse is split by a beam splitter into two beams: a pump beam (strong) and a probe beam (weak). The pump beam illuminates the THz emitter (e. g. a photoconductive-antenna emitter or an optical-rectification emitter). The generated radiation is a short electromagnetic pulse with a duration on the order of one picosecond. so the frequency is in the terahertz range. The radiation normally consists of one cycle or a small number of cycles, so it is broadband. The THz beam is focused by a pair of parabolic mirrors onto an EO crystal. The beam modifies the index ellipsoid of the EO crystal transiently. via the Pockels effect. The linearly polarized probe beam copropagates inside the crystal with THz beam, and its phase is modulated by the refractive-index change induced by the electric field of the THz pulse. This phase change is converted to an intensity

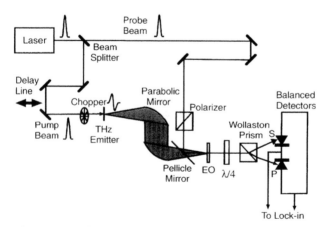

Fig. 4. Typical setup for free-space electro-optic sampling (FS-EOS)

change by a polarization analyzer (here a Wollaston prism). Usually a pair of balanced detectors are used to suppress the common laser noise. This also doubles the measured signal (see the analysis below). A mechanical delay line is used to change the time delay between the THz pulse and the probe pulse, and the THz electric-field waveform can be obtained by scanning this time delay and performing a repetitive sampling measurement. To increase the sensitivity, the pump beam is modulated by a mechanical chopper, and the THz-field-induced modulation of the probe beam is extracted by a lock-in amplifier.

3.1   Measurement Principle

The principle of EO sampling can be explained as follows. Suppose that the probe beam is propagating in the z direction, and x and y are the crystal axes of the EO crystal (see Fig. 5). When an electric field is applied to the EO crystal, the electrically induced birefringence axes x' and y' are at an angle of $45°$ with respect to x and y. If the input beam is polarized along x, then the output light can obtained from the following expression:

$$
\begin{pmatrix} E_x \\ E_y \end{pmatrix} = \begin{pmatrix} \cos\dfrac{\pi}{4} & -\sin\dfrac{\pi}{4} \\ \sin\dfrac{\pi}{4} & \cos\dfrac{\pi}{4} \end{pmatrix} \begin{pmatrix} \exp(i\delta) & 0 \\ 0 & 1 \end{pmatrix} \begin{pmatrix} \cos\dfrac{\pi}{4} & \sin\dfrac{\pi}{4} \\ -\sin\dfrac{\pi}{4} & \cos\dfrac{\pi}{4} \end{pmatrix} . \tag{7}
$$

where $\delta = \Gamma_0 + \Gamma$ is the phase difference between the x' and y' polarizations. including both the dynamic (Γ, THz induced) and the static (Γ_0, from the intrinsic or residual birefringence of the EO crystal and the compensator) phase difference. Following (7), the light intensities in the x and y polarizations are

$$
\begin{cases} I_x = |E_x|^2 = I_0 \cos^2 \dfrac{\Gamma_0 + \Gamma}{2} \\ I_y = |E_y|^2 = I_0 \sin^2 \dfrac{\Gamma_0 + \Gamma}{2} \end{cases} \tag{8}
$$

where $I_0 = E_0^2$ is the input intensity. It can be seen that I_x and I_y are complementary, i.e. $I_x + I_y = I_0$. This is the result of energy conservation,

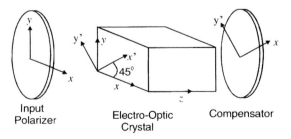

Input
Polarizer

Electro-Optic
Crystal

Compensator

Fig. 5. Coordinates for EO sampling

and follows from the fact that absorption in the crystal has been ignored. To extract the light in the x and y polarizations separately, a Wollaston prism is usually used.

The static phase term Γ_0 (also called the optical bias) is often set equal to $\pi/2$ for balanced detection. For an EO crystal without intrinsic birefringence (such as ZnTe), a quarter-wave plate is often used to provide this optical bias. Because $|\Gamma| \ll 1$ in most cases in EO sampling, we have

$$\begin{cases} I_x = \dfrac{I_0}{2}(1 - \Gamma) \\ I_y = \dfrac{I_0}{2}(1 + \Gamma) \end{cases} \tag{9}$$

In the two beams, the signals have the same magnitudes but opposite signs. For balanced detection, the difference between I_y and I_x is measured, giving the signal

$$I_\mathrm{s} = I_y - I_x = I_0 \Gamma . \tag{10}$$

The signal is proportional to the THz-field-induced phase change Γ, and Γ, in turn, is proportional to the electric field of the THz pulse. For a $\langle 110 \rangle$-oriented ZnTe crystal, the following relation holds:

$$\Gamma = \frac{\pi d n^3 \gamma_{41}}{\lambda} E . \tag{11}$$

where d is the thickness of the crystal, n is the refractive index of the crystal at the wavelength of the probe beam, λ is the probe wavelength, γ_{41} is the EO coefficient, and E is the electric field of the THz pulse.

Figure 6 shows a typical waveform of a freely propagating THz pulse and its spectral distribution measured by free-space EO sampling. In this case, a large-aperture photoconductive antenna was used as the emitter and a 1 mm thick $\langle 110 \rangle$-oriented ZnTe crystal was used as the sensor. FS-EOS gives very

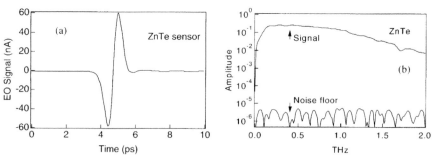

Fig. 6. (a) Temporal electro-optic signal of a 1 ps THz pulse measured by a ZnTe sensor. (b) Amplitude spectrum after FFT shows a signal-to-noise ratio (SNR) greater than 100 000 between 0.1 THz and 1.2 THz

good dynamic range and spectral bandwidth. With a Coherent RegA 9000 laser amplifier (820 nm central wavelength, 4 μJ pulse energy, 250 fs pulse duration, 250 kHz repetition rate), a dynamic range as high as 1.8×10^6 was obtained [59].

It should be emphasized that the above discussion on EO sampling is based on the assumption of a DC electric field. For a transient electric field such as a THz pulse, phase matching should be considered. When the probe pulse has a group velocity different from that of the THz pulse (the so-called group velocity mismatch, or GVM), the probe does not always sample the same position on the THz pulse. Instead, it scans across the THz pulse as the two propagate through the crystal, leading to broadening of the measured waveform. GVM can be discussed in the time domain [64], as well as in the frequency-domain [61]. The frequency-domain treatment is more accurate because it more naturally accounts for the dispersion of the dielectric constant in the THz range. The frequency response function for detection is the same as for generation, so (5) and (6) can be used. Figure 7b plots the frequency response function of a gallium phosphide (GaP) sensor for several different crystal thicknesses. Figure 7a plots the dispersion of the group index of the probe beam and the dielectric constant. Clearly, thinner sensor crystals provide broader frequency response functions. Therefore, once the material is specified, it is essential to use a thin crystal to obtain a broad frequency bandwidth. However, a small thickness means a short interaction distance, and therefore a small sensitivity. The trade-off depends on the specific application. Figure 8 shows the thickness dependence of the measured signal. For a thick sensor (2.57 mm), the measured waveform is severely distorted, and the spectral distribution is substantially lower than that measured with thinner sensors.

EO sampling requires very little probe power, and thus exhibits very good linearity. Figure 9 plots the measured EO signal versus the THz electric field and the probe power. The excellent linearity extends for more than 6 orders

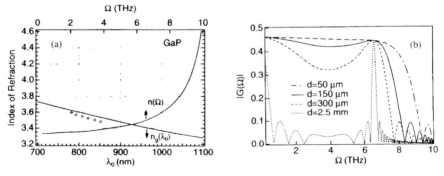

Fig. 7. (a) Refractive indices for THz and near-infrared light in GaP. (b) Frequency response functions for different sensor crystal thicknesses

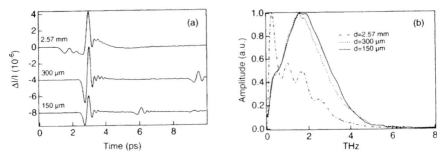

Fig. 8. (a) THz transients detected with 2.57 mm, 300 μm, and 150 μm thick ⟨110⟩-oriented GaP sensors. (b) The spectra of these waveforms

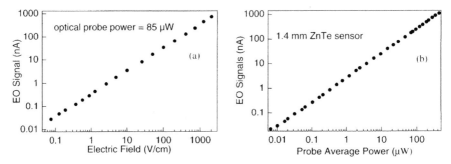

Fig. 9. A demonstration of the linearity of FS-EOS. (a) Electro-optic signal versus THz field. (b) Electro-optic signal versus power in the optical probe beam

of magnitude. This facilitates the application of EO sampling for parallel measurements, described below.

Many different materials have been used for EO detection. Table 2 lists some common materials for EO measurements. Owing to its good velocity match and relatively large EO coefficient, ZnTe is one of the best EO materials for free-space THz measurements from the sub-THz range to several tens of THz (see Fig. 2), and has been widely used [65,66].

Table 2. Comparison of electro-optic materials for THz measurement

	ZnTe	GaAs	InP	GaP	ZnS
V_π ($d = 1$ mm) (kV/cm)	89.0	161	153	252	388
Field sensitivity (mV/cm $\sqrt{\text{Hz}}$)	3.20	5.80	5.51	9.07	12.2
NEP (10^{-16} W $\sqrt{\text{Hz}}$)	0.27	0.89	0.80	2.2	5.2
$\sqrt{\varepsilon}$ (300 μm)	3.18	3.63	3.54	3.34	2.88
n (800 nm)	2.85	3.63	3.53	3.18	2.32

3.2   Measurement of Coherent Mid-Infrared Fields

The frequency response of a photoconductive antenna is generally limited by the resonance frequency of the antenna structure and by the lifetime of the photon-induced carriers. These considerations limit the usable spectral range to lower than 6 THz [68]. On the other hand, the electro-optic effect is nonresonant, and therefore very fast. As a result, the frequency response of an EO sampling system is limited by the laser pulse duration and group velocity mismatch. Since ultrashort laser pulses below 10 fs are widely available, the only limiting factor is the group velocity mismatch. To obtain the broadest bandwidths, one therefore typically chooses EO materials that have the best group velocity mismatch, and then uses the thinnest possible crystal. ZnTe is the best known EO material for this purpose. With a thin ⟨110⟩-oriented ZnTe crystal as both emitter and sensor, the frequency response has been pushed to about 40 THz [27–29] and recently to over 70 THz [65]. Figure 10 gives a typical temporal waveform and the corresponding spectrum obtained using ZnTe as emitter and sensor.

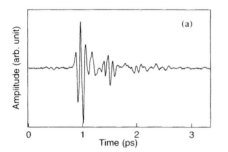 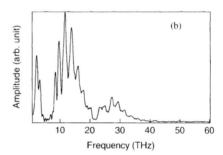

Fig. 10. Detection of a mid-infrared pulse. (**a**) A typical temporal waveform, and (**b**) its amplitude spectrum. For these measurements, the emitter was a 30 μm thick ZnTe crystal, and the detector was a 27 μm thick ZnTe crystal

3.3   Parallel Measurement: Chirped-Pulse Measurement

Conventional time-domain optical measurements, such as THz time-domain spectroscopy, use a mechanical translation stage to vary the optical path between the pump and probe pulses [69–71]. The intensity or polarization of the optical probe beam, which carries information generated by the pump beam, is repetitively recorded for each sequential time delay. In general, this data acquisition in the temporal scanning measurement is a serial acquisition; the signal is recorded during sampling by the probe pulse from a very small part of the THz waveform (a temporal window defined by the pulse duration of the optical probe beam). Therefore, the data acquisition rate in this single-channel detection scheme is limited to less than 100 Hz for a temporal scan

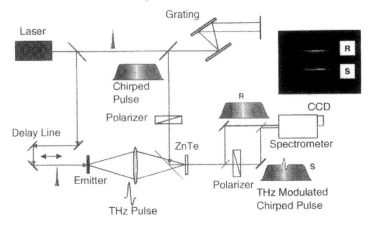

Fig. 11. Setup for chirped-pulse measurement

on the order of tens of picoseconds [72]. Clearly, this relatively slow acquisition rate cannot meet the requirements for real-time measurements, such as time-domain THz spectroscopy of fast-moving objects or flame analysis. To increase the acquisition rate, parallel data acquisition or multichannel detection is required. One possible method is to extend the novel design of "real-time picosecond optical oscilloscopes" [73,74] for local-field characterization so that it can be applied to freely propagating THz fields.

Figure 11 schematically illustrates the experimental setup for an electro-optic measurement with a chirped optical probe beam. The geometry is similar to that of the conventional free-space electro-optic sampling setup, except for the use of a grating pair for chirping and stretching the optical probe beam, and a spectrometer with a detector array for the measurement of the spectral distribution [73]. An amplified Ti:sapphire laser (Coherent RegA 9000) with an average power of 0.9 W and a pulse duration of 200 fs at 250 kHz was used. The center wavelength of Ti:sapphire lasers is about 820 nm, with a spectral bandwidth from 10 nm (Rega) to 17 nm (Tsunami). The THz emitter used in the setup described here is an 8 mm wide GaAs photoconductor. A polylens of focal length 5 cm focuses the THz beam onto a 4 mm thick ⟨110⟩ ZnTe crystal. The optical probe pulse is frequency chirped and time stretched from subpicosecond to over 30 picoseconds by the grating pair. Owing to the negative chirp of the grating (the pulse has a decreasing frequency versus time), the blue component of the pulse leads the red component. The fixed delay line is used only for the positioning of the THz pulse within the duration of the synchronized probe pulse (acquisition window) and for temporal calibration.

When the chirped optical probe pulse (approximately 30 ps long) and a THz pulse copropagate in the ZnTe crystal, the polarizations of different wavelength components of the chirped pulse are rotated by different portions

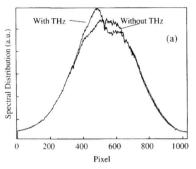 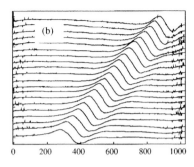

Fig. 12. Experimental results: (**a**) spectral distributions with and without THz modulation, and (**b**) difference signal with various THz-probe delays

of the THz pulsed field through the Pockels effect. The degree and direction of rotation are proportional to the THz field strength and polarity. After the optical analyzer, the polarization modulation is converted to an amplitude modulation on the probe pulse spectrum. The spectrometer is used to disperse and focus the collimated probe beam on a CCD camera. The electro-optic modulation must be operated near zero optical transmission to avoid detector saturation due to the finite dynamic range of the detector array. The residual birefringence from the ZnTe crystal provides a moderate background. The net change of the probe beam is small compared with the background light, so the electro-optic measurement is close to linear operation even though a crossed analyzer is used [75].

Figure 12a shows the spectra with and without THz modulation. Here, the difference between these two spectra is proportional to the THz electric field. When the time delay between the THz beam and the probe beam is varied, the THz pulse modulation is shifted on the spectrum (Figure 12b), showing the useful temporal window and linearity.

Acquiring the spectrum requires only one dimension of the CCD camera, so it is possible to realize 1D spatial and 1D temporal imaging [63]. To accomplish this, the setup of Fig. 11 needs some modifications. The probe beam is first expanded and then focused by a cylindrical lens to a thin line on the EO crystal; after the EO crystal, the probe beam is reconverted into a circular beam by another cylindrical lens. In this way, a 1D spatial distribution can be measured together with the temporal waveform. Figure 13 plots the spatiotemporal distribution emitted by a photoconductive antenna imaged by a polyethylene lens, at three transverse positions. The wavefront curvature is clearly visualized in this imaging scheme.

Because a measurement made with a chirped pulse is a parallel measurement, a single pulse contains all the information. Thus, it is obviously possible to perform a single-shot measurement. Figure 14 shows a single-shot spatiotemporal image (corresponding to Fig. 13b).

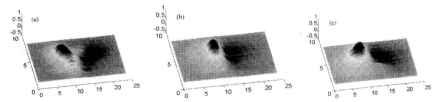

Fig. 13. Spatiotemporal imaging using the chirped-pulse technique

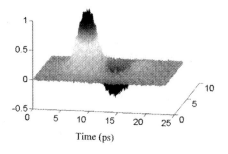

Fig. 14. Single-shot spatiotemporal image

For single-point measurements, the reference spectrum (R) without THz modulation can be sent to the spectrometer and CCD camera at the same time, and used as a dynamic reference. Therefore a true single-shot measurement can be obtained in a single pulse. This also provides a better signal-to-noise ratio because the laser fluctuations are corrected dynamically. An example is shown in Fig. 15.

The parallel-sampling property of the chirped-pulse measurement technique provides some unique features, including single-shot capability and ultrafast measuring speed. The technique can therefore be used in areas where conventional sampling techniques cannot be used. Possible applications include the study of unrepeatable events, such as emitter breakdown, spatiotemporal imaging of non-THz signals, unsynchronized microwave and other unsynchronized fast phenomena, and nonlinear effects.

3.4   Parallel Measurement: Terahertz Streak Camera

The chirped-pulse measurement technique makes it possible to study unrepeatable events on a single-shot basis. However, a theoretical analysis shows that the temporal resolution of a chirped-pulse measurement is given by $\Delta T = \sqrt{T_c T_0}$, where T_c and T_0 are the durations of the chirped and unchirped optical probe pulse, respectively [76]. Therefore $\Delta T \approx 5\,\mathrm{ps}$ when $T_c = 100\,\mathrm{ps}$ and $T_0 = 0.25\,\mathrm{ps}$. One of the factors limiting the temporal resolution is the spectral bandwidth of the laser pulse, because the chirped-pulse technique is a frequency-domain technique. A THz pulse modulates the chirped probe pulse in the time domain, and the signal is extracted in the frequency do-

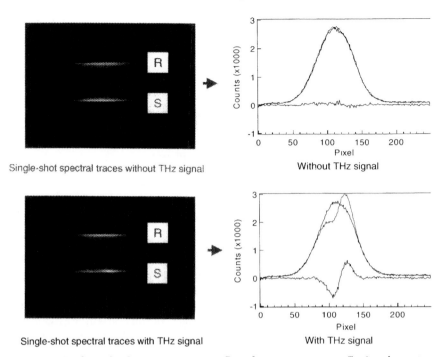

Fig. 15. Real single-shot measurement. R: reference spectrum, S: signal spectrum

main. Hence the temporal resolution is limited because of the well-known time–frequency relation.

An optical streak camera measures ultrafast light pulses in the time domain. For a direct measurement, the photon energy of the measured light must be greater than the cathode work function so as to photoemit electrons. Owing to the limitations of cathode materials, conventional streak cameras are only suitable for short wavelengths such as X-ray, UV, visible, and near-infrared radiation [77].

Much effort has been put into extending the measurable wavelength range. Recently, a far-infrared streak camera using high-lying Rydberg-state atoms has been reported. This extended the measurable wavelength range to 100 μm [78–80]. In this instrument, an ultraviolet laser source is needed to pump electrons into excited states, and the gas atoms must be in a vacuum chamber. In about 1990, a streak camera was used to measure radio-frequency and microwave radiation indirectly, with an EO modulator as a converter. The radio-frequency or microwave signals were converted into intensity modulations of a continuous-wave He–Ne laser. The highest measurable frequency was limited to about 40 GHz by the bandwidth of the EO modulator [81,82]. As discussed above, with the development of free-space EO sampling, the bandwidth has been extended to over 40 THz. The speed of the EO mod-

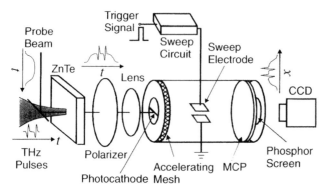

Fig. 16. Principle of operation of the electro-optic THz streak camera

ulator is no longer a limiting factor. In addition, the temporal resolution of a state-of-the-art streak camera is better than 200 fs. Therefore it is feasible to cover the previously inaccessible frequency range by combining an electro-optic device and an optical streak camera.

The measuring principle of such a camera is shown in Fig. 16. The THz pulses and a linearly polarized long probe pulse copropagate in an EO crystal. The polarization of the probe beam is modulated by the electric field of the THz pulses via the Pockels effect. and this polarization modulation is converted into an intensity modulation by a polarization analyzer. When the modulated probe pulse hits the photocathode of the streak tube, photoelectrons are generated. These photoelectrons are accelerated toward a microchannel plate (MCP). At the same time. they are deflected by a properly synchronized sweeping voltage provided by the sweep electrodes. Therefore. the electrons generated at different times hit different locations on the MCP. As the electrons pass through the MCP, they are multiplied several thousand times, and then impact on a phosphor screen to produce a visible trace. The image of the phosphor screen is acquired by a CCD camera and sent to a computer.

Figure 17 shows a double THz pulse measured by the streak camera. Figures 17a and 17b show an averaged and a single-shot measurement, respectively, of a THz pulse. The CCD traces are also shown above the curves. The THz streak camera provides an alternative method for single-shot THz pulse measurement. THz pulses usually have a bipolar temporal structure. However, in order to obtain a good contrast ratio, an optical bias near zero has been used in Figure 17, resulting in a unipolar temporal structure. Nevertheless, the bipolar property of the THz pulses can still be observed by means of the interference of two THz pulses. When the time delay between two THz pulses is scanned, constructive and destructive interference are obtained as a result of the bipolar temporal structure (see Fig. 18). The temporal resolution of the THz streak camera is mainly limited by that of the optical streak camera. The sensitivity is estimated to be 6 V/cm.

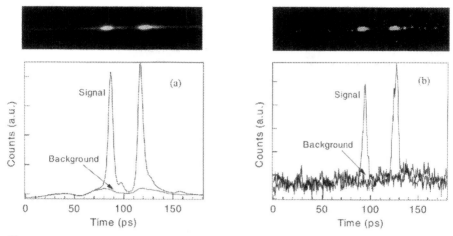

Fig. 17. Temporal waveforms of THz pulses measured by the electro-optic THz streak camera, with a ZnTe crystal as the converter: (a) averaged and (b) single-shot results

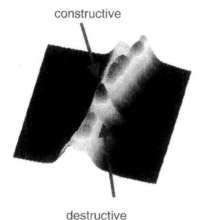

constructive

destructive

Fig. 18. Interference of two THz pulses measured by an optical streak camera

3.5   Parallel Measurement: 2D Imaging

Since the first demonstration of THz imaging in 1995, THz imaging has attracted much attention [83–86,55,59,60]. Despite the advantages of high signal-to-noise ratio, good uniformity, and large imaging area, the 2D scanning imaging system has the disadvantage of low speed. The speed is limited by the 2D mechanical scanning. Depending on the number of pixels, a picture may take minutes, or even hours. Alternative methods are required if the object under test is changing rapidly or evolving nonrepeatably.

To increase the speed, an array device is desired. For photoconductive antennas, it is technically difficult to make an array with a large number

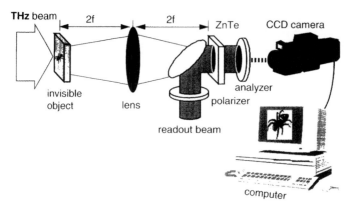

THz beam | 2f | 2f | ZnTe | CCD camera

invisible object | lens | readout beam | polarizer | analyzer

computer

Fig. 19. 2D real-time THz imaging system

of elements, because the wiring and alignment are problematic. However, it is very straightforward to realize two-dimensional measurements with EO sampling. A system that does this is shown in Fig. 19. Here, a lens images the object, which is illuminated by an expanded and collimated THz beam, onto the EO crystal (ZnTe); the probe (readout) beam is also expanded and copropagates with the THz beam through the crystal. The polarization state of the probe beam is modulated by the local THz electric field, and this modulation is converted into an intensity modulation by an analyzer. A CCD camera is used to measure the two-dimensional distribution of the probe beam.

This CCD-based 2D imaging system is completely parallel, so there is no wiring problem. The speed is ultimately limited by the frame rate of the CCD camera. For a fast CCD camera, the frame rate can be over 100 frames per second. Therefore it is possible to realize real-time THz imaging. Moreover, even single-shot imaging is possible, provided that there is enough THz modulation on the probe beam.

However, a lock-in amplifier, which is used in the single-detector system and suppresses noise dramatically, cannot be used with the CCD camera. So this system suffers a dramatically reduced signal-to-noise ratio. Nevertheless, the phase-sensitive detection principle of the lock-in amplifier can be adopted by using dynamic subtraction. The principle of operation is as follows (Fig. 20): the pump beam (and thus the THz beam) is modulated by an EO modulator, which is synchronized to the CCD camera. With the modulation frequency set equal to half of the CCD frame rate, the THz signal appears in the probe beam in alternate frames. The difference between two consecutive frames gives the signal image. More frame accumulation increases the SNR. The lower part of Fig. 20 shows the timing between the CCD camera and EO modulator. While the in-phase signal is not changed, the out-of-phase noise is decreased dramatically. Experiments with a Coher-

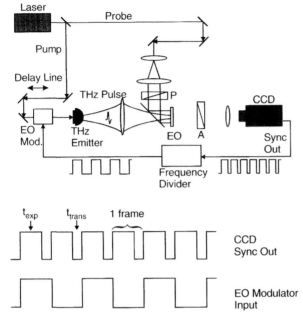

Fig. 20. 2D real-time THz imaging system with dynamic subtraction

ent RegA 9000 laser showed that the signal-to-noise ratio was increased by two orders of magnitude by dynamic subtraction of 1000 frames, as compared with simply averaging 1000 times without dynamic subtraction. The detectable fractional modulation can be as small as 5×10^{-5}.

Figure 21 shows some THz signals obtained by use of the CCD-based imaging system with dynamic subtraction. The emitter was a 2 mm thick $\langle 110 \rangle$ ZnTe crystal, and the optical rectification effect was used. The THz beam was focused by two pairs of parabolic mirrors (see Fig. 4). As mentioned above, the radiation from an optical-rectification source is much weaker than that from a biased photoconductive-antenna emitter. The modulation depth is only a few percent, of the same order as the laser fluctuations. It would not be possible to obtain a signal with good SNR without using dynamic subtraction. Figure 21a shows the 2D distribution at the main peak, and Fig. 21b shows the measured waveform, indicating good SNR. Figures 21c,d show the spatiotemporal distribution when the EO sensor is located at the focal plane or 1.5 cm away from the focal plane. The curved wavefront in Fig. 21d means that the beam is divergent. The CCD-based THz imaging system is a powerful tool for visualizing THz beams. Using this technique, the beam can be directly observed on a computer screen in real time. This is very valuable in optimizing the alignment of the optics [87,88].

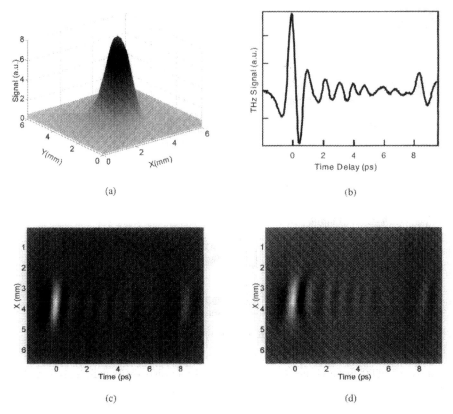

Fig. 21. Electric-field distribution, measured with dynamic subtraction, from an optical-rectification emitter. (**a**) 2D distribution at peak signal, (**b**) temporal waveform at the main peak, (**c**) 1D spatial and 1D temporal distribution at the focal plane, (**d**) distribution 1.5 cm from the focal plane

3.6   Near-Field Terahertz Imaging

Near-field measurements are an effective way to improve the spatial resolution of THz imaging [89,90]. It is obviously not possible with FS-EOS to realize near-field imaging by putting the sample in the near field of an evanescent THz wave, because of the requirement that the THz and probe beams copropagate in the detector crystal. Therefore a reflection geometry is used. A system of this kind is shown in Fig. 22; the THz beam is incident from the left side of the EO crystal, whereas the probe beam is incident from the right side and is reflected by the left facet of the EO crystal. It has been shown experimentally, as well as theoretically, that the THz waveform measured in this reflection geometry is the same as that measured in the more conventional transmission geometry. The reason is that the contribution from the counterpropagation part is negligible. With this modification, the object can

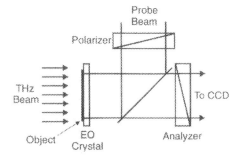

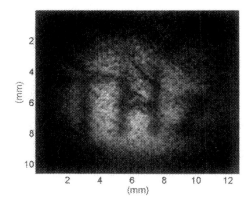

Fig. 22. 2D near-field THz imaging system

Fig. 23. THz image of a mask with the three letters "THz", taken using the near-field geometry shown in Fig. 22

be located on the surface of the EO crystal, and near-field measurements can be realized. We used a mask with the three letters "THz", in which the width of the metal letters was 0.5 mm and the size of the word was 1 cm × 0.5 cm. Figure 23 shows the imaged result. The spatial resolution is improved by at least five times compared with that in the $2f$–$2f$ imaging system, and the image can be displayed on a computer screen in real time. This result also indicates a capability to image moving and living objects [87].

3.7 Detection Geometry and Working Conditions

In single-point measurements with a lock-in amplifier, EO modulation is usually implemented at the linear-optical bias point that gives the largest dynamic range. However, in the case of parallel measurements, such as 2D real-time imaging and chirped-pulse measurements, a 2D array device (usually CCD camera) has to be used, instead of a lock-in amplifier. To obtain the highest SNR, it is essential to generate the highest THz modulation depth. Therefore a crossed-polarizer geometry has to be used, and the EO modulation works at an optical bias point near zero. Here, we analyze this somewhat different situation.

The transmitted light can be written as

$$I = I_0 \left[\eta + (\Gamma_0 + \Gamma)^2 \right] , \tag{12}$$

where I_0 is the input light intensity. η the contribution from scattering. Γ_0 the optical bias, and Γ the electric-field-induced contribution to the birefringence. Note that at near-zero optical bias. $|\Gamma_0| \ll 1$ and $|\Gamma| \ll 1$.

The modulation depth γ. defined as the signal-to-background ratio. can be readily obtained as

$$\gamma \equiv \frac{I_{\Gamma \neq 0} - I_{\Gamma = 0}}{I_{\Gamma \neq 0} + I_{\Gamma = 0}} = \frac{2\Gamma_0 \Gamma + \Gamma^2}{2\eta + \Gamma_0^2 + (\Gamma_0 + \Gamma)^2} . \tag{13}$$

The optimal working point that gives the maximum modulation depth is given by

$$\Gamma_0^{\mathrm{m}} = -\frac{\Gamma}{2} \pm \sqrt{\left(\frac{\Gamma}{2}\right)^2 + \eta} \approx \pm\sqrt{\eta}. \tag{14}$$

and the maximum depth of modulation is

$$\gamma_{\mathrm{max}} \approx \frac{\Gamma}{2\sqrt{\eta}} . \tag{15}$$

Figure 24 shows the experimental and calculated results for the modulation depth versus the optical bias. The excellent agreement between the experiment and calculation proves that this analysis is valid.

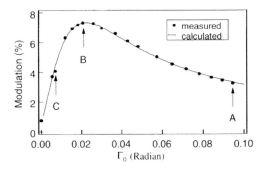

Fig. 24. Modulation depth versus the optical bias. B is the optimal working point

One problem with operating near zero optical bias is the nonlinear response if the signal Γ is comparable to the optical bias Γ_0. Figure 25 shows two waveforms. measured at two optical-bias points (A and C in Fig. 24). While the distortion in curve A is negligible, curve C is severely distorted. Fortunately, this distortion can be recovered, provided that Γ_0 is known. The corrected signal is given by

$$\Gamma = \begin{cases} -\Gamma_0 + \sqrt{\Gamma_0^2 + I_{\mathrm{s}}/I_0}. & \Gamma_0 > 0 \\ -\Gamma_0 - \sqrt{\Gamma_0^2 + I_{\mathrm{s}}/I_0}. & \Gamma_0 < 0 \end{cases} \tag{16}$$

where I_{s} is the distorted signal. The corrected signal is also plotted in Fig. 25. and is in good agreement with the undistorted curve A.

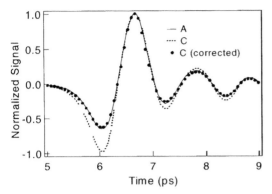

Fig. 25. Normalized THz waveforms at different optical biases. The two curves correspond to the two Γ_0 positions A and C marked in Fig. 24. The *filled circles* are the corrected data obtained from curve C using (16), which overlaps well with curve A

3.8 Comparison Between Photoconductive Sampling and EO Sampling

Both a photoconductive (PC) antenna and EO sampling can be used for the measurement of freely propagating THz pulses. It is of interest to compare the performances of these two methods [54.91–93]. For low-frequency THz signals (less than 3 THz) and a low chopping frequency (kHz), PC antennas have much better SNR (by about two orders). However, with special high-frequency chopping techniques, the noise can be greatly reduced with EO sampling (by about two orders), leading to comparable SNR for both methods [91]. For frequencies above several THz, the performance of PC antennas decreases dramatically, but EO sampling still has high sensitivity. Figure 26 plots the THz signal from the same emitter, obtained with PC and EO sampling. The waveform obtained with the EO sensor is obviously narrower than that obtained with the antenna, showing the larger bandwidth attainable with EO sampling.

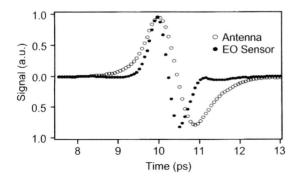

Fig. 26. Comparison between photoconductive-antenna and EO sampling

EO sampling requires substantially less power for the probe beam than does PC sampling. This makes it possible to realize parallel measurements with EO sensors, as described above. The implementation of the parallel measurement both spatially and temporally is straightforward with the EO technique; however, it is very hard to realize parallel measurement with PC antennas. The problems include the manufacture of the antenna array, wiring, and probe beam power. The alignment and stability of EO sensors are easier and better than those of PC antennas. However, the EO technique is more sensitive to laser noise.

3.9   EO Sampling for Continuous-Wave Terahertz Beams

The sampling techniques take advantage of the coherent property of the radiation. While the coherent signal is accumulated, the incoherent noise is averaged out owing to its random phase. Therefore it is possible to achieve a noise-equivalent power (NEP) much smaller than the thermal background. For pulsed THz beams, the duty cycle is usually much smaller than 1, so even if the averaged THz power is small, the peak electric field is still high. The gating effect of the sampling techniques eliminates most of the background noise.

As can be seen, the key to these advantages is to have the sampling pulse and THz pulse synchronized in time. For a conventional pulsed THz system, this synchronization is automatic because the THz pulse is induced by the same laser that generates the probe pulse. In contrast, the detection of a CW THz beam using these techniques is not trivial. For an unsynchronized THz beam, EO sampling gives zero output because the contributions from the positive and negative halves of the electric field cancel each other. Although it is possible to obtain a rectification output at an optical-bias point near zero, most of the signal is canceled, leading to very poor sensitivity. Nevertheless, EO sampling measurements of a CW THz beam can be realized by beating two laser beams with a frequency difference in THz range. This acts as a pulsed laser beam, and is used both to generate and to sample the CW THz signal [94].

4   Applications

4.1   Dynamics of Interaction of Lattice with Infrared Photons

A conventional technique to determine the dielectric function of materials is to use a CW IR source for a transmission or reflection measurement. In this form of wavelength-domain spectroscopy, however, the details of the dispersion function are often not clearly discernible. Especially in the reststrahl band, where the dispersion function is sensitively determined by factors such as the oscillator strength and the broadening of phonons, a subtile detail of

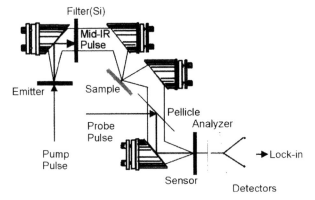

Fig. 27. Experimental setup for mid-infrared spectroscopy

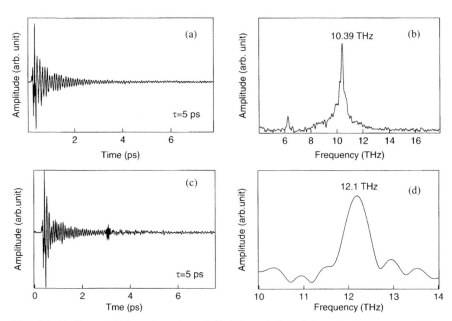

Fig. 28. Reflected THz pulses from InP (**a**) and GaP (**c**), and the corresponding spectra ((**b**) and (**d**))

the dispersion function can give rise to an unexpected time-domain response for IR spectroscopy. With mid-infrared EO sampling, it is possible to directly study the dielectric response of strongly infrared-active phonon resonances of polar compound materials by observing the coherent dynamics of the reflection in the reststrahl frequency band. An experimental setup for this purpose is shown in Fig. 27. Figure 28 shows reflected THz pulses from InP and GaP and the corresponding spectra [29].

4.2   Spatiotemporal Coupling of Few-Cycle Pulses

The propagation and diffraction of few-cycle, even sub-cycle THz pulses has attracted much attention recently, both theoretically and experimentally. Many interesting phenomena have been reported, including pulse shaping, spectral shifts, Gouy phase shifts, and apparent superluminal effects [95–104]. THz pulses are good candidates for studying these propagation and diffraction effects, because the methods for generation and detection of few-cycle, or even sub-cycle THz pulses is well established. Also, the measurement of THz pulses are coherent, i. e. both the amplitude and the phase are measured simultaneously.

Nevertheless, although the techniques for detection of the THz electric field are mature, most available techniques are based on single-point measurements, i. e., spatially, only one point is measured at a time. If an image of a 2D distribution is needed, it is necessary to scan the detection system. This scanning involves mechanical movement, and therefore is very time-consuming and cannot produce results in real time. Besides, it is difficult to maintain an the exact timing between the pump and probe beams. The lack of adequate experimental measurements prevents one from obtaining deep insight into the complicated propagation of few-cycle pulses.

However, an electro-optic, CCD-based THz imaging system is an ideal tool for studying the propagation of few-cycle THz pulses, using the dynamic subtraction techniques described above. The 2D electric-field distribution of the THz pulses can be mapped with unprecedented speed and SNR without mechanical movement, except for the time delay. Figure 29 shows a setup used to study the propagation and spatiotemporal coupling of few-cycle THz pulses. This is a simple one-lens system, with the emitter (optical rectification) located at a distance d_1 from the lens, and the EO sensor at d_2 on the other side. The lens is made of polyethylene, with a 5 cm focal length. In the experiment, the distance $d_2 = 10$ cm is fixed. Time zero is set at the main peak when $d_1 = 10$ cm. Figure 30 plots the 2D distribution at three emitter-to-lens distances: 31, 15, and 10 cm. The circular pattern arises from

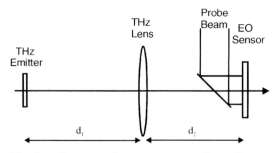

Fig. 29. Geometry for spatiotemporal coupling experiment

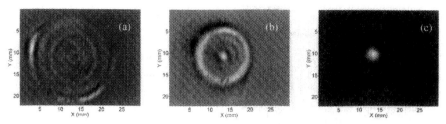

Fig. 30a–c. 2D spatial distribution for three emitter positions

the curvature of the wavefront. Roughly speaking, assuming R is the radius of curvature radius of the wavefront, the measured field is

$$E(r) = A(r) \cos\left(\frac{kr^2}{R}\right) . \tag{17}$$

The cosine term leads to the observed ring structure. Since R depends on the distance d_1, and the spacing between two adjacent rings is determined by R and r, when d_1 is scanned, the ring spacing changes. Figure 30c corresponds to the $2f$–$2f$ geometry, where the EO sensor is at the conjugate plane of the emitter. A small, very strong spot is observed.

However, the bright rings in Fig. 30a,b cannot be explained simply by the geometric wavefront. When d_1 is fixed at 10 cm and the time delay between the THz and probe pulses is scanned, the spatiotemporal distribution is obtained. The image in Fig. 31 resembles ordinary beam focusing, except that the horizontal axis in this figure is not the propagation distance, but rather the time delay. The measured X-shaped spatiotemporal distribution demonstrates dramatic spatiotemporal coupling: unlike the case in Figs. 21c,d, the beam shape is no longer time-invariant and the spatial and temporal coordinates are no longer separable.

There are two reasons for this strong spatiotemporal coupling. One is the ultrabroad spectral bandwidth of the THz pulses. For few-cycle pulses, the

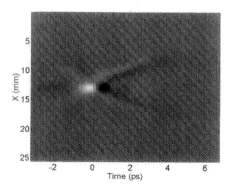

Fig. 31. The X-shaped spatiotemporal distribution obtained with the $2f$–$2f$ imaging system, showing strong space-time coupling in the propagation of few-cycle pulses

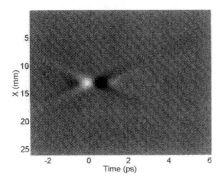

Fig. 32. Simulated spatiotemporal distribution (see Fig. 31 for the corresponding experimental result)

spectral bandwidth is comparable to the central frequency. Another reason is the limited diameter of the imaging lens. Because the wavelength of the THz beam is very large (0.3 mm for 1 THz), diffraction is a much more significant effect than it is for visible or near-infrared beams. Therefore the THz beam is truncated by the finite aperture of the lens, with longer wavelengths experiencing more severe truncation. This wavelength-dependent effect leads to the strong spatiotemporal coupling seen in Fig. 31.

This phenomenon can be simulated by means of diffraction integrals, taking the finite size of the lens into consideration. Two diffraction integrals are needed: one of them calculates the propagation from the emitter to the lens plane, and the other the propagation from the lens plane to the sensor plane. To consider the aperture effect of the lens, the integral is taken over an area corresponding to the size of the lens. Either time-domain or frequency-domain diffraction integrals can be used. Figure 32 shows the simulated result. The main feature (i.e. the X-shaped spatiotemporal structure) is in good agreement with the experimental result in Fig. 31.

The above results show that, in contrast to the case of long pulses, where the spatial distribution is time-independent, the spatial distribution of few-cycle pulses shows dramatic changes with time. A similar phenomenon should exist in the optical frequency range as well. These techniques provide a means to manipulate and study the spatiotemporal structure of ultrashort pulses, and may find applications in nonlinear optics, optical information processing, and optical communication. It is worth noting that the X-shaped spatiotemporal structure is mainly due to the finite size of the lens. That is, the "mask" is only a circular aperture. It is expected that if a more complicated mask were used, as in optical pulse shaping [106], one would have more freedom to control the spatiotemporal structure of an ultrashort pulse.

4.3 Point Scanning Terahertz Imaging

As discussed above, the EO system has a high signal-to-noise ratio and large bandwidth, and it is easy to align. These features make it an attractive alternative method for THz imaging. Figure 33 shows a schematic of a scanning

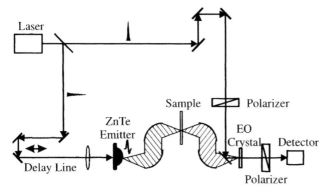

Fig. 33. Schematic of a scanning THz imaging system using an EO emitter and sensor

THz imaging system using an EO emitter and sensor. A piece of EO crystal (e. g. $\langle 110 \rangle$ ZnTe) is used to generate THz radiation, using the optical rectification effect. A silicon ball may be used to increase the collection efficiency. The first pair of parabolic mirrors is used to focus the THz beam down to a small spot, and the second pair of parabolic mirrors is used to refocus the beam onto the EO sensor crystal. The detection system is a typical example of EO sampling as described above. The sample is located at the focal plane between two pairs of parabolic mirrors, and is raster-scanned to obtain the 2D distribution of the absorption and/or refractive index.

Figure 34 shows THz images of a mammographic phantom obtained with the setup shown in Fig. 33. This phantom is one used to evaluate x-ray imaging systems. It contains five nylon fibers, five masses and five groups of specks.

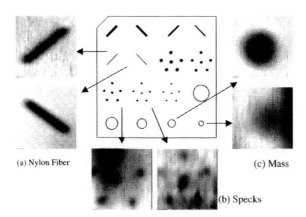

Fig. 34. Mammographic phantom, and THz images of portions of it. The central diagram shows the structure, and the surrounding pictures were obtained with the THz imaging setup of Fig. 33

All these structures are invisible to the naked eye. A good x-ray imaging system can see the third speck group, the fourth fiber, and the third mass. The scanning THz imaging system can resolve the fifth fiber, the fifth mass, and the fourth speck group, indicating very good spatial resolution. Since the EO system provides much larger bandwidth than does a photoconductive antenna, it is possible to perform THz imaging using much higher-frequency components, and therefore increase the spatial resolution greatly. With 12 fs laser pulses, and thin EO crystals (on the order of 10 µm) as both emitter and sensor, the usable frequency bandwidth is over 40 THz. Figure 35 shows a mid-infrared image of an array of 50 µm holes, in which the spatial resolution is better than 10 µm. With the increased spatial resolution, it is even possible to image biological cells using mid-infrared radiation. Figure 36 shows an optical and a THz image of onion cells. Note that the optical image is a reflection image, whereas the THz image is a transmission image. These are similar but not identical, indicating that THz image contains different information about the object [107].

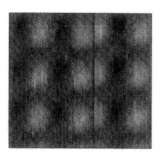

Fig. 35. Mid-infrared image of an array of 50 µm in diameter

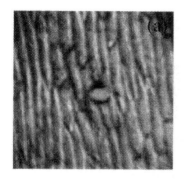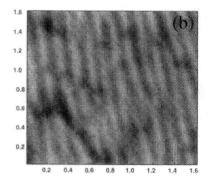

Fig. 36. Images of onion cells obtained by optical (**a**) and THz (**b**) imaging

4.4   Electro-Optic Terahertz Transceiver

In a conventional experimental setup for THz time-domain spectroscopy, separate THz transmitters and THz receivers are used for the generation and detection of the THz signal. However, electro-optic detection is the reverse process of generation by rectification. This suggests that the transmitter and the receiver can be the same crystal. It is thus feasible to realize an EO THz transceiver, which alternately transmits pulsed electromagnetic radiation (optical rectification) and receives the returned signal (electro-optic effect) [106]. Figure 37 shows an experimental setup that achieves this. A pair of synchronized optical pulses (pump and probe) generated by a Michelson interferometer illuminates the ZnTe crystal. The first optical pulse generates a THz pulse by optical rectification. The THz pulse is collimated by a parabolic mirror and reflected by a metallic mirror. A mechanical chopper modulates the THz beam. The second optical pulse samples the returned THz signal via the electro-optic effect in the same crystal, with a lock-in amplifier. The polarization directions of the optical pump and probe beams are parallel. Theoretical calculation predicts that the optimum orientation of the pump beam polarization is 26° counterclockwise from the z axis of $\langle 110 \rangle$ ZnTe.

Figure 37b shows a set of waveforms measured by moving the metallic mirror along the THz propagation direction in 1 mm step (6.6 ps round-trip time). The first signal is the THz reflection from the metallic chopper blade, which is roughly perpendicular to the propagation direction of the THz beam, and whose time position is fixed. The second peak is the reflected THz sig-

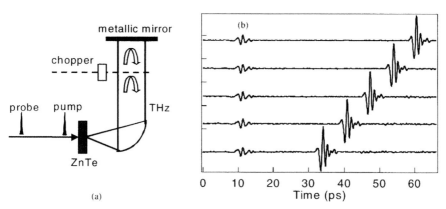

Fig. 37. (a) Schematic experimental setup of an electro-optic THz transceiver. The THz signal is generated and detected by the same ZnTe crystal. (b) Temporal waveforms of a THz signal reflected from the metallic mirror, with 1 mm displacement along the THz propagation direction at each step. The first, inverted signal is the reflection from the metallic chopper blade, and the second peak is the returned signal from the metallic mirror

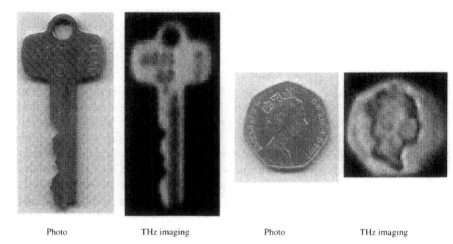

Photo   THz imaging   Photo   THz imaging

Fig. 38. Images obtained using the EO THz transceiver. Photos are shown for comparison

nal from the metallic mirror, whose time position can be changed by the location of the mirror. The time delay between the two THz signals is the round-trip time of a THz pulse traveling between the chopper and the metallic mirror. The reflection from the chopper blade automatically serves as a reference marker for calibration of the system. Similarly to the demonstration of a photoconductive THz transceiver [109]. there is a phase difference of π between the phases of the reflected signals from the chopper and the metallic mirror. Therefore these two signals, when measured with a lock-in amplifier, show opposite polarities. This THz transceiver is a good candidate for tomography applications because the complexity of the system is greatly reduced. Figure 38 shows some reflection images obtained using the EO THz transceiver.

4.5   Compact System

THz waves can be used in time-domain spectroscopy, tissue imaging, chemical-reaction analysis, environmental monitoring, and material and package inspection. However, a THz system is normally complicated, large, and expensive. A simple, compact, and user-friendly THz system would be valuable [108]. Figure 39 shows such a compact THz system, which uses EO sampling for detection. The entire system is smaller than a fax machine. In this system, a very small femtosecond fiber laser is used, and is located beneath the optics of the EO system. After frequency doubling, the fiber laser delivers 30 mW of 130 fs, 800 nm pulses at 80 MHz repetition rate. An acousto-optic modulator is used to modulate the pump beam. The emitter is a biased, large-aperture PC antenna. A 2 mm thick $\langle 110 \rangle$ ZnTe crystal is used as the sensor. A dy-

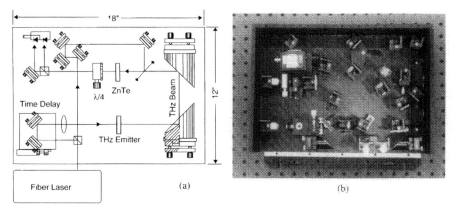

Fig. 39. Compact THz system using EO sampling for detection. The whole system is smaller than a fax machine. (a) Layout, (b) photograph

namic range of over 1000 is achieved. The system can be used in time-domain spectroscopy, THz imaging, and other applications where space limitations and portability are important.

References

1. G. Mourou, C. V. Stancampiano, D. Blumenthal, "Picosecond microwave pulse generation", *Appl. Phys. Lett.* **38**, 470 (1981).
2. C. Waschke, H. G. Roskos, R. Schwedler, K. Leo, H. Kurz, K. Kohler, "Coherent submillimeter-wave emission from Bloch oscillations in a semiconductor superlattice", *Phys. Rev. Lett.* **70**, 3319 (1993).
3. T. Dekorsy, P. Leisching, K. Kohler, H. Kurz, "Electro-optic detection of Bloch oscillations", *Phys. Rev. B* **50**, 8106 (1994).
4. T. Dekorsy, H. Auer, C. Waschke, H. J. Bakker, H. G. Roskos, H. Kurz, "THz-wave emission by coherent optical phonons", *Physica B* **219–220**, 775 (1996).
5. R. Kersting, K. Unterrainer, G. Strasser, H.F. Kauffmann, E. Gornik, "Few-cycle THz emission from cold plasma oscillations", *Phys. Rev. Lett.* **79**, 3038 (1997).
6. M. Bass, P. A. Franken, J. F. Ward, G. Weireich, "Optical Rectification", *Phys. Rev. Lett.* **9**, 446 (1962).
7. G. A. A'skaryan, "Cherenkov radiation and transition radiation from electromagnetic waves", *Sov. Phys. JETP* **15**, 943 (1962).
8. D. H. Auston, M. C. Nuss, "Electro-optic generation and detection of femtosecond electrical transients", *IEEE J. Quantum Electron.* **24**, 184 (1988).
9. Y. R. Shen, "Far-infrared generation by optical mixing", *Prog. Quantum Electron.* **4**, 207 (1976).
10. J. Morris, Y. R. Shen, "Far-infrared generation by picosecond pulses in electrooptic materials", *Opt. Commun.* **3**, 81 (1971).
11. K.P. Yang, P.L. Richards, Y.R. Shen, "Generation of far-infrared radiation by picosecond light pulses in LiNbO$_3$", *Appl. Phys. Lett.* **19**, 320 (1971).

12. T. K. Gustafson, J.-P. E. Taran, P. L. Kelley, R. Y. Chiao, "Self-modulation of picosecond pulses in Electro-optic crystals", *Opt. Commun.* **2**, 17 (1970).

13. D. Bagasaryan, A. Makaryan, P. Pogosyan, "Cherenkov radiation from a propagating nonlinear polarization wave", *JETP Lett.* **37**, 595 (1983).

14. D. H. Auston, "Subpicosecond electrooptic shockwaves", *Appl. Phys. Lett.* **43**, 713 (1983).

15. D. A. Leinman, D. H. Auston, "Theory of electrooptic shockwave radiation in nonlinear optic media", *IEEE J. Quantum Electron.* **20**, 964 (1983).

16. D. H. Auston, K. P. Cheung, J. A. Valdmannis, D. A. Kleinman, "Cherenkov radiation from femtosecond optic pulses in electrooptic media", *Phys. Rev. Lett.* **53**, 1555 (1984).

17. C. Fattinger, D. Grischkowsky, "A Cherenkov source for freely-propagating terahertz beams", *IEEE J. Quantum Electron.* **25**, 2608 (1989).

18. B. B. Hu, X.-C. Zhang, D. H. Auston, P. R. Smith, "Free Space Radiation from Electro-Optic Crystal", *Appl. Phys. Lett.* **56**, 506 (1990).

19. X.-C. Zhang, X. F. Ma, Y. Jin, T.-M. Lu, E. P. Boden, P. D. Phelps, K. R. Stewart, C. P. Yakymyshyn, "THz optical rectification from highly nonlinear organic crystals", *Appl. Phys. Lett.* **61**, 3080 (1992).

20. X.-C. Zhang, Y. Jin, X.F. Ma, "Coherent measurement of THz Optical rectification from electro-optic crystals", *Appl. Phys. Lett.* **61**, 1764 (1992).

21. X.-C. Zhang, D. H. Auston, "Transition radiation from femtosecond optical pulses in electro-optic materials", *Appl. Phys. Lett.* **61**, 1784 (1992).

22. H. J. Bakker, G. C. Cho, H. Kurz, Q. Wu, X.-C. Zhang, "Distortion of THz pulses in electro-optic sampling", *J. Opt. Soc. Am. B* **15**, 1795 (1998).

23. X.-C. Zhang, B. B. Hu, J. T. Darrow, D. H. Auston, "Generation of femtosecond electromagnetic pulses from semiconductor surfaces", *Appl. Phys. Lett.* **56**, 1011 (1990).

24. T. J. Carrig, G. Rodriguez, T. Sharp Clement, A. J. Taylor, K. R. Stewart, "Generation of terahertz radiation using electro-optic crystal mosaics", *Appl. Phys. Lett.* **66**, 10 (1995).

25. T. J. Carrig, G. Rodriguez, T. Sharp Clement, A. J. Taylor, K. R. Stewart, "Scaling of terahertz radiation via optical rectification in electro-optic crystals", *Appl. Phys. Lett.* **66**, 121 (1995).

26. F. G. Sun, X.-C. Zhang, W. Ji, "Two-Photon-Absorption Induced Saturation of THz Radiation in ZnTe," in *Conference on Lasers and Electro-Optics*, OSA Technical Digest (Washington DC 2000), p. 479.

27. Q. Wu, X.-C. Zhang, "Free-Space Electro-Optic Sampling of Mid-Infrared Pulses", *Appl. Phys. Lett.* **71**, 1285 (1997).

28. P.Y. Han, X.-C. Zhang, "Coherent, broadband mid-infrared terahertz beam sensors", *Appl. Phys. Lett.* **73**, 3049 (1998).

29. P. Y. Han, G. C. Cho, X.-C. Zhang, "Broad band mid-infrared THz pulse: measurement technique and applications", *J. Nonlin. Opt. Phys. Mater.* **8**, 89 (1999).

30. G. Mourou, C. V. Stancampiano, D. Blumenthal, "Picosecond microwave pulse generation", *Appl. Phys. Lett.* **38**, 470 (1981).

31. D. H. Auston, K. P. Cheung, P. R. Smith, "Picosecond photoconducting Hertzian dipoles", *Appl. Phys. Lett.* **45**, 284 (1984).

32. A. P. DeFonzo, M. Jarwala, C. R. Lutz, "Transient response of planar integrated optoelectronic antennas", *Appl. Phys. Lett.* **50**, 1155 (1987).

33. C. Johnson, F. J. Low, A. W. Davidson, "Germanium and germanium-diamond bolometers operated at 4.2 K, 2.0 K, 1.2 K, 0.3 K, and 0.1 K", *Opt. Eng.* **19**, 255 (1980).

34. R. C. Johns, "The ultimate sensitivity of radiation detectors", *J. Opt. Soc. Am.* **37**, 879 (1947).

35. C. Fattinger, D. Grischkowsky, "Point source terahertz optics", *Appl. Phys. Lett.* **53**, 1480 (1988).

36. J. A. Valdmanis, G. Mourou, C.W. Gabel, "Electrical transient sampling system with two picosecond resolution", in *Picosecond Phenomena III. Proceedings of the Third International Conference on Picosecond Phenomena*, ed. by K. B. Eisenthal, R. M. Hochstrasser, W. Kaiser, A. Laubereau (Springer, Berlin, Heidelberg, 1982), p. 101.

37. J. A. Valdmanis, G. A. Mourou, C. W. Gabel, "Picosecond electro-optic sampling system", *Appl. Phys. Lett.* **41**, 211-212 (1982).

38. J. A. Valdmanis, G. A. Mourou, C. W. Gabel, "Subpicosecond electrical sampling", *IEEE J. Quant. Electron.* **QE-19**, 664 (1983).

39. B. H. Kolner, D. M. Bloom, P. S. Cross, "Electrooptic sampling with picosecond resolution", *Electro. Lett.* **19**, 574 (1983).

40. D. R. Dykaar, T. Y. Hsiang, G. A. Mourou, "An application of picosecond electro-optic sampling to superconducting electronics", *IEEE Trans. Magn.* **21**, 230 (1985).

41. J. A. Valdmanis, G. Mourou, "Subpicosecond electrooptic sampling: principles and applications", *IEEE J. Quant. Electron.* **QE-22**, 69 (1986).

42. B. H. Kolner, D. M. Bloom, "Electrooptic sampling in GaAs integrated circuits", *IEEE J. Quant. Electron.* **QE-22**, 79 (1986).

43. H. Takahashi, S. Aoshima, Y. Tsuchiya, "Sampling and real-time methods in electro-optic probing system", *IEEE Trans. Instrum. Meas.* **44**, 965 (1995).

44. T. Itatani, T. Nakagawa, F. Kano, K. Ohta, Y. Sugiyama, "Electrooptic vector sampling-measurement of vector components of electric field by the polarization control of probe light", *IEICE Trans. Electron.* **E78-C**, 73 (1995).

45. D. Jacobs-Perkins, M. Currie, C.-C. Wang, C. A. Williams, W. R. Donaldson, R. Sobolewski, T. Y. Hsiang, "Subpicosecond imaging system based on electrooptic effect", *IEEE J. Sel. Top. Quant. Electron.* **2**, 729 (1996).

46. T. Pfeifer, H.-M. Heiliger, T. Loffler, C. Ohlhoff, C. Meyer, G. Lupke, H. G. Roskos, H. Kurz, "Optoelectronic on-chip characterization of ultrafast electric devices: Measurement techniques and applications", *IEEE J. Sel. Top. Quant. Electron.* **2**, 586 (1996).

47. K. Yang, G. David, S. V. Robertson, J. F. Whitaker, L. P. B. Katehi, "Electrooptic mapping of near-field distributions in integrated microwave circuits", *IEEE Trans. Microwave Theory Tech.* **46**, 2338 (1998).

48. Q. Wu, X.-C. Zhang, "Free-space electro-optic sampling of terahertz beam", *Appl. Phys. Lett.* **67**, 3523 (1995).

49. A. Nahata, D. H. Auston, T. F. Heinz, C. Wu, "Coherent detection of freely propagating terahertz radiation by electro-optic sampling", *Appl. Phys. Lett.* **68**, 150 (1996).

50. P. Uhd Jepsen, C. Winnewisser, M. Schall, V. Schya, S. R. Keiding, H. Helm, "Detection of THz pulses by phase retardation in lithium tantalate", *Phys. Rev. E* **53**, 3052 (1996).

51. Q. Wu, X.-C. Zhang, "Electro-optic sampling of freely propagating THz field", *Opt. Quantum Electron.* **28**, 945 (1996).

52. Q. Wu. X.-C. Zhang, "Ultrafast electro-optic field sensors", *Appl. Phys. Lett.* **68**. 1604 (1996).
53. Q. Wu. M. Litz. X.-C. Zhang, "Broadband detection capability of electro-optic field probes". *Appl. Phys. Lett.* **68**, 2924 (1996).
54. Q. Wu. F. G. Sun. P. Campbell. X.-C. Zhang. "Dynamic range of an electro-optic field sensor and its imaging applications". *Appl. Phys. Lett.* **68**. 3224 (1996).
55. Q. Wu. T. D. Hewitt, X.-C. Zhang. "Electro-optic imaging of terahertz beams". *Appl. Phys. Lett.* **69**, 1026 (1996).
56. X.-C. Zhang, Q. Wu. T. D. Hewitt. "Electro-optic imaging of terahertz beams". *Ultrafast Phenomena X*, in *Springer Series in Chemical Physics* ed. by P. F. Barbara. J. G. Fujimoto. W. H. Knox. W. Zinth (Springer, Berlin. Heidelberg. 1996). p. 54.
57. Q. Wu. M. Litz. X.-C. Zhang, "Free-space electro-optic samplers", *Ultrafast Phenomena X*, in *Springer Series in Chemical Physics* ed. by P. F. Barbara, J. G. Fujimoto, W. H. Knox. W. Zinth (Springer, Berlin, Heidelberg. 1996), p. 60.
58. Q. Wu. X.-C. Zhang, "Design and characterization of traveling-wave electro-optic THz sensors". *IEEE J. Sel. Top. Quantum Electron.* **3**, 693 (1996).
59. Z. G. Lu. P. Campbell. X.-C. Zhang. "Free-space electro-optic sampling with a high-repetition-rate regenerative amplified laser". *Appl. Phys. Lett.* **71**, 593 (1997).
60. Z. Jiang. X.-C. Zhang, "THz imaging via electro-optic effect". *IEEE Trans. Microwave Theory Tech.* **47**. 2644 (1999).
61. Q. Wu. X.-C. Zhang. "7 THz ultrabroadband GaP electro-optic sensors". *Appl. Phys. Lett.* **70**. 1784 (1997).
62. Z. Jiang, X.-C. Zhang, "Electro-optic measurement of THz pulses with a chirped optical beam". *Appl. Phys. Lett.* **72**, 1945 (1998).
63. Z. Jiang. X.-C. Zhang, "Single-shot spatial-temporal THz field imaging". *Opt. Lett.* **23**. 1114 (1998).
64. C. Winnewisser. P. Uhd Jepsen. M. Schall. V. Schyja, H. Heim, "Electro-optic detection of THz radiation in LiTaO₃. LiNbO₃ and ZnTe", *Appl. Phys. Lett.* **70**. 3069 (1997).
65. A. Leitenstorfer. S. Hunsche. J. Shah. M. C. Nuss, W. H. Knox, "Detectors and sources for ultrabroadband electro-optic sampling: experiment and theory". *Appl. Phys. Lett.* **74**, 1516 (1999).
66. G. Gallot. D. Grischkowsky. "Electro-optic detection of terahertz radiation". *J. Opt. Soc. Am. B* **16**. 1204 (1999).
67. P. Y. Han. G. C. Cho. X.-C. Zhang. "Time-domain transillumination of biological tissues with terahertz pulses". *Opt. Lett.* **25**. 242 (2000).
68. N. Katzenllenbogen. H. Chan. D. Grischkowsky. in *Quantum Electronics and Laser Science Conference*, OSA Technical Digest Series (Optical Society of America, Washington. DC. 1993). vol. 3, p. 155.
69. For example, see: J. Shah. *Ultrafast Spectroscopy of Semiconductors and Semiconductor Nanostructures*. Springer Series in Solid-State Sciences, vol. 115 (Springer Berlin, Heidelberg, 1999).
70. P. R. Smith, D. H. Auston, M. C. Nuss, "Subpicosecond photoconducting dipole antennas". *IEEE J. Quantum. Electron.* **24**, 255 (1988).
71. Ch. Fattinger. D. Grischkowsky. "Terahertz beam", *Appl. Phys. Lett.* **54**. 490 (1989).

72. D. M. Mittleman, R. H. Jacobsen, M. C. Nuss, "T-ray imaging", *IEEE J. Sel. Top. Quantum Electron.* **2**, 679 (1996).

73. J. A. Valdmanis, "Real Time Picosecond Optical Oscilloscope", in *Proceedings of Ultrafast Phenomena V*, ed. by G. R. Fleming, A. E. Siegman, (Springer, Berlin, Heidelberg, 1996), p. 82.

74. J. A. Valdmanis, *Solid State Technol./Test & Measurement World* **6**, S40 (1986).

75. Z. Jiang, F. G. Sun, Q. Chen, X.-C. Zhang, "Electro-optic sampling near zero optical transmission point", *Appl. Phys. Lett.* **74**, 1191 (1999).

76. F. G. Sun, Z. Jiang, X.-C. Zhang, "Analysis of THz pulse measurement with a chirped probe beam", *Appl. Phys. Lett.* **73**, 2233 (1998).

77. Z. Jiang, F.G. Sun, X.-C. Zhang, "Terahertz pulse measurement with an optical streak camera", *Opt. Lett.* **24**, 1245 (1999).

78. M. Drabbels, L. D. Noordam, "Streak camera operating in the mid infrared", *Opt. Lett.* **22**, 1436 (1997).

79. M. Drabbels, G. M. Lankhuijzen, L. D. Noordam, "Demonstration of a far-infrared streak camera", *IEEE J. Quantum Electron.* **34**, 2138 (1998).

80. M. Drabbels, L. D. Noordam, "Infrared imaging camera based on a Rydberg atom photodetector", *Appl. Phys. Lett.* **74**, (1999).

81. J. Chang, C. N. Vittitoe, "An electro-optical technique for measuring high frequency free space electric field", in *Fast Electrical and Optical Measurement*, ed. by J. E. Thompson, L. H. Luessen, NATO ASI Series, Series E: Applied Science, No. 108 (1983), p. 57.

82. S. Williamson, G. Mourou, "Picosecond electro-electron optic oscilloscope", in *Picosecond Electronics and Optoelectronics*, ed. by G. A. Mourou, D. M. Bloom, C.-H. Lee (Springer, Berlin, Heidelberg, 1985), p. 58.

83. B. B. Hu, M. C. Nuss, "Imaging with terahertz waves", *Opt. Lett.* **20**, 1716 (1995).

84. D. M. Mittleman, S. Hunsche, L. Boivin, M. C. Nuss. "T-ray tomograhy", *Opt. Lett.* **22**, 904 (1997).

85. D. M. Mittleman, M. Gupta, R. Neelamani, R. G. Baraniuk, J. V. Rudd, M. Koch, "Recent advances in terahertz imaging", *Appl. Phys. B: Lasers Opt. B* **68**, 1085 (1999).

86. D. M. Mittleman, R. H. Jacobsen, M. C. Nuss, "T-ray imaging", *IEEE J. Sel. Top. Quantum Electron.* **2**, 679 (1996).

87. Z. Jiang, X.-C. Zhang, "Improvement of terahertz imaging with phase sensitive technique", *Appl. Opt.* **39**, 2982 (2000).

88. Z. Jiang, X.-C. Zhang, "2D measurement and spatio-temporal coupling of few-cycle THz pulses", *Opt. Express* **5**, 243 (1999).

89. S. Hunsche, M. Koch, I. Brener, M. C. Nuss, "THz near-field imaging", *Opt. Commun.* **150**, 22 (1998).

90. K. Wynne, D. A. Jaroszynski, "Superluminal terahertz pulses", *Opt. Lett.* **24**, 25 (1999).

91. Y. Cai, I. Brener, J. Lopata, J. Wynn, L. Pfeiffer, J. B. Stark, Q. Wu, X. C. Zhang, J. F. Federici, "Coherent terahertz radiation detection: direct comparison between free-space electro-optic sampling and antenna detection", *Appl. Phys. Lett.* **73**, 444 (1998).

92. S.-G. Park, M. R. Melloch, A. M. Weiner, "Comparison of terahertz waveforms measured by electro-optic and photoconductive sampling", *Appl. Phys. Lett.* **73**, 3184 (1998).

93. S.-G. Park, M. R. Melloch, A. M. Weiner, "Analysis of terahertz waveforms measured by photoconductive and electrooptic sampling", *IEEE J. Quantum Electron.* **35**, 810 (1999).

94. A. Nahata, J. T. Yardley, T. F. Heinz, "Free-space electro-optic detection of continuous-wave terahertz radiation", *Appl. Phys. Lett.* **75**, 2524 (1999).

95. D. You, P. H. Bucksbaum, "Propagation of half-cycle far infrared pulses", *J. Opt. Soc. Am. B* **14**, 1651 (1997).

96. A. E. Kaplan, "Diffraction-induced transformation of near-cycle and subcycle pulses", *J. Opt. Soc. Am. B* **15**, 951 (1999).

97. S. Feng, H. G. Winful, R. W. Hellwarth, "Gouy shift and temporal reshaping of focused single-cycle electromagnetic pulses", *Opt. Lett.* **23**, 385 (1998): "Erratum". *Opt. Lett.* **23**, 1141 (1998).

98. S. Hunsche, S. Feng, H. G. Winful, A. Leitenstorfer, M. C. Nuss, E. P. Ippen, "Spatiotemporal focusing of single-cycle light pulses", *J. Opt. Soc. Am. A* **16**, 2025 (1999).

99. E. Budiarto, P. Nen-Wen, J. Seongtae, J. Bokor, "Near-field propagation of terahertz pulses from a large- aperture antenna", *Opt. Lett.* **23**, 213 (1998).

100. J. Bromage, S. Radic, G. P. Agrawal, C. R. Stroud, Jr., P. M. Fauchet, R. Sobolewski, "Spatiotemporal shaping of terahertz pulses", *Opt. Lett.* **22**, 627 (1997).

101. J. Bromage, S. Radic, G. P. Agrawal, C. R. Stroud, Jr., P. M. Fauchet, R. Sobolewski, "Spatiotemporal shaping of half-cycle terahertz pulses by diffraction through conductive apertures of finite thickness", *J. Opt. Soc. Am. B* **15**, 1953 (1998).

102. A. Nahata, T. F. Heinz, "Reshaping of freely propagating terahertz pulses by diffraction", *IEEE J. Sel. Top. Quantum Electron.* **2**, 701 (1996).

103. A. B. Ruffin, J. V. Rudd, J. F. Whitaker, S. Feng, H. G. Winful, "Direct observation of the Gouy phase shift with single-cycle terahertz pulses", *Phys. Rev. Lett.* **83**, 3410 (1999).

104. P. Kuzel, M. A. Khazan, J. Kroupa, "Spatiotemporal transformations of ultrashort terahertz pulses", *J. Opt. Soc. Am. B* **16**, 1795 (1999).

105. A. M. Weiner, "Femtosecond optical pulse shaping and processing", *Prog. Quantum Electron.* **19**, 161 (1995).

106. Q. Chen, Z. Jiang, M. Tani, X.-C. Zhang, "Electro-optic Terahertz transceiver", *Electron. Lett.* **36**, 1298 (2000).

107. M. Tani, Z. Jiang, X.-C. Zhang, "Photoconductive terahertz transceiver", *Electron. Lett.* **36**, 804 (2000).

108. M. Li, X.-C. Zhang, G. Sucha, D. Harter, "Portable THz system and its applications", *Proc. SPIE* **3616**, 126 (1999).

Photomixers for Continuous-Wave Terahertz Radiation

Sean M. Duffy, Simon Verghese and K. Alex McIntosh

Abstract. Optical heterodyne conversion, or photomixing, is a frequency-agile technique that generates continuous-wave radiation at THz frequencies using thin films of low-temperature-grown (LTG) GaAs. Optimizing photomixers for maximum output power requires careful design of the epitaxial growth sequence, and detailed analyses of the RF circuitry and of the optical feed. Control of the LTG GaAs epitaxy leads to a material with short photocarrier lifetime and robustness to thermal failure. Key aspects of the photocarrier lifetime, quantum efficiency, and thermal performance are described for the photomixer. Trade-offs for optimizing the RF and optical design for THz output power are discussed. Promising applications for photomixers include local oscillators for THz heterodyne detectors based on superconductors, and high-resolution spectrometers useful for rotational spectroscopy of airborne molecules.

1 Introduction

Photomixers are compact, all-solid-state sources that use a pair of single-frequency, tunable diode lasers to generate a THz difference frequency by photoconductive mixing in low-temperature-grown (LTG) GaAs [1–3]. The output frequency can be tuned over several THz by temperature or current detuning of the two diode lasers by a few nanometers in wavelength. As sources, photomixers have been used for local oscillators with cryogenic THz heterodyne detectors [4] and for high-resolution gas spectroscopy in conjunction with liquid-helium-cooled bolometers [5,6]. On the most basic level, optical heterodyne conversion with a photoconductive switch (photomixing) is analogous to the operation of a transistor amplifier. In a transistor, a small RF signal applied to the gate modulates the conductance of a switch under a relatively large DC bias. The output power is drawn from the source providing the DC bias. In photomixing, the photoconductance is modulated by the optical beating of the two laser diodes and the output THz power is generated from the source that provides the DC bias between the photoconductor electrodes. Photomixing is fundamentally different from optical difference-frequency generation using $\chi^{(2)}$ processes in a material such as ZnTe or LiNbO$_3$. In this type of difference-frequency generation, the output THz power is generated from the optical photons, and only one THz photon can be created per pair of optical photons. The fact that two hot photons are required to create one cold photon results in an efficiency penalty that makes photomixing more favorable than $\chi^{(2)}$ mixing at lower THz frequencies. At frequencies above several THz, however, the $\chi^{(2)}$ mixing process is more efficient than photomixing, since photomixers suffer from parasitic impedances that limit their bandwidth. This trade-off is discussed in more detail by Brown et al. [7].

1.1   Demonstrated CW Technology for Terahertz Generation

Many sources are available for generating coherent continuous-wave THz radiation, including backward-wave oscillators, molecular lasers, Schottky diode upconverters, and parametric downconverters. Some of these are discussed in the chapter by De Lucia, in this volume. If we restrict the discussion to solid-state devices, the most widely used sources are varactor multipliers [8], fundamental sources such as FETs and negative-resistance diodes, and, more recently, high-electron-mobility transistors (HEMTs) and heterojunction base transistors (HBTs). In general, electronic technologies that use electron transport for THz generation are limited by parasitic resistance and capacitance as frequencies exceed 1 THz. A developing technology is the THz quantum cascade laser, this technology represents a fundamentally different approach to THz generation from the electronic technologies [9,10]. Quantum-effect devices such as Bloch oscillators and quantum cascade lasers are so far limited to operation at cryogenic temperatures since the thermal energy, kT, at 300 K otherwise smears out the quantum levels.

Photomixers can be thought of as combining some of the attributes of electronic technologies and optical technologies to generate widely tunable coherent THz radiation at room temperature. Although their output power is modest (1 μW typical), it is suitable or nearly suitable for application in local oscillators for cryogenic heterodyne receivers and for spectrometers. Two of the most attractive features of photomixers are the wide tuning range (25–3000 GHz), and room temperature operation.

2   Photomixers: Principle of Operation

The system configuration of a typical photomixer setup is shown in Fig. 1. The outputs of two diode lasers at frequencies v_1 and v_2 slightly higher than the band gap of GaAs, are combined in a single-mode fiber. The fiber is aligned so as to illuminate an interdigitated electrode region constructed on an LTG GaAs active layer. An applied bias collects the photogenerated carriers (here, the photocurrent is assumed to be dominated by the higher-mobility electrons rather than the low-mobility holes) and causes radiation from a planar antenna. The radiation is collected using a substrate lens into a quasi-optical system.

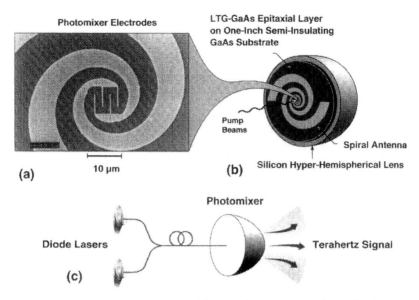

Fig. 1. Photomixer implementation. (**a**) Set of photomixer electrodes integrated with a log-spiral antenna on LTG GaAs, (**b**) the LTG GaAs is mounted on a silicon hyper-hemispherical lens that collects the radiation (**c**) The output of two diode lasers are combined in a fiber and illuminate the electrodes so as to generate a terahertz signal

In this section, an overview of the basic physics and operation of photomixers is presented that relates the optical power, photocurrent, and carrier lifetime to the THz output power. The growth conditions and bias affect the short lifetime (0.2 ps) of LTG GaAs that enables terahertz operation of the photomixer. The maximum output power of the photomixer is affected by the thermal performance of the substrate materials. The inherently low thermal conductivity of LTG GaAs is a barrier to achieving high output powers, and methods of improving the thermal design are discussed. Finally, an overview of the possible trade-offs between some of the physical parameters is presented.

2.1 Overview of Operation

A photomixer constructed from biased interdigitated electrode fingers, with an incident optical signal, is shown in Fig. 2. The photomixer behaves similarly to a photoconductor [1]. In a photoconductor, the incident light modulates the conductivity through generation of photocarriers. The photomixing process occurs during illumination of an electrode region with two single-

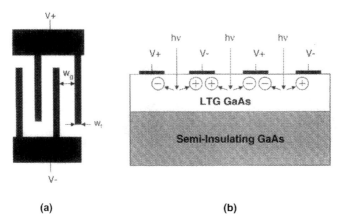

Fig. 2. Electrode geometry: (a) top view of electrode geometry with gap width w_g and finger width w_f defined: (b) side view of electrode geometry. LTG GaAs epitaxially grown on GaAs. An optical signal $h\nu$ generates carriers in the LTG GaAs. which are collected as a photocurrent

mode lasers with average powers P_1 and P_2 and frequencies ν_1 and ν_2. The instantaneous optical power incident on the photomixer is given by [3]

$$P_i = P_1 + P_2 + 2\sqrt{P_1 P_2}\left[\cos 2\pi(\nu_1 - \nu_2)t + \cos 2\pi(\nu_1 + \nu_2)t\right]. \tag{1}$$

where $\nu_1 - \nu_2$ is the difference frequency and can easily be adjusted from the microwave to submillimeter-wave frequency band. The $\nu_1 + \nu_2$ term of the mixing product is at twice the optical frequency and is not coupled well to the metal circuitry.

Interdigitated electrode fingers are used in the design shown here to apply the electric field to the photogenerated carriers and collect the mixed signal. Other electrode structures are possible. such as gap-coupled lines: however. for efficient transfer of the optical to the electrical signal, it is desirable to have an electrode region equal in size to the laser spot on the substrate. The electric field is highly nonuniform for interdigitated fingers: however. a simple approximation that works well for our purposes is

$$E_o = \frac{V_B}{w_g} \tag{2}$$

where V_B is the bias voltage between the electrodes and w_g is the spacing between fingers. The breakdown field in typical LTG GaAs is approximately 5×10^5 V/cm [3].

The DC photocurrent. I_o, is the amount of current generated by the optical power:

$$I_o = \eta_c \left(\frac{e}{h\nu}\right) P_o \tag{3}$$

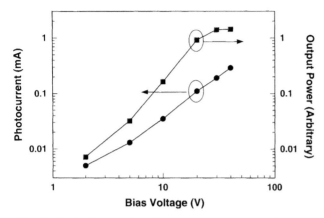

Fig. 3. Typical current–voltage curves and output power for a photomixer at low optical power ($P_{\rm o} = 30\,{\rm mW}$) ($w_{\rm f} = 0.2\,\mu{\rm m}$, $w_{\rm g} = 1.8\,\mu{\rm m}$, electrode region $20\,\mu{\rm m}$ $\times 20\,\mu{\rm m}$)

where $e/h\upsilon$ is a constant equal to 0.69 for an optical wavelength of 850 nm, $P_{\rm o}$ is the incident optical power, and $\eta_{\rm e}$ is the external quantum efficiency. The measured DC photocurrent for a typical photomixer is plotted versus applied bias voltage in Fig. 3.

The amount of THz signal generated can be found using the solution of the current-continuity equation. For the photomixer, this yields the small-signal result of optical heterodyne theory [11]:

$$P_\omega = \frac{1}{2}\frac{I_{\rm o}^2}{(1+\omega^2\tau_{\rm c}^2)}\eta_{\rm ant}{\rm Re}(Z_{\rm load}),\qquad(4)$$

where $\omega = 2\pi f$ is the angular frequency, $\tau_{\rm c}$ is the lifetime of the carriers, $\eta_{\rm ant}$ is the antenna efficiency and ${\rm Re}(Z_{\rm load})$ is the real part of the load impedance formed by the antenna impedance and the capacitance of the interdigitated-finger electrode. The current at the THz frequency as a fraction of the DC photocurrent is given by

$$|I_\omega| = \frac{I_{\rm o}}{(1+\omega^2\tau_{\rm e}^2)^{1/2}}.\qquad(5)$$

In essence, the reduction in the current contributing to the THz signal at high frequencies is a result of the response time of the carriers compared with the modulation frequency. Because the response of the carriers decreases with increasing modulation frequency, the current decreases. Measured DC photocurrents and output powers for a typical photomixer are plotted in Fig. 3. Several nonlinear features of the measured data are discussed in the following sections.

The equivalent-circuit model of the photomixer is shown in Fig. 4. The low conductance G of the photomixer, typically $(15\,{\rm k}\Omega)^{-1}$, leads to the circuit

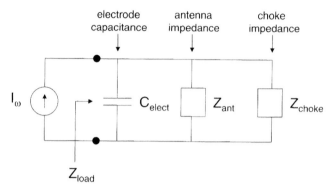

Fig. 4. Equivalent-circuit model for photomixer. The photoconductance is low $(15\,\mathrm{k\Omega})^{-1}$ and is assumed to be a current source. The load impedance incorporates electrode capacitance, antenna impedance, and choke impedance

approximation that it is a current source with a shunting capacitor caused by the interdigitated electrode fingers [1]. The antenna impedance is, in general, in a shunt connection to the current source. Therefore, the electrode capacitance is included in the load impedance and causes design difficulty for terahertz operation. At high frequencies, the electrode capacitance tends to "short out" the antenna resistance, thus limiting the achievable output power. These concepts and issues will be discussed in detail in later sections of this chapter.

In particular, we focus in the next few sections on three issues that must be considered in order to achieve a greater understanding of the photomixing process and are necessary for accurately characterizing these devices. First, the lifetime of the carriers, τ_c in equations (4) and (5), is field dependent, which affects the high-bias behavior. Second, the external quantum efficiency η_c is important for understanding the conversion of the optical to the electrical signal. Finally, the thermal limitations of the material determine the power-handling capacity of the device.

2.2   Lifetime of Carriers

Carriers generated in the LTG GaAs owing to absorption of an optical photon exist for a period of time before recombining. In the presence of an applied electric field, these carriers contribute to a photoinduced current. In general, the lifetime of the material depends on the details of the growth and annealing conditions, as shown in Fig. 5 [12]. A minimum carrier lifetime of approximately 0.2 ps occurs at growth temperatures around 195°C, with annealing at 600°C in an arsenic flux for 10 minutes. LTG GaAs grown at temperatures higher than 195°C incorporates less excess arsenic and has a lower density of arsenic defects [12]. The time it takes for photogenerated carriers to be captured by arsenic defects would therefore be expected to increase as the

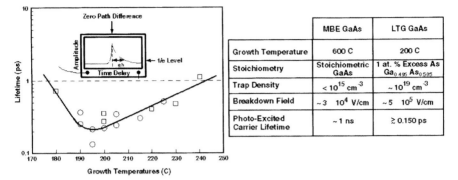

Fig. 5. Optical pump–probe measurements of the photocarrier lifetime for various LTG GaAs wafers. Also shown are typical values of electronic properties of LTG GaAs compared with conventional epitaxial GaAs

growth temperature increases, as seen in Fig. 5. At growth temperatures below 195°C, the strain induced by the greater excess As incorporation in the GaAs begins to cause the formation of extended defects, which also acts to increase the lifetime [12].

Typical I–V curves (see Fig. 3) of photomixer devices show a superlinear bias dependence for the photocurrent at high bias levels. The simple photoconductive model [1] with a field-independent carrier lifetime does not explain this behavior; instead, this behaviour has been found to be a result of a field dependence of the carrier lifetime in the LTG GaAs [13,14]. This field dependence arises because the effective cross section of Coulombic attractive electron traps decreases with increasing electric field, owing to electron heating and Coulomb-barrier lowering [15]. The electron capture time is inversely proportional to their cross section, and the field dependent lifetime can be written as

$$\tau_e(E_o) = \tau_{e0} \left(1 + \frac{\lambda e E_o}{kT}\right)^{3/2} \left(\frac{r(E=0)}{r(E_o)}\right)^3 , \qquad (6)$$

where T is the lattice temperature, r is the radius of the well in the direction of the applied field, k is Boltzmann's constant, and λ is the optical-phonon mean free path. The field-independent lifetime, τ_{e0}, can be found using a time-resolved photoreflectance measurement [12]. The second factor in (6) is an electron-heating term, and the third factor is due to Coulomb-barrier lowering [13,14]. The radius corresponding to the capture cross section can be found as follows. The electric potential $V(z)$ due to ionized donors at $z = 0$ and $z = z_o$ and an applied field E in the positive z direction is

$$V(z) = \frac{-e}{4\pi\varepsilon|z|} - Ez - \frac{e}{4\pi\varepsilon|z - z_o|}. \qquad (7)$$

$V(z)$ has a maximum between 0 and $z_0/2$. The effective capture radius r of the potential well centered at $z = 0$ is such that $V(r) = V_{max} - 2kT/e$. The potential well centered at $z = z_0$ accounts for the effect of a nearest-neighbor potential on barrier lowering. The parameters λ and z_0 are used to match the measured result. The optical-phonon mean free path is found to be 2 nm. somewhat lower than that of normal GaAs. for which $\lambda = 5.8$ nm. The factor z_0 is 7 nm and is consistent with the estimated ionized-donor density of 10^{19} cm^{-3} [13,14].

A numerical solution for the lifetime. and measurements made using an autocorrelation technique [13,14] are shown in Fig. 6. The field-dependent lifetime is one of the reasons that a superlinear bias dependence [3] is seen in the measured photocurrents at high bias voltage for typical photomixers. Beyond a certain voltage (electric field), the lifetime starts to increase. causing a higher quantum efficiency and higher photocurrent.

A curve-fitting approximation is used in the following calculations in this chapter. The fitted curve fit equation shown in Fig. 6 does not provide the physical insight of the above solution but is based on the measurements shown in Figs. 6 and 9 and of others [14]. The field-dependent lifetime in this approximation is

$$\tau_e(E_o) = \tau_{e0} \left[1 + \left(\frac{E_o}{\chi \times 0.6 \times 10^5} \right)^3 \right]^{0.5}. \tag{8}$$

where E_o is in V/cm and χ is a factor that accounts for the nonuniformity of the electric field. In Fig. 6. τ_{e0} is assumed to be 0.25 ps, on the basis of the growth temperature. and χ is 0.679. The scaling factor χ has the gap width dependence seen in typical I V curves (see Fig. 9). An expression for

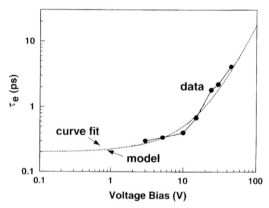

Fig. 6. Measured values of photocarrier lifetime inferred from the decay time obtained from correlating two femtosecond laser pulses on an LTG GaAs photoconductor. Also shown are the predictions of a model [13,14] and a fitted curve given by (8)

χ is derived in the next section. An important feature of this approximation is a sharp transition region from a bias-independent lifetime to a functional dependence of $E^{3/2}$, as seen in Fig. 6.

2.2.1   Lifetime Modified by Electric Field Profile

The photocarrier lifetimes in Fig. 6 were calculated and measured for a gap width of 1.8 µm, assuming that the electric field is uniform throughout the gap [13,14]. However, the conversion from applied voltage to electric field should account for the nonuniformity of the field if it is to have general applicability to different gap spacings.

A solution of the Poisson equation for two thin strips of width w_f and gap spacing w_g gives the magnitude of the electric field in the gap as

$$|E| = \frac{V_B}{2 \ln[(2w_g + w_f)/w_f]} \frac{w_g}{\sqrt{(w_g/2 + x)^2 + y^2} \sqrt{(w_g/2 - x)^2 + y^2}}. \qquad (9)$$

This equation has been solved for the midpoint between the fingers and is plotted as a function of depth into the LTG active layer in Fig. 7 for a gap width of 0.8 µm and a bias of 16 V. The equation shows that different gap widths lead to different electric fields for a constant value of V_B/w_g. Therefore, to apply the results of [13,14] for the field-dependent lifetime to other gap widths, a scaling factor χ must be included. The scaling is determined using the midpoint between the fingers, leading to

$$\chi = \frac{2}{\ln\left[(2w_g + w_f)/w_f\right]}. \qquad (10)$$

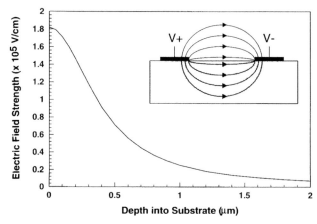

Fig. 7. Electric field strength in the substrate due to the electrodes at the midpoint between fingers ($w_f = 0.2$ µm, $w_g = 0.8$ µm, $V_B = 16$ V). Also shown is an approximate representation of the electric field between fingers

This factor is included in (8) (and could be applied to (6)). In essence, the increase in lifetime becomes significant at different uniform electric fields (as defined in (2)) for different gap widths. This is important when one is considering the external quantum efficiency and DC photocurrent, since a 1.8 µm gap is affected by the increase in lifetime at a lower bias voltage than a 0.8 µm gap.

2.3   External Quantum Efficiency

The external quantum efficiency captures the physics of converting the optical signal to an electrical signal. It describes the amount of DC current generated as a result of an applied bias and optical power. The intrinsically low value of the external quantum efficiency for a photomixer on LTG GaAs also demonstrates one of the limitations of this technology.

The external quantum efficiency is defined as the integral of an absorption factor f times a collection factor g; f is the fraction of incident photons that generate carriers, and g is the number of electrons induced in the external circuit per absorbed photon [16]. Following the derivation by Brown [16]. the external quantum efficiency is

$$\eta_e = \int_V fg \, \mathrm{d}v/V = \frac{\alpha \eta_{\mathrm{opt}}}{w_g} \int_{-w_g/2}^{w_g/2} \int_0^{d_{\mathrm{LTG}}} g \, e^{-\alpha y} \, \mathrm{d}x \, \mathrm{d}y. \tag{11}$$

where α is the absorption coefficient ($1.5\,\mu m^{-1}$) [17,18], d_{LTG} is the thickness of the LTG GaAs region, V is the volume of the active area, and the laser spot size is assumed to match the active region.

The photoconductive gain g is given by the ratio of the recombination lifetime to the transit time of the active area [16,19], i.e. $g = \tau_e/t_e$, where t_e is the transit time determined by the distance between the electrodes and the velocity of the carriers. The photoconductive gain can also be viewed as the fraction of the transit distance covered by the carriers before they are trapped. For transit times much greater than the lifetime. the photoconductive gain can be written as

$$g = \frac{v_e \tau_e}{L_e}, \tag{12}$$

where v_e is the velocity of the carriers, τ_e is the lifetime of the carriers, and L_e is the length of the carrier path from electrode to electrode. Consideration of the external quantum efficiency for a typical photoconductor with a parallel-plate region of constant electric field leads to the conclusion that closer finger separation will achieve higher quantum efficiencies. Although this occurs at low bias, the field dependence of the lifetime leads to a more complicated behavior.

The velocity of the carriers is affected by saturation effects and is given by [13,14,19]

$$v_e = \frac{\mu_e E}{1 + (\mu_e E/v_{\mathrm{sat}})} \tag{13}$$

where μ_e is the mobility of the carriers, v_{sat} is the saturation velocity, and E is given by (9). Typical values of μ_e and v_{sat} in LTG GaAs are $164\,\text{cm}^2/\text{V s}$ and $4.4 \times 10^6\,\text{cm/s}$ [13,14,19], respectively, for normal growth conditions [12]. Therefore, carrier velocity is also a function of position in the interdigitated region owing to the nonuniformity of the electric field.

The carrier path for the photoconductive gain in (12) is a curved path along the electric-field lines of force (shown in Fig. 7). For simplicity, the length of the path is assumed to be the average of two easily found paths, as given by [19]

$$L_e = \frac{1}{2}\left[\sqrt{(w_g/2 + x)^2} + \sqrt{(w_g/2 - x)^2} + w_g + 2y\right]. \tag{14}$$

Path 1 is formed by straight lines connecting a point in the layer directly to the electrodes and path 2 is formed by traveling vertically down to the appropriate depth, horizontally across the device through the point and then vertically back up to the other electrode.

Finally, the optical coupling efficiency η_{opt} accounts for the reflection loss of the optical wave incident from the air into the active region

$$\eta_{opt} = \frac{w_g}{w_f + w_g}\frac{4n_{LTG}}{(n_{LTG} + 1)^2}. \tag{15}$$

The refractive index for a wavelength of $850\,\text{nm}$, $n_{LTG} = 3.62$ [17], of the LTG GaAs causes a reduction in the amount of optical power that enters the dielectric. Two methods have been used to increase η_{opt}: adding an antireflective (AR) coating to the surface and constructing a Distributed Bragg Reflector (DBR) inside the substrate. The DBR is discussed in the next section and is shown in Fig. 11. When either of these structures is used, the reflection from the dielectric is eliminated and $\eta_{opt} = w_g/(w_f + w_g)$ can be used. Reflection of the laser radiation from the electrode fingers is included in this definition, although proper choice of the polarization of the laser relative to the fingers is necessary to maximize the coupling into the active region [20,21].

The external quantum efficiency can be determined numerically and has been calculated for three different gap spacings, 0.3, 0.8 and $1.8\,\mu\text{m}$, as shown in Fig. 8 for low-bias operation. For field-independent lifetime behavior ($E < 10^4\,\text{V/cm}$), these η_e calculations agree qualitatively with solutions which ignore this effect [16,19]. In this regime, smaller gap spacings lead to higher quantum efficiency.

The bias dependence of the external quantum efficiency shown in Fig. 9 is an important result for understanding photomixer operation. In this figure, the measurements and calculations for gap widths of 0.8 and $1.8\,\mu\text{m}$ show a sublinear bias dependence at low voltages due to the saturating velocity of the carriers, and then a superlinear bias dependence at high voltages after field-dependent-lifetime effects become important [2,14,19,22]. The electric

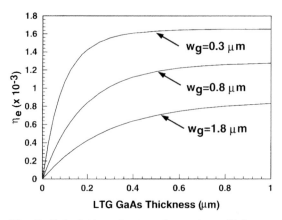

Fig. 8. Calculation of external quantum efficiency at low bias ($E_o = 10^4$ V/cm) for varying LTG GaAs thickness and three gap widths 0.3, 0.8 and 1.8 μm. $w_f = 0.2$ μm. $\tau_{e0} = 0.27$ ps

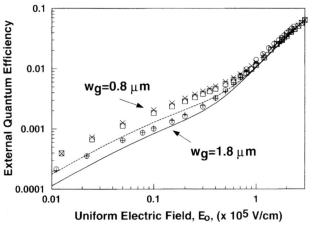

Fig. 9. Measured and calculated external quantum efficiency at varying bias levels for two different gap widths, 0.8 and 1.8 μm ($w_f = 0.2$ μm). Two sets of measured data are shown for each gap width. $\tau_{e0} = 0.27$ ps, AR coating, $d_{\mathrm{LTG}} = 1$ μm

field (as given by V_B/w_g) at which the external quantum efficiency becomes superlinear is higher for the smaller gap spacing seen in Fig. 9, consistent with the above discussion of the nonuniformity of the electric field. The external quantum efficiencies for different gap spacings have almost the same value when V_B/w_g is greater than approximately 10^5 V/cm. However, a large gap spacing achieves an increase in external quantum efficiency at constant E_o by producing a longer carrier lifetime. The implication for output power is investigated in Sect. 5. For accurate characterization of the photocurrent, the devices used to obtain the results shown in Fig. 9 had an AR coating, which eliminates a possible standing-wave problem between the fiber end and the electrodes. This and other measurement details are discussed in more depth in Sect. 5

2.4 Thermal Limits

The thermal dissipation of the photomixer limits the achievable THz power by constraining the optical and bias power levels. A combination of ohmic and laser heating at sufficiently high levels causes catastrophic failure in photomixers [19,23]. Also, the thermal conductivity of LTG GaAs is significantly poorer than even the relatively low thermal conductivity of normal GaAs, further limiting the THz power [19,24].

The thermal conductivities of several materials are shown in Table 1. The thermal conductivity of LTG GaAs was investigated as a function of growth and anneal conditions in [19,24]. LTG GaAs grown at 240°C with an anneal at 640°C for 30 s leads to a result approximately 40% of that of normal GaAs. However, the thermal conductivity is a function of the growth and annealing temperatures and is as low as 23% of that of normal GaAs for anneal temperatures below 550°C. For anneal temperatures around 600°C, thermal conductivity of LTG is approximately 0.35 that of normal GaAs.

Table 1. Thermal conductivity of typical photomixer materials

Material	Thermal conductivity (W/m K)
LTG GaAs	15
GaAs	44
AlAs	91
Au	297

The thermal power dissipated in the electrode region is a function of the laser and ohmic heating in the substrate [19,23]. The thermal power is defined as

$$P_{\text{therm}} = P_{\text{o}}A_{\text{th}} + V_{\text{B}}I_{\text{o}}. \tag{16}$$

The laser heating is $P_{\text{o}}A_{\text{th}}$ where A_{th} is the effective optical absorption in the active layer. The optical absorption can be written as $A_{\text{th}} = \eta_{\text{opt}}(1 - e^{-\alpha d_{\text{LTG}}})$, which for an absorption coefficient of $1.5\,\mu\text{m}^{-1}$, an active-layer thickness of $1\,\mu\text{m}$, a gap width of $0.8\,\mu\text{m}$, a finger width of $0.2\,\mu\text{m}$, and without an AR coating or DBR, gives 0.42. Unfortunately, this definition of A_{th} does not agree with measurements in [19] that demonstrate that a $7 \times 5\,\mu\text{m}$ active region had an A_{th} of 0.25 independent of active-layer thickness. It has been postulated that highly localized heating near the electrodes, due to the larger electric field, leads to a more rapid failure in those regions [19]. Therefore, laser heating deeper in the active layer contributes less to failure and a constant value of A_{th} is obtained. A second issue is the lower value, 0.25 obtained for A_{th}. The rectangular region of $7 \times 5 \mu m$ might be expected to provide a lower value of A_{th} owing to the spillover loss from the laser. Therefore, a value for A_{th} of 0.3 has been used in calculations for square electrode regions. P_{therm} has been reported for a number of photomixers in the literature [1,19,25], and these results are listed in Table 2 and plotted in Fig. 10.

From Fig. 10, the maximum thermal power at the thermal-failure limit is found to be related to the square root of the area of the electrode region by a constant:

$$\frac{P_{therm,max}}{\sqrt{A_e}} = \theta_c. \tag{17}$$

where A_e is the area of the electrode region and θ_c is a material-dependent constant. θ_c is defined as a power per unit length that can be dissipated from the electrode region. Experimental results from published data show a value of θ_c of approximately $7.8\,mW/\mu m$ for a $1\,\mu m$ layer of LTG GaAs on normal GaAs. This value is plotted in Fig. 10 with the values given in Table 2. A close fit to the measured data is obtained with equation (17). This result is somewhat unexpected since the thermal dissipation should scale with area [26]. However, there is other data that shows large electrode regions ($400\,\mu m^2$) to have a failure limit scaling much less than the square root of the area [23]. It should also be remembered that the thermal conductivity of the substrate is dependent on the growth and anneal conditions of the LTG GaAs [19,24]. Some of the growth data is unavailable for the cases listed in Table 2. Therefore, it is important to view equation (17) as only a rough guide for

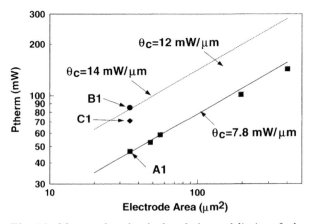

Fig. 10. Measured and calculated thermal limits of photomixers for electrode regions of varying size and three substrate topologies

estimating thermal limits for electrode regions of varying size. Future work should resolve some of the uncertainties about the details of thermal failure, so that a more accurate limit on thermal power can be attained.

Table 2. Thermal failure limits of 1 μm thick LTG on GaAs

Area ($\mu m \times \mu m$)	P_o (mW)	V_B (V)	w_g (μm)	VI (mW)	$P_{\mathrm{therm,max}}$ (mW)	References
7×5	120	15	0.8	18	48	[19]
8×6	63	30	1.8	34	53	[25]
8×7	105	15	0.8	24	56	[25]
14×14	120	30	1.8	62	98	[19][a]
20×20	170	36	1.0	86	137	[1]

[a] Current for ohmic power derived from low-frequency output power result and assuming $P = (1/2)|I|^2 R$

2.4.1   Improved Thermal Designs: Thin LTG GaAs on AlAs

A significant improvement in thermal dissipation can be obtained by thinning the 1 μm LTG GaAs layer. This does not necessarily lead to a large decrease in η_e, as Fig. 8 demonstrates. However, a better approach, taken by Jackson [19] and shown in Fig. 11, uses a buried AlAs layer below the LTG GaAs that utilizes the fact that thermal conductivity of AlAs is more than double that of GaAs. The use of a thin (0.3 μm) active layer of LTG GaAs with a thick (2.5 μm) layer of AlAs leads to an increase in the maximum themal

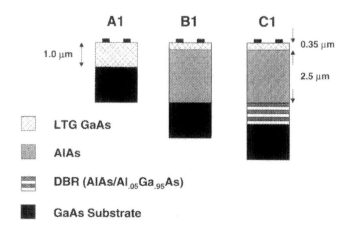

Fig. 11. Substrate topologies. A1: 1 μm LTG GaAs on GaAs, B1: 0.35 μm LTG GaAs on 2.5 μm AlAs on GaAs, C1: 0.35 μm LTG GaAs on 2.5 μm AlAs on GaAs

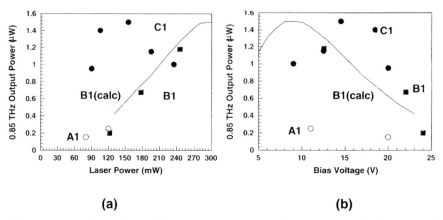

Fig. 12a,b. Scatter plots of laser power and bias voltage versus output power at 0.85 THz for the structures A1. B1. and C1 taken from [19]. Also. shown is calculated result for B1: output power has been normalized for comparison

power $P_{\text{therm,max}}$ of 80% for identical electrode structures [19]. The value of θ_c for a 0.35 µm LTG GaAs layer on a 2.5 µm AlAs layer, as shown in Fig. 10. is roughly 14 mW/µm on the basis of the scatter plots of powers shown in Fig. 12. It has been assumed that this substrate topology scales with the square root of the electrode area as in equation (17), and tests to confirm this behavior are needed.

In obtaining the results shown in Fig. 12. measured output powers at failure were obtained at varying the bias and laser power conditions for the three substrate configurations [19]. The predicted output power for thermal design B1 has been calculated and is also shown on Fig. 12 (the calculated output power is normalized in the figure). A peak in the output power is observed when the bias is high enough that sufficient external quantum efficiency is obtained but low enough that the field-enhanced lifetime does not limit the output power. The calculation of maximum power seems to lead to an optimum bias voltage slightly lower than that achieved experimentally. but clearly a peak in output power versus voltage is observed [19].

2.4.2   Improved Thermal Designs: Optically Resonant Cavity (DBR)

A further improvement in the overall efficiency can be obtained by reducing the necessary optical drive power. This can be accomplished by adding a mirror to increase the absorption in the active LTG region [16]. This mirror can be realized using a distributed Bragg reflector constructed from alternating layers of $Al_{0.05}Ga_{0.95}As$ and AlAs. However. the low thermal conductivity of the DBR (0.48 of normal GaAs) limits the thermal loading and would yield lower overall output powers [19]. The thermal conductivity can be improved by incorporating a thick layer (2.5 μm) of AlAs [19]. This approach also utilizes the larger band gap of AlAs, since the optical signal is not absorbed in this region. The configuration is shown schematically in Fig. 11.

The derived θ_c shown in Fig. 10 for a 0.35 μm LTG GaAs layer on a 2.5 μm AlAs layer with a DBR is roughly 12 mW/μm, on the basis of the scatter plots of powers shown in Fig. 12. The reduction in thermal limits is consistent with predictions showing a lowered thermal conductance with a DBR. A DBR with ten periods was calculated to have a reflectivity greater than 90% [19]. The increase in optical coupling efficiency is consistent with the near two times reduction in optical power for constant THz power shown in Fig. 12.

The significantly improved output power results (up to six times) from a thin active layer of LTG GaAs backed by a thick layer of AlAs can be considered an important step in photomixer development for moderate power levels (~ 1 μW) in the 1–3 THz frequency range. The incorporation of the DBR, although compromising the thermal performance somewhat, offers a two times reduction in optical power for an equivalent THz power, enabling the use of lower-power and lower-cost diode lasers.

2.5   Trade-offs for Enhanced Output Power

Three trade-offs for achieving higher output power are discussed here. First, the use of a short-lifetime material and the trade-off versus optical power are investigated. Second, consideration of the breakdown electric field for varying gap spacing should account for the nonuniform electric field. Finally, the optimum LTG GaAs active-layer thickness is considered: the result shows that the external quantum efficiency can be traded off for improved thermal conductivity to arrive at higher output power.

Rewriting the output power equation (4) to emphasize the carrier lifetime behavior gives

$$P_\omega \propto \frac{\tau_e^2 P_o^2}{1 + \omega^2 \tau_{e^2}} \mathrm{Re}(Z_{\mathrm{load}}), \tag{18}$$

which, for sufficiently high frequencies (> 0.8 THz at low bias, > 0.1 THz at high bias) is roughly

$$P_\omega \propto \frac{P_o^2}{\omega^2} \mathrm{Re}(Z_{\mathrm{load}}). \tag{19}$$

From this equation, we find the 6 dB/octave decrease in output power with increasing frequency observed experimentally at high frequencies [19,27]. Therefore, the high-frequency, high-bias output power is independent of the lifetime, although it still remains dependent on the optical power and load resistance. The lifetime affects the thermal loading of the device since longer lifetimes cause higher external quantum efficiency and therefore more photocurrent. In terms of thermal limits.

$$P_{\text{therm.max}} = P_{\text{o}} \left[A_{\text{th}} + V_B \eta_e \left(\frac{e}{h\upsilon} \right) \right] = \theta_c \sqrt{A_c}. \tag{20}$$

and $P_{\text{therm.max}}$ is constant for a given electrode design and material. Therefore, the area of the electrode region and the material thermal load can be written simply as a constant function, $\beta = \theta_c \sqrt{A_c}$, and the DC bias portion $V_B \eta_e (e/h\upsilon)$ can be rewritten as $\gamma \tau_e$ since η_e is directly related to τ_e, and γ is constant for a given design and operating bias point. The relation between the lifetime and optical power is

$$\frac{\beta}{P_{\text{o}}} - \gamma \tau_e = A_{\text{th}} = \text{constant}. \tag{21}$$

As this equation demonstrates, there is a trade-off between optical power and lifetime. For low-frequency applications, the lifetime can be increased. allowing an identical photocurrent with a lower optical power. However, at high frequencies, the fraction of photocurrent generating the terahertz signal is reduced for long lifetimes (see (5)). The increase in photocurrent due to the lifetime adds to the thermal loading while not contributing to the output power. Therefore, the best approach for obtaining high terahertz power levels is to use short-lifetime material.

The electric breakdown field, E_{break}, of the LTG GaAs epitaxial layer is typically quoted as 5×10^5 V/cm [1]. However, the nonuniformity of the electric field due to the interdigitated electrode fingers affects the breakdown. One would expect

$$V_{\text{break}} = \chi E_{\text{break}} w_g \tag{22}$$

with χ given by (10). Table 3 shows breakdown voltages (22) with varying gap spacings and assuming $E_{\text{break}} = 3.7 \times 10^5$ V/cm. The uniform electric field solution for the breakdown voltage (2) is also shown. It is seen that accounting for the nonuniformity of the electric field leads to an accurate prediction of dielectric breakdown for varying gap widths.

A third consideration in understanding the physical limitations is the thermal conductivity (or loading) of the device. The following development of

Table 3. Measured and calculated breakdown voltages for varying gap spacings. Calculated results normalized to 0.8 μm gap

w_g (μm)	w_f (μm)	$V_B(V)$ (2)	$V_{\text{break}}(V)$ (22)	$V_{\text{break}}(V)$ (measured)	Reference
0.3	0.2	11	16	18	[22]
0.8	0.2	30	27	27	[19,25]
1.8	0.2	67	45	44	[24,25]

Table 4. Finite-element calculations of thermal conductance for varying LTG GaAs thicknesses, normalized to $1\,\mu$m thickness

d_{LTG} (μm)	θ^{-1}
0	1.55
0.125	1.5
0.25	1.45
0.375	1.35
0.5	1.25
0.75	1.125
1	1
2	0.875

the thermal limitations is taken from Jackson [19]. Including heating effects, the thermal power can be rewritten from (3) and (16) as $P_{\text{therm}} = P_oA_{\text{th}} + V_B\eta_eP_o(e/h\upsilon)$. The temperature rise ΔT caused by this heating source is determined by the estimated thermal impedance θ of the structure, $\Delta T = \theta P_{\text{therm}}$. The thermal impedance found from finite element calculations [19] is shown in Table 4. If it is assumed that the maximum incident optical pump power is limited by the maximum allowable temperature rise of the device, then

$$P_o^{\text{max}} = \frac{\Delta T_{\text{max}}}{\theta[P_oA_{\text{th}} + V_B\eta_eP_o(e/h\upsilon)]}. \tag{23}$$

Inserting this equation into the solution for the THz power (4) gives the following expression:

$$P_{\text{max}} \propto \frac{\eta_e^2}{\theta^2[A_{\text{th}} + V_B\eta_e(e/h\upsilon)]^2} \frac{1}{[1 + \omega^2\tau_e^2]}. \tag{24}$$

This equation is plotted in Fig. 13 for varying LTG GaAs layer thicknesses, with a gap spacing of $0.8\,\mu$m biased at $16\,$V. A peak in the maximum power is observed where the external quantum efficiency is sufficiently high to generate adequate photocurrent and the thermal impedance is low enough to provide good thermal dissipation. The optimal thickness for a $0.8\,\mu$m gap is approximately $0.5\,\mu$m.

2.6   Proven High-Power Methods

Several strategies have been used to achieve higher output powers from photomixers. The first strategy is to improve the thermal loading capability of the substrate. One of these approaches, discussed above, uses an AlAs layer as a thermal conductor below a thin active region of LTG GaAs and leads to improvements in output power of four times [19]. A second approach to improving the thermal loading is to cool the photomixer to $77\,$K to exploit the

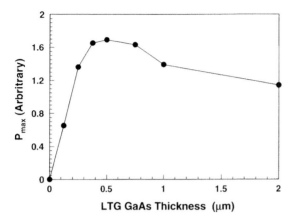

Fig. 13. Calculated output power versus LTG GaAs thickness, with $w_g = 0.8\,\mu m$, $w_f = 0.2\,\mu m$, normalized to $1\,\mu m$ LTG GaAs

higher thermal conductivity of materials at lower temperatures [23]. Cooling the photomixer to 77 K also provides an improvement in output power of aproximately four times. A second strategy for achieving high output power is to improve the impedance-matching design by increasing the antenna resistance and tuning out the electrode capacitance [27]. This is discussed in detail in Sect. 3. This strategy has increasing benefits at higher frequencies, and improvements in output power of six to ten times have been obtained, compared with log-spiral antennas. However, the small size of the electrode region still limits the output power owing to the thermal limitations. A third strategy to achieve high output powers is to use traveling-wave designs [28–30] which utilize a large active-electrode region. Finally, work is ongoing to improve the output power by optimizing the electrode structure. Early work showed close to milliwatt power levels (at 200 MHz) using narrow gap spacings [22]. However, a small gap spacing increases the electrode capacitance and can actually cause lower output power at high frequencies (see Fig. 14).

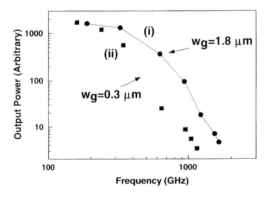

Fig. 14. Measured data for log-spiral antenna photomixers using $20\,\mu m \times 20\,\mu m$ electrode regions with gap widths of $0.3\,\mu m$ and $1.8\,\mu m$. $P_{opt} = 30\,mW$, $\tau_{e0} = 0.27\,ps$

3   Planar Antennas and Circuitry

One of the real advantages of the photomixer concept and its implementation as described in this chapter is the use of a single layer of metal on the LTG GaAs. This provides a flat, uniform substrate in which to construct planar circuitry and antennas. More complex heterostructures or multiple metal layers may someday provide higher powers owing to improved saturation velocities and/or shorter transit lengths for the carriers [31–33]. However, as we shall see in Sect. 6, structures using LTG GaAs with Au metal has achieved tunable, usable power levels.

In this section, we start with the electromagnetic behavior of the inter-digitated-electrode finger geometry and describe an equivalent circuit. The electrode fingers possess a capacitance that presents a key design challenge for providing high output power and accurate frequency response. Log-spiral antennas are then discussed since these designs provide several THz of tuning bandwidth [1–3]. Recent work has focused on resonant antennas [4,27], since there are two primary advantages over the wideband designs: the electrode capacitance is tuned out, which eliminates one of the pair of 6 dB/octave rolloffs in output power, and the radiation resistance is increased. Dipole and slot antennas, used as single and dual elements, have significant power advantages over log-spiral designs when large bandwidths are not necessary. Finally, traveling-wave designs are discussed, since the correspondingly large active-electrode region can handle a larger optical power and thus produce more THz power [28–30].

3.1   Electrode Capacitance

The interdigitated electrodes of the finger region possess a capacitance that strongly affects the behavior of the photomixer at terahertz frequencies. The photomixer can be modeled as a modulating conductance G with a shunting capacitance due to the interdigitated fingers of the electrodes, as shown in Fig. 4. The typical conductance is low and the photomixer is assumed to act as a current source. The shunt capacitance causes the antenna impedance to short at high frequencies and represents a design challenge for achieving high power and accurate prediction of the frequency response of the photomixer.

As shown above, high output power from a photomixer is achieved by having a high external quantum efficiency (closely spaced electrode fingers) and a high thermal limit (large electrode area). However, this conflicts with high-frequency operation since the capacitance is directly proportional to the area and inversely related to the spacing of the fingers. The quasi-static electrode capacitance due to the interdigitated fingers can be found using a conformal mapping technique [34] as

$$C = \frac{K(k)}{K(k')} \frac{\varepsilon_0 (1 + \varepsilon_r) A}{(w_g + w_f)} \tag{25}$$

where A is the interdigitated area, ε_0 is the permittivity, ε_r is the relative permittivity of LTG GaAs (ε_r equal to 12.8), and $K(k)$ is the complete elliptical integral of the first kind. Here

$$k = \tan^2 \frac{\pi w_f}{4(w_g + w_f)} \tag{26}$$

and

$$k' = \sqrt{1 - k^2}. \tag{27}$$

This solution is expected to provide adequate agreement with the real situation for lengths of the electrode region of up to 0.1–0.15λ_c [27]. The wavelength of interest for planar circuitry is the guided wavelength, which, for the quasi-TEM structures considered here, is given by

$$\lambda_e = \frac{\lambda_o}{\sqrt{\varepsilon_{\text{eff}}}}. \tag{28}$$

where λ_o is the free-space wavelength. The effective permittivity is [35]

$$\varepsilon_{\text{eff}} = \frac{\varepsilon_r + 1}{2}. \tag{29}$$

The prediction of the electrode capacitance (25) has been shown [27] to be generally valid for log-spiral and dipole antennas but less accurate for slot antennas. The improvement and ubiquity of commercial electromagnetic (EM) simulation tools over the past decade has made the widespread use of full-wave moment-method and finite-element solutions available to most physicists and engineers. Therefore, (25) can be considered a first pass design with more detailed designs being performed with full-wave EM codes.

An important step towards improved prediction of the frequency response is recognizing the finite height of the fingers. The thickness of the metal of the electrodes causes additional capacitance of the electrode region beyond that corresponding to the infinitely thin metal assumed in the above calculations. Typical fingers have a cross-sections with a width of 0.2 μm and a height of 0.15 μm. Results from a finite element solver[1] and measurements demonstrate that the 0.2 μm wide and 0.15 μm high finger can be modeled as a zero-height finger of width 0.23–0.24 μm. It should be mentioned that the use of an antireflective coating for improving the coupling of the laser radiation, into the LTG GaAs increases the capacitance of the electrode region slightly, owing to the increased effective permittivity, and can reduce the high-frequency performance.

---

[1] Agilent HFSS, version. 5.5, Agilent Corporation.

3.2   Log-Spiral

Early photomixer work integrated the electrodes with log-spiral antennas [1,2]. A photograph of one of these designs is shown in Fig. 1. One of the primary disadvantages of the photomixer is the low output power at high frequencies. This is partly a result of the $-6\,\mathrm{dB/octave}$ roll-off arising from the lifetime of the carriers (19), but is also a result of the reduction in load impedance due to the RC time constant (resulting in another $-6\,\mathrm{dB/octave}$ roll-off). A second reason for the low output power is the relatively low radiation resistance of the log-spiral antenna. From the equivalent-circuit model (see Fig. 4), the real part of the load impedance for use in (4) is

$$\mathrm{Re}(Z_{\mathrm{load}}) = \mathrm{Re}\left[(G_{\mathrm{spiral}} + \mathrm{j}\omega C_{\mathrm{elect}})^{-1}\right] = \frac{R_{\mathrm{spiral}}}{1 + \omega^2 R_{\mathrm{spiral}}^2 C_{\mathrm{elect}}^2}, \qquad (30)$$

where $R_{\mathrm{spiral}} = (G_{\mathrm{spiral}})^{-1}$ is equal to $60\pi\sqrt{\varepsilon_{\mathrm{eff}}}$ or $72\,\Omega$ for a complimentary log spiral on a GaAs half-space [35]. Therefore, the log-spiral photomixer design is subject to a $-12\,\mathrm{dB/octave}$ reduction in output power at high frequencies. It is important to note that in (30) the resistance presented to the current source reduces as a function of frequency, although the inherent resistance of the spiral does not change. In effect, the capacitor is "shorting out" the resistance of the spiral. As the frequency becomes infinite, the resistance presented to the current source becomes zero.

The tuning bandwidth obtained using a log-spiral antenna is shown for two different electrode structures in Fig. 14. Curve (i) shows data for a $20\,\mu\mathrm{m}$ region with $0.2\,\mu\mathrm{m}$ wide fingers spaced at $2\,\mu\mathrm{m}$ intervals and curve (ii) shows data for a $20 \times 20\,\mu\mathrm{m}$ region with $0.2\,\mu\mathrm{m}$ wide fingers spaced at $0.5\,\mu\mathrm{m}$ intervals. The larger capacitance of the structure corresponding to curve (ii) leads to a pole at a lower frequency, and thus a reduction in signal relative to the lower-capacitance geometry.

3.3   Single Full-Wave Dipoles

Another class of antennas, single and dual dipoles and slots, uses the approach of having a high radiation resistance (at the expense of bandwidth) and using inductive tuning to cancel out the electrode capacitance at the desired frequency. This is a useful approach because, for many applications, several terahertz of bandwidth is not necessary, but only a metallic waveguide band (less than an octave) is required. Therefore, resonant antenna elements may be used for improved output power for these applications [27].

The process of obtaining an optimum impedance match relies on accurate predictions of the THz impedance of the various elements depicted in Fig. 15. The input impedance of a full-wave dipole with the interdigitated electrodes at the drive point was calculated. The excitation source considered was a coplanar-strip (CPS) mode that is a standard calibrated port

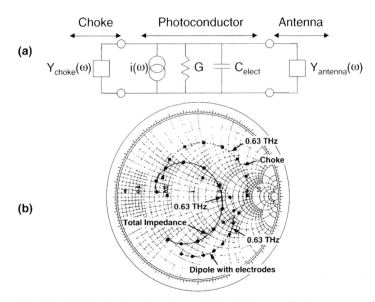

Fig. 15. (a) Equivalent-circuit model for full-wave dipole antenna. (b) calculated impedances for full-wave dipole antenna

available from a commercial method-of-moments simulator[2]. The dc bias of the photomixer requires planar lines. since contacts are applied near the edge of the substrate. THz leakage down the bias lines was minimized by presenting a high impedance at the antenna terminals, using a choke made from alternating quarter-wave sections of high- and low-impedance CPS lines. In this design, the first section of high-impedance line near the antenna is operated at less than a quarter wavelength and presents an inductive susceptance that tunes out the capacitance from the photomixer. Fig. 15b shows the choke impedance. To a good approximation. the individual elements. whose S-parameters were calculated with the EM simulator, can then be connected together in a simple circuit simulator to predict the total embedding impedance. also shown in Fig. 15b. The resonant resistance of the dipole with the photomixer and choke is 174 Ω. A comparison of the ratio of the resistance of the dipole to that of the spiral (71 Ω) with the result shown in Fig. 16 shows that the expected 2.45 times improvement in output power is obtained.

3.4   Dual-Antenna Elements

3.4.1   Dual-Dipole Elements

Dual-antenna elements have several advantages over a single-antenna design. These include more symmetric beam patterns, leading to higher Gaussian

---

[2] Agilent Momentum. Agilent Corporation.

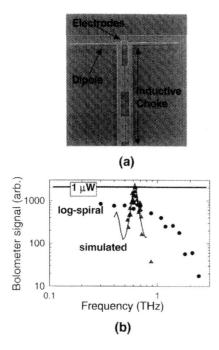

(a)

(b)

Fig. 16. (a) Scanning electron micrograph of full-wave dipole antenna, (b) measured versus calculated performance. *Circles*: log-spiral antenna measurements, *triangles*: full-wave dipole antenna measurements

beam efficiency [36,37] and higher radiation resistance. The dual-dipole design shown in Fig. 17 uses an electrode region centered between two dipoles and connected via CPS transmission lines, whose lengths are chosen to resonantly tune out the electrode capacitance. The primary advantage of the dual dipole is the ease in design of the photomixer for high-frequency operation.

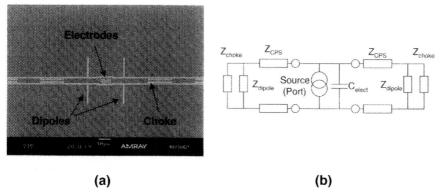

(a)   **(b)**

Fig. 17. (a) Scanning electron micrograph of dual-dipole antenna, (b) equivalent-circuit model of dual dipole

The behavior of the dual dipole is illustrated by the equivalent-circuit model in Fig. 17. The choke and dipoles are connected in parallel with the electrode capacitance and photoconductance. Therefore, the admittance at the photoconductor is

$$Y_{\text{dipole}} = 2G_{\text{trans}} + 2jB_{\text{trans}}, \tag{31}$$

where G_{trans} and B_{trans} are the conductance and susceptance, respectively, of a single dipole transformed through the CPS transmission line. Therefore, two important characteristics are displayed: (1) the input resonant resistance from a single dipole arm is reduced by half, and (2) only half of the inductive tuning needs to come from each dipole arm. Therefore, the maximum attainable resistance from a dual dipole is limited (~ 200–$300\,\Omega$). However, the reduced amount of tuning required for the dual dipoles allows operation at higher frequencies, with examples up to 2.7 THz having been demonstrated [27]. Also, electrode structures with many closely spaced fingers (higher capacitance) may be used.

The problem in applying the equivalent-circuit model of Fig. 17 is finding accurate and meaningful port definitions for use with EM simulators. To deal with this problem, the electrodes can be solved using an EM simulator and then converted to a lumped circuit model. The antenna, CPS transmission line, and choke can also be calculated using the EM simulator and then combined in a circuit simulation where the drive port is across the photoconductor capacitance, as above. A convenient transmission line for dual dipoles uses coplanar strips, as shown in Fig. 17. The CPS acts as an impedance transformer connecting the photomixer to the dipole antennas and provides DC bias to the photomixer region with the bias contact pads placed away from the antenna near the edge of the substrate.

3.4.2   Graphical Design Procedure

A graphical procedure is presented here for dual-dipole designs. The technique ignores loss, mutual coupling, and the choke but gives a starting point for estimating the resistance at the photoconductor and the peak operating frequency. It also should provide practitioners without access to full-wave EM simulators a means for achieving high-power dual-dipole designs. The design discussed in this section is a typical approach used in microwave engineering [38] and the reader may wish to review the techniques of matching using a Smith chart.

The input impedance of a thin half-wave dipole in air is well known to be $73 + j42.5\,\Omega$ [39]. On a half-space of GaAs, the input resistance is expected to be approximately $73/\sqrt{\varepsilon_{\text{eff}}}$, or $28\,\Omega$ [35]. The reactance can be approximated as zero since the bend to the CPS feed adds some capacitance and the high-impedance line makes the reactance close to the real axis of a Smith chart. The electrode capacitance can be found using the solution shown above in (23).

The characteristic impedance of a CPS is [40]

$$Z_{cps} = \frac{120\pi}{\sqrt{\varepsilon_{eff}}} \frac{K(k)}{K(k')},\tag{32}$$

where in this case

$$k = a/b,\tag{33}$$

and a is the separation from the inner edge of the conductors and b is the separation from the outer edge of the conductors.

The design goal for using resonant elements maximizes the resistance presented across the capacitor of the photoconductor at the desired frequency. The parallel connection of the dual dipoles doubles the admittance of an individual dipole at the electrode, making the desired, goal susceptance (B_{goal}) of one arm of the dual dipole equal to half the electrode susceptance (B_{elect}), i. e.

$$B_{goal} = -B_{elect}/2,\tag{34}$$

where B_{elect} is found from

$$B_{elect} = 2\pi f C_{elect} Z_{cps}\tag{35}$$

(normalized to the line impedance). This transformation lowers the goal susceptance closer to the real axis of a Smith chart, allowing operation at higher resistances and frequencies.

As an example, a 1.05 THz design (D2) has been calculated using this method and compared with results in [27]. The design is shown in Fig. 21 in Sect. 5 with dimensions in Table 5 (and results in Table 6 and Fig. 22). The line made of 1 μm width strips with a center-to-center spacing of 6 μm ($a = 5$ μm. $b = 7$ μm) is calculated with (32) to have a characteristic impedance of 145 Ω. An eight finger 7×5 μm electrode region with the characteristics given in Table 5 is calculated with (25) to have a capacitance of 1.24 fF. Therefore, $B_{elect} = j1.19$ (normalized to 145 Ω) and $B_{goal} = -j0.59$; these are plotted in Fig. 18. A constant Γ (reflection coefficient) circle is shown connecting Y_{dipole} and B_{goal} yielding the transformed admittance $Y_{trans} = 0.26 - j0.59$. The conductance at the photoconductor is $G_{rad} = 2G_{trans} = 0.46$ which gives a radiation resistance of 315 Ω. The connecting line length found in Fig. 18 is $0.164\lambda_e$ or 17.9 μm, for a single dipole arm. Therefore, the total CPS length between dipoles is calculated to be 35.8 μm, which should be compared with the fabricated 44.2 μm. The predicted resistance of 315 Ω modified by a radiation efficiency of 67% (see Table 6) is 211 Ω, which can be compared with the value of 214 Ω in Table 6 found using a full EM approach [27]. The graphical approach provides a useful starting point for the more accurate modeling described in [27] or for useful designs when precise frequency operation is not necessary.

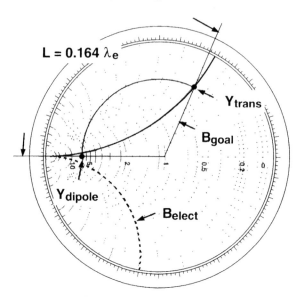

Fig. 18. Graphical design procedure for a dual dipole shown on 145 Ω Smith chart. B_{elect} is the electrode capacitance. B_{goal} is the required susceptance a single dipole arm must present for resonant operation. Y_{dipole} is the dipole admittance, and Y_{trans} is the admittance at the photoconductor of a single dipole arm after transforming through a 145 Ω CPS line of length $0.164\lambda_e$.

3.4.3   Dual-Slot Elements

The dual-slot antenna is illustrated in the SEM photograph in Fig. 19. A coplanar waveguide (CPW) is used as the connecting transmission line, and bias is applied across the electrodes. The primary advantage of the dual slot is the very high output power that can be attained at lower frequencies. Additionally. the Gaussian beam efficiency is increased over that of a single-slot element owing to the reduced beam width in the E-plane. However. dual-slots are best suited to low-frequency designs (< 1 THz) because it is more difficult to tune out the electrode capacitance than with the dual dipole.

The behavior is illustrated in the equivalent-circuit model in Fig. 19. The slots are connected in series. making the input impedance twice that of a single-slot arm:

$$Y_{\text{slot}} = G_{\text{trans}}/2 + jB_{\text{trans}}/2. \tag{36}$$

Therefore. twice the resonant resistance is obtained with a dual slot as with a single-slot arm, and four times as much as with a dipole system. This creates the possibility of a very high resistance ($> 800\,\Omega$) being presented to the photoconductor. However, the inductive tuning is halved, making it necessary for each slot arm to present twice as much susceptance as does a single-slot arm. and four times as much as dipole arms. This means that

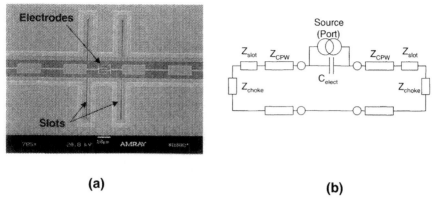

(a)   **(b)**

Fig. 19. (a) Scanning electron micrograph of dual-slot antenna, (b) equivalent-circuit model of dual slot

the electrode capacitance must be kept small by using a wide gap spacing and/or small electrode area, which reduces the external quantum efficiency and thermal limits possibly leading to overall lower-power designs.

3.5 Distributed Photomixers

The inherent problem, but primary advantage from a modeling viewpoint, of the log-spirals, single and dual dipoles and slots discussed so far is the use of the photomixer as a lumped element. There are two reasons for this restriction to small devices. First, the capacitance must remain small enough that inductive tuning is possible. Second, for large electrode regions, partial cancellation of photocurrents due to phase mismatch across the active region can occur. Therefore, the difficulty in achieving higher powers is related to the small area of the electrode region. For example, a $0.15\lambda_e \times 0.15\lambda_e$ active region for use at 2.7 THz is only 40 μm^2 in area and the maximum capacitance is 0.91 fF (these limitations are discussed in Sect. 5). The power-handling capability of such a small electrode region is limited.

An alternative approach to photomixer design is the distributed photomixer. These devices can potentially improve the bandwidth and output power of photomixers operating above 1 THz. The relevant analogy is the success of traveling-wave transistor amplifiers and nonlinear transmission lines in enhancing the high-frequency performance of transistors and Schottky diodes, respectively. Two approaches look encouraging. A surface-illuminated approach has been developed by Matsuura et al. [28,29] that has demonstrated record output power above 1 THz. This approach has the advantage of relatively simple fabrication. Velocity matching of the optical group velocity to the THz phase velocity is achieved by controlling the angle of incidence of the optical waves. A second approach has been described by Duerr et al.

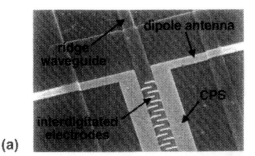

(a)

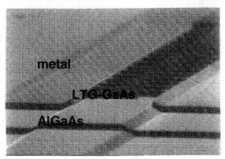

(b)

Fig. 20. Scanning electron micrographs of a distributed photomixer. (a) Top view shows the AlGaAs ridge waveguide. the electrodes and coplanar strips and the full-wave dipole antenna: (b) side view shows the LTG GaAs film on top of the Al-GaAs waveguide and the metal electrodes

[30] that is similar to the design of high-power photodiodes used for analog optical communication links.

Figure 20 shows two SEM images of a distributed photomixer terminated by a full-wave dipole antenna [30]. Here. an AlGaAs ridge waveguide couples the optical beat signal to a thin layer of LTG GaAs photoconductor that is patterned on its surface. Interdigitated electrodes collect the photocurrent onto a CPS and add just enough capacitance per unit length to velocity-match the THz and optical waves. In theory, this device should outperform lumped-element photomixers above $\sim 1.5\,\text{THz}$. In practice. engineering details such as efficient thermal management. optical facet coatings. and RF antenna design need to be optimized before the device realizes its potential.

4   Hyperhemisphere Lens

The hyperhemisphere lens [41] shown in Fig. 1 provides a convenient and efficient means of collecting the radiation from a planar antenna and focusing it into a Gaussian beam for use with quasi-optics. At the frequencies of interest (1–3 THz), quasi-optics is a preferred means of transferring and routing signals, owing to low loss and dispersion. Classic millimeter-wave guided-wave structures such as waveguide are too lossy and difficult to manufacture in this frequency region.

A silicon lens (refractive index close to that of GaAs) formed into a hyperhemisphere or elliptical lens is a good candidate for use with photomixers

[41–45]. It uses the high refractive-index contrast of the GaAs or silicon substrate relative to air and the fact that the radiation fields around the antenna prefer to stay inside the dielectric. The ratio of the radiation intensity into the dielectric U_{diel} to that into the air U_{air} is [35]

$$\frac{U_{\mathrm{diel}}}{U_{\mathrm{air}}} = \varepsilon_{\mathrm{r}}^{1.5} \tag{37}$$

for spirals, dipoles, and slots. Therefore, the backlobe (or power not collected in the lens) is $-16.6\,\mathrm{dB}$ below the main lobe into the lens. Ignoring diffraction, the fraction of power that radiates into the lens out of the total is approximately 96%.

The reflection from the lens into air has been analyzed carefully in [44,45]. In general, however, it was found that a simple mismatch loss calculated using $\eta_{\mathrm{mismatch}} = 4n_{\mathrm{lens}}/(n_{\mathrm{lens}} + 1)^2$, predicts this loss sufficiently well. For a silicon hemisphere with $n_{\mathrm{lens}} = 3.43$, this equation adds 1.55 dB loss due to reflection. A quarter-wave matching layer, defined as one with a dielectric $\varepsilon_{\mathrm{match}} = \sqrt{\varepsilon_{\mathrm{lens}}}$, would recover nearly all of the lost power over a finite bandwidth ($\sim 25\%$).

The dielectric attenuation constant of the lens can be calculated by assuming a plane wave in a lossy dielectric [38]:

$$\alpha_{\mathrm{d}} = 27.288 \frac{\sqrt{\varepsilon_{\mathrm{r}}} f \tan \delta}{c}, \tag{38}$$

where $\tan \delta$ is the loss tangent of the dielectric, c is the speed of light, and f is the frequency. However, the complex conductivity of materials up to several terahertz has a complicated behavior. Rewriting (38) using the approach of [47] leads to

$$\alpha_{\mathrm{d}} = 27.288 \frac{\sqrt{\varepsilon_{\mathrm{r}}} f}{c} \frac{\sqrt{\varepsilon_{\mathrm{r}}} \alpha_{\mathrm{m}}(f) c}{2 \pi f} = 4.343 \varepsilon_{\mathrm{r}} \alpha_{\mathrm{m}}(f), \tag{39}$$

where α_{m} is the power absorption coefficient and is a function of frequency. The power absorption coefficient of float-zone high-resistivity silicon ($10\,\mathrm{k\Omega}$ cm) is approximately $0.04\,\mathrm{cm}^{-1}$ from 0.1 to 2 THz [46]. This leads to a lens loss of $\alpha_{\mathrm{d}} l_{\mathrm{lens}}$, where l_{lens} is the length of the lens (the length is assumed to be the radius of the hyperhemisphere). For a 0.5 cm silicon lens, the calculated loss is a constant 1.0 dB over the 0.1–2 THz frequency range. Other measurements suggest this is true well past 4 THz [48]. On the other hand, the GaAs substrate has a value of α_{m} that increases from $0.5\,\mathrm{cm}^{-1}$ at 1 THz to $2.4\,\mathrm{cm}^{-1}$ at 2 THz [46], although this value may vary with the quality of the semi-insulating GaAs. The loss due to a typical GaAs (500 µm) substrate can become significant [27,48], equal to 1.2 dB at 1 THz and 6.0 dB at 2 THz. Minimizing this thickness to reduce this effect for terahertz operation may be desirable.

A further issue is the use of a low-loss attachment of the GaAs substrate to the silicon lens. Typically this is done with a silicone grease for laboratory

measurements or a thin-bond-line glue for permanent applications. Air gaps
in the bond can cause significant mismatch and lead to significant power
reduction. However, a high-quality bond layer provides a low-loss path to the
output beam ($\sim 0.2\,$dB). Typical bonds provide a poor thermal path [19],
and a potential problem is a reduced thermal load as a result of thinning the
GaAs.

Finally, the use of quasi-optics is required to collect the output beam from
the hyperhemisphere. For power measurements, a Gaussian beam-coupling
efficiency of 80–90% is possible [37]. However, proper quasi-optical techniques
should be employed to ensure maximum power coupling into the measuring
device.

5   Photomixer Design and Examples

5.1   Dual-Dipole Results

A number of dual-dipole designs (as illustrated in Fig. 21), with dimensions
given in Table. 5, have been fabricated in the 0.85–2.7 THz range. Their
THz output characteristics were measured as described in [19]. Fig. 22 shows
the measured and calculated results. The calculated results were obtained
by taking a ratio of the resistance of that of a dipole design to that of a

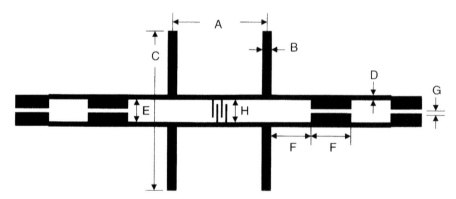

Fig. 21. Geometry of dual dipoles, with parameters given in Table 5

Table 5. Dimensions for dipole designs in μm

Name	A	B	C	D	E	F	G	H	N_f	w_f	w_g
D1	55.4	2.5	68.4	1.0	5.0	33.7	1.0	5.0	8	0.2	0.8
D2	46.7	2.5	54.6	1.0	5.0	27.7	1.0	5.0	8	0.2	0.8
D3	28.8	1.5	34.3	1.0	5.0	17.9	1.0	5.0	8	0.2	0.8
D4	20.0	1.0	20.7	1.0	4.0	10.6	1.0	4.0	6	0.2	0.8
S1									8	0.2	0.8

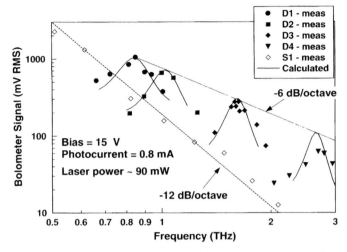

Fig. 22. Results from a bolometer for dual dipoles D1, D2, D3, and D4 and a spiral S1. Calculated results are shown by *solid lines*. The −6 dB/octave curve demonstrates the predicted roll-off for a dipole and the −12 dB/octave curve demonstrates the predicted roll-off for a spiral

spiral design. For identical electrode structures, this eliminates the effects of heating, carrier lifetime, and losses in coupling to the Gaussian beam optics. The calculated characteristics are also tabulated in Table 6. The radiation efficiency accounts for the losses in the metal circuitry [27]. The basic dual-dipole design uses half-wavelength dipoles spaced a half-wavelength apart (minus the length required to tune out the electrode capacitance). This basic design was used for D1, D2, D3 and D4 for operation at 0.85, 1.05, 1.6, and 2.7 THz, respectively. The 2.7 THz design has measured values lower than predicted, as demonstrated in Fig. 22. Two reasons may account for this. The first is that D4 has a smaller electrode region than the other designs have, possibly allowing more spillover from the optical fiber. Second, additional THz power may have been lost owing to substrate absorption in GaAs, as discussed above. Results measured using a spiral antenna with an identical electrode geometry of $7 \times 5\,\mu m$ with 8 fingers are also shown in Fig. 22. Measurable power levels are obtainable to 2 THz before noise degrades the accuracy.

The output power behavior of the dual dipoles in Fig. 22 demonstrates the predicted 6 dB/octave roll-off due to the lifetime of the carriers. This is in contrast to the 12 dB/octave roll-off for the log-spiral. Therefore, for operation at frequencies above 1 THz, the improved high-output-power behavior of resonantly tuned antennas becomes more pronounced. The improvement in output power of 6–10 dB for the dual dipoles over the spiral is noted in Table 6. The corresponding output powers of the D1, D2, D3 and D4 designs

Table 6. Calculated predictions for photomixer designs

Name	Frequency (THz)	Radiation efficiency (%)	Dipole resistance (Ω)	Spiral load resistance (Ω)	Dipole/spiral (dB)
D1	0.85	72	217	58	5.7
D2	1.05	67	214	53	6.0
D3	1.6	64	203	39	7.2
D4	2.7	60	217	21	10.1

on the best material are 3, 2, 0.8 and 0.2 μW, respectively [19], and represent state-of-the-art photomixer results.

5.2   Practical Measurement Issues and Power Calibration Difficulties

A number of measurement issues and difficulties arise in obtaining quantitatively accurate results with photomixers. These problems include optimizing the optical signal entering the electrode region, efficiently collecting the THz signal using quasi-optics, and the elimination of errors arising from the measuring device (bolometer).

One of the difficulties with a photomixer is maximizing the optical signal into the electrode region. The fiber should be cleaved cleanly to ensure a proper Gaussian output of the optical signal. Alignment with the electrode region is typically done with micrometers but small vibrations can cause some variation in photocurrent. Also, standing waves between the end of the fiber and the electrode region can affect the amount of optical power entering the LTG GaAs. Maximizing this power may cause a simple analysis of the optical coupling efficiency to be less accurate. It is also important to use a fiber polarization controller to maximize the coupling into the electrode region since grating effects occur [20,21]. Properly aligned, polarization maintaining fibers can also be used.

Accurate determination of photomixer output power is accomplished through careful calibration of the detector used to measure the power, and knowledge of the free-space coupling efficiency into the detector. Diffraction and interference phenomena within the detector, and between the detector and external elements, contribute to producing a complicated spectral and spatial dependence of the bolometer responsivity. For example, the bolometer sensitivity has been shown to vary by almost an order of magnitude within some frequency bands. For this reason, calibration of the detector with broadband thermal sources can be somewhat misleading. A more accurate determination can be made with narrowband sources such as solid-state oscillators or gas lasers, although such sources typically are fixed frequency or are limited to a narrow tuning band. The quasi-optical coupling efficiency from the photomixer to the detector can be calculated by taking into account some measured output beam parameters for the photomixer and using appropriate Gaussian-beam overlap integrals [37] and the Friis equation [39].

5.3 Maximum Power Limitations

In this section, the maximum power obtainable from a photomixer is investigated. For simplification, the antenna choices are limited to the dual dipole, owing to its advantages over the other antenna types. This study will also focus only on the small-area (lumped element) electrode region. The larger size of the distributed photomixers will undoubtedly lead to higher power levels than those obtainable with the lumped-element versions. However, this approach is still under development, with a clear design guide not yet available.

There are a few fundamental limitations that are assumed. For the lumped element photomixer design, the length of the electrode region should remain smaller than $0.15\lambda_e$. For larger regions, a single lumped element no longer describes the behavior adequately [27]. Also, since the photocurrent in the electrodes is generated in phase, for large electrode regions, the current starts to add out of phase, thus reducing the overall current gain. The maximum area as determined from electrical considerations is

$$A_{\mathrm{max}} = (0.15\lambda_e)^2. \tag{40}$$

The minimum area of the electrode region is determined by the size of the laser spot on the substrate. For a single-mode fiber at $0.85\,\mu\mathrm{m}$, this minimum area is assumed to be $20\,\mu\mathrm{m}^2$.

A second consideration is the amount of capacitance that can be tuned out without losing antenna resistance. As demonstrated in the graphical design procedure, when the reactance of the electrode region increases, the radiation resistance decreases. For a dual dipole to achieve a sufficiently large resistance, it is assumed that

$$B_{\mathrm{elect}} = \omega C_{\mathrm{elect}} Z_{\mathrm{cps}} \leq 2. \tag{41}$$

This means that the normalized goal susceptance B_{goal} of each dipole arm is 1, not far from the example given above.

The limitations on the area and electrode capacitance expressed by (40) and (41) are shown in Fig. 23. The area limitation (40) is a vertical line in the figure and the capacitance limitation (41) is a horizontal line. The allowable region is inside the box defined by these lines (to the lower left). Superimposed on this region is the electrode capacitance obtained using ((25)–(27)) for varying gap widths (constant $0.2\,\mu\mathrm{m}$ finger width with zero height is assumed).

The region of maximum output power for each frequency is also shown in Fig. 23. The optimal output power for an unlimited optical supply is found

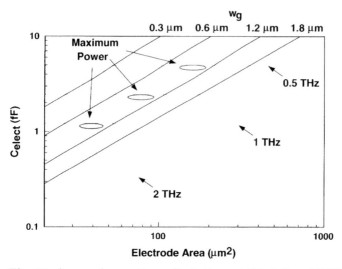

Fig. 23. Area and capacitance limitations at 0.5. 1.0. and 2.0 THz. Also shown is the calculated capacitance of the electrodes with the corresponding areas and gap widths. The maximum-power locations are given for unlimited optical power

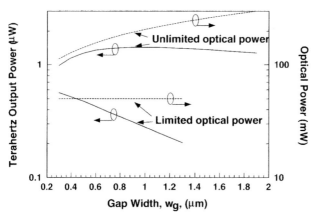

Fig. 24. Optical and output power for varying gap width when using an unlimited optical supply and a limited optical supply (50 mW). The calculated data is given in Tables 7 and 8

when the gap width is varied. as demonstrated in Fig. 24. This result was obtained by finding the maximum possible area for a given electrode capacitance (2.45 fF at 1 THz). The maximum output power versus voltage was found (see Fig. 12). and then the maximum thermal power (19) was calculated for 1 μm LTG GaAs on GaAs. The solution used (4) for the calculation of output power at 1 THz with an antenna efficiency of 0.33 (1.55 dB for lens reflection, 0.2 dB from the backlobe. 1.0 dB from the silicon in lens. 1.2 dB from the GaAs substrate, 0.2 dB from the bond layer, 0.7 dB from the Gaussian beam coupling). A tabulation of the calculated values is given in Table 7, with the output power and optical power versus gap width shown in Fig. 24. A peak in output power is seen with a maximum for a gap width of approximately 0.9 μm. However, the power is fairly insensitive to gap width, being within 10% of the maximum from 0.5 to 1.9 μm. The necessary optical power, on the other hand, is much less for the small gap widths (owing to the higher external quantum efficiency). A compromise of optical power versus optimal output power yields best gap widths of around 0.8 μm to 1 μm for a finger width of 0.2 μm.

The second set of curves in Fig. 24 is an example for limited optical supply. In many cases, large optical powers are not available; therefore an example with 50 mW of optical signal has been considered. Results of the calculation are tabulated in Table 8. This example uses 1 μm LTG GaAs on GaAs with an AR coating to maximize coupling into the active region. For limited optical power, it is important to make η_{opt} high by using either an AR coating or a

Table 7. Output power at failure limit for varying gap widths and unlimited optical power. Frequency: 1 THz, $\tau_{e0} = 0.25\,\mathrm{ps}$, $d_{\mathrm{LTG}} = 1\,\mu\mathrm{m}$, $C_{\mathrm{elect}} = 2.45\,\mathrm{fF}$, $A_{\mathrm{max}} = 295\,\mu\mathrm{m}^2$, $w_{\mathrm{f}} = 0.2\,\mu\mathrm{m}$, $\tau_e = 0.8\,\mu\mathrm{ps}$

w_{g} (μm)	A_e (μm^2)	V_{bias} (V)	η_e	P_{therm} (mW)	P_{o} (mW)	I_{o} (mA)	P_ω (mW)
0.3	25	5.5	0.0113	39	113	0.885	0.983
0.5	41	7.0	0.0102	50	143	1.005	1.27
0.7	59	8.5	0.00896	60	170	1.054	1.40
0.9	78	9.8	0.00794	69	195	1.070	1.44
1.1	98	11.1	0.00709	77	218	1.068	1.44
1.3	119	12.4	0.00638	85	240	1.057	1.40
1.5	141	13.6	0.00579	92	261	1.043	1.37
1.7	163	14.8	0.0053	100	281	1.027	1.32
1.9	185	16.0	0.00487	106	300	1.008	1.28

Table 8. Output power at failure for varying gap widths and a limited optical power of 50 mW. Frequency: 1 THz, $\tau_{e0} = 0.25\,\mathrm{ps}$, $d_{\mathrm{LTG}} = 1\,\mu\mathrm{m}$, $C_{\mathrm{elect}} = 2.45\,\mathrm{fF}$, $A_{\mathrm{max}} = 295\,\mu\mathrm{m}^2$, $w_{\mathrm{f}}0.2\,\mu\mathrm{m}$, AR coating

w_{g} (μm)	A_e (μm^2)	P_{therm} (mW)	V_{break} (V)	V_{bias} (V)	η_e	τ_c (ps)	I_{o} (mA)	P_ω (mW)
0.3	25	39	17.0	11.8	0.0583	2.44	2.01	0.565
0.5	41	50	22.0	16.7	0.0611	2.79	2.11	0.475
0.7	59	60	25.6	21.5	0.0608	3.09	2.10	0.383
0.9	78	69	28.3	26.4	0.0593	3.35	2.05	0.312
1.1	98	77	30.5	30.5	0.0549	3.45	1.90	0.251
1.3	119	85	32.4	32.4	0.0461	3.22	1.59	0.204

DBR. The limitation in this case is primarily caused by dielectric breakdown due to high electric fields. The gap widths are varied. while the maximum area is obtained. The voltage breakdown restricts how much the bias can be increased (since below the thermal-failure limit increasing the bias creates more output power). For limited optical power. the optimal output power is obtained when small gap spacings. with the resulting inherently higher external quantum efficiency. are used.

6   Applications

6.1   Local Oscillators

Recent advances in superconducting heterodyne detectors promise a significant scientific payoff in submillimeter-wave astronomy for the study of planetary atmospheres. Hot-electron bolometers (HEBs) and superconductor-insulator-superconductor (SIS) mixers can achieve almost quantum-limited sensitivity at frequencies above 600 GHz. Such systems presently use local oscillators (LOs) consisting of either frequency-multiplied diode oscillators. vacuum tubes, or molecular lasers. The photomixer is an interesting alternative LO technology. It can be assembled with no moving parts. has relatively low power consumption. has a wide tuning range. and is based on commercial lasers and a custom photoconductor. For array applications. multiple photomixer LOs can be located remotely from the lasers. thereby easing signal distribution to remote antennas.

The main challenges in making a viable LO out of a photomixer are generating a coherent tone with minimal frequency jitter and amplitude noise. and generating sufficient power to overcome diplexer and other losses incurred when coupling an LO to a cryogenic heterodyne detector. An experiment was performed which demonstrated the use of a photomixer as a local oscillator at 630 GHz [4]. The heterodyne detector was a Nb SIS mixer mounted in a waveguide mixing block. The photomixer was a single-element full-wave dipole. Using the photomixer as an LO. approximately 0.2 µW of RF power was coupled to the SIS mixer. and the resulting double-sideband noise temperature was 331 K (see Fig. 25)   in good agreement with the 323 K noise temperature obtained when a multiplied Gunn oscillator generating 0.25 µW was substituted for the photomixer. The twin-element photomixers described above have higher output power and operate at higher frequencies. These devices should be adequate for laboratory experiments that integrate HEB mixers with photomixer LOs. The photomixer technology should become a reliable LO technology as it becomes more robust against thermal failure and as the LO power consumption of the HEB mixers drops.

6.2   Transceivers

Photomixers are also attractive sources for high-resolution spectroscopy of gases. Pine et al. used a photomixer in conjunction with a liquid-helium-

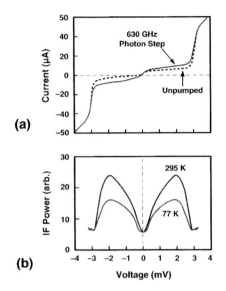

(a)

(b)

Fig. 25. (a) Pumped and unpumped current–voltage characteristic for an SIS junction exhibiting a 630 GHz photon step. (b) IF power for hot and cold black-body signals. The lowest noise temperature was 331 K double sideband

cooled bolometer to measure the rotational-broadening parameters of sulfur dioxide (SO_2) [5]. Chen et al. measured the rotational spectrum of acetonitrile (CH_3CN) [6]. Gas spectrometers based on photomixers and bolometers exhibit high sensitivity, as well as a frequency resolution that is unmatched by systems such as Fourier transform spectrometers that do not use coherent sources. A practical disadvantage of fielded systems, however, is the requirement of a helium-cooled bolometer. The photomixer transceiver addresses this limitation [49].

The transceiver uses a pair of photomixers that are pumped by the same pair of diode lasers, as shown in Fig. 26. One photomixer is a coherent transmitter of THz radiation. Its input consists of a DC bias voltage and the

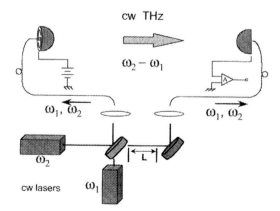

Fig. 26. Schematic of the photomixer transceiver. A pair of diode lasers generates a THz signal used for spectroscopy

optical beat signal produced by the combined output of the two lasers. Its output is a coherent THz beam. The THz beam passes through a gas cell and is detected by the second photomixer. The second photomixer is a coherent homodyne receiver of THz radiation. Its input consists of a THz waveform and the optical beat signal. Its output is a DC current that is proportional to the amplitude of the incident THz electric-field strength. Twin-slot antennas with a center frequency of 1.4 THz (see Fig. 19) were used to perform a demonstration of high-resolution spectroscopy on water vapor [50]. This particular set of photomixers had sufficient THz power to measure the water vapor spectrum from 1.15 to 1.5 THz, as given in Fig. 27. Figure 28 shows successive measurements of the transmission at 1.411 THz through a 50 cm long gas cell containing water vapor at pressures varying between 0.1 and 1 Torr (from top to bottom). Note the narrowing of the line that occurs as the pressure is reduced, and the ability of the transceiver to resolve it.

Recently, an on-chip photomixer transceiver was demonstrated, where the two photomixers were coupled by a coplanar waveguide rather than through free space [51]. An on-chip transceiver is especially useful for studying the frequency dependence of the behaviour of the photoconductor and interdigitated electrodes, since propagation along a well-designed transmission line is mostly linear phase and can be de-embedded more accurately than with quasi-optical systems. Analysis of the data from the on-chip transceiver confirmed the predictions of the model described by Zamdmer et al. for the electric-field dependence of the photocarrier lifetime [13].

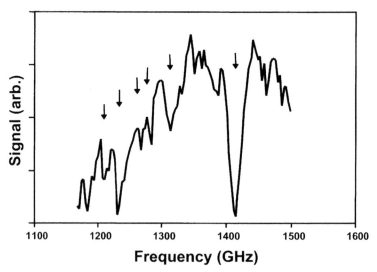

Fig. 27. Signal measured by the transceiver with a 65 cm air path between transmitter and receiver. *Arrows* mark location of absorption lines due to water molecules

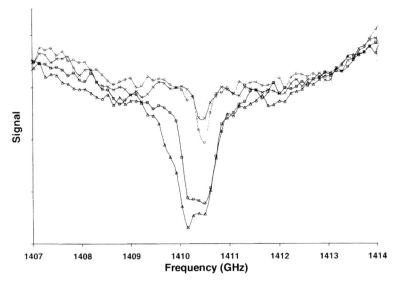

Fig. 28. Spectra measured with a water vapor pressure varying from 0.1 Torr to 1 Torr in a 50 cm-long gas cell

Acknowledgments

This work was supported under Contract No. F19628-00-C-0002. Opinions, interpretations, conclusions, and recommendations are those of the authors and are not necessarily endorsed by the United States Air Force.

References

1. E. R. Brown, F. W. Smith, K. A. McIntosh, "Coherent millimeter-wave generation by heterodyne conversion in low-temperature-grown GaAs photoconductors", *J. Appl. Phys.* **73**, 1480, (1993).
2. K. A. McIntosh, E. R. Brown, K. B. Nichols, O. B. McMahon, W. F. Dinatale, T. M. Lyszczarz, "Terahertz photomixing with diode lasers in low-temperature-grown GaAs", *Appl. Phys. Lett.* **67**, 3844, (1995).
3. S. Verghese, K.A. McIntosh, E.R. Brown, "Highly tunable fiber-coupled photomixers with coherent terahertz output power", *IEEE Tran. Microwave Theory Tech.* **45**, 1301, (1997).
4. S. Verghese, E. K. Duerr, K. A. McIntosh, S. M. Duffy, S. D. Calawa, C.-Y. E. Tong, R. Kimberk, and R. Blundell, "A photomixer local oscillator for a 630-GHz heterodyne receiver". *IEEE Microwave Guided Wave Lett.* **9**, 245, (1999).
5. A. S. Pine, R. D. Suenram, E. R. Brown, and K. A. McIntosh, "A terahertz photomixing spectrometer: Application to SO_2 self broadening", *J. Mol. Spectrosc.* **175**, 37, (1996).

6. P. Chen, G. A. Blake, M. C. Gaidis, E. R. Brown, K. A. McIntosh, S. Y. Chou, M. I. Nathan, F. Williamson, "Spectroscopic applications and frequency control of submillimeter-wave photomixing with distributed-Bragg-reflector diode lasers in low-temperature-grown GaAs", *Appl. Phys. Lett.* **71**, 1601, (1997).

7. E. R. Brown, S. Verghese, K. A. McIntosh, "Terahertz photomixing in low-temperature-grown GaAs", *Proc. SPIE* **3357**, 132, (1998).

8. N. Erickson, "Diode frequency multipliers for THz local oscillator applications", *Proc. SPIE* **3357**, 75, (1998).

9. B. Xu, Q. Hu, M. R. Melloch, "Electrically pumped tunable terahertz emitter based on intersubband transition", *Appl. Phys. Lett.* **71**, 440, (1997).

10. R. Köhler, A. Tredicucci, F. Beltram, H. E. Beere, E. H. Linfield, A. G. Davies, D. A. Ritchie, R. C. Iotta, and F. Rossi, "Terahertz semiconductor heterostructure laser", *Nature*, **417**, 156 (2002).

11. R.H. Kingston, *Detection of Optical and Infrared Radiation* Springer, Berlin, Heidelberg, (1978). vol. 10.

12. K. A. McIntosh, K. B. Nichols, S. Verghese, E. R. Brown, "Investigation of ultrashort photocarrier relaxation times in low-temperature-grown GaAs", *Appl. Phys. Lett.* **70**, . 354, (1997).

13. N. Zamdmer, Q. Hu, K. A. McIntosh, S. Verghese, "Increase in response time of low-temperature-grown GaAs photoconductive switches at high voltage bias", *Appl. Phys. Lett.* **75**, 2313, (1999).

14. N. Zamdmer, *The Design and Testing of Integrated Circuits for Submillimeter-Wave Spectroscopy*, PhD dissertation, Massachusetts Institute of Technology, (1999).

15. T. H. Ning, "High-field capture of electrons by Coulomb-attractive centers in silicon dioxide", *J. Appl. Phys.* **47**, 3203, (1976).

16. E. R. Brown, "A photoconductive model for superior GaAs THz photomixers", *Appl. Phys. Lett.* **75**, 769, (1999).

17. S. U. Dankowski, D. Streb, M. Ruff, P. Kiesel, M. Kneissl, B. Knupfer, G. H. Dohler, U. D. Keil, C. B. Sorenson, A. K. Verma, "Above band gap absorption spectra of the arsenic antisite defect in low temperature grown GaAs and AlGaAs", *Appl. Phys. Lett.* **68**, 37, (1996).

18. C. J. Johnson, G. H. Sherman, R. Weil, "Far infrared measurement of the dielectric properties of GaAs and CdTe at 300 K and 8 K", *Appl. Opt.* **8**, 1667, (1969).

19. A. W. Jackson, *Low-Temperature-Grown GaAs Photomixers Designed for Increased Terahertz Output Power*, PhD dissertation, University of California, Santa Barbara, (1999).

20. J. J. Kuta, H. M. van Driel, D. Landheer, J. A. Adams, "Polarization and wavelength dependence of metal-semiconductor-metal photodetector response", *Appl. Phys. Lett.* **64**, 140, (1994).

21. E. Chen, S. Y. Chou, "Polarimetry of thin metal transmission gratings in the resonance region and its impact on the response of metal-semiconductor-metal photodetectors", *Appl. Phys. Lett.* **72**, 2673, (1997).

22. E. R. Brown, K. A. McIntosh, F. W. Smith, K. B. Nichols, M. J. Manfra, C. L. Dennis, J. P. Mattia, "Milliwatt output levels and superquadratic bias dependence in a low-temperature-grown GaAs photomixer", *Appl. Phys. Lett.* **64**, 3311, (1994).

23. S. Verghese, K. A. McIntosh, E. R. Brown, "Optical and terahertz power limits in low-temperature-grown GaAs photomixers", *Appl. Phys. Lett.* **71**, 2743, (1997).
24. A. W. Jackson, J. B. Ibbetson, A. C. Gossard, U. K. Mishra, "Reduced thermal conductivity in low-temperature-grown GaAs", *Appl. Phys. Lett.* **74**, 2325, (1999).
25. E. K. Duerr, K. A. McIntosh, S. M. Duffy, unpublished results.
26. R. E. Simons, V. W. Antonetti, W. Nakayama, S. Oktay, "Heat transfer in electronic packages", in *Microelectronics Packaging Handbook* ed. R.R. Tummala, E. J. Rymaszewski, and A. G. Klopkenstein, pp. 314–398, 2nd ed. (Chapman & Hall, New York, 1997).
27. S. M. Duffy, S. Verghese, K. A. McIntosh, A. W. Jackson, A. C. Gossard, S. Matsuura, "Accurate modeling of dual dipole and slot elements used with photomixers for coherent terahertz output power", *IEEE Trans. Microwave Theory Tech.*, vol. 49, pp. 1032–1038, June 2001.
28. S. Matsuura, G. A. Blake, R. A. Wyss, J. C. Pearson, C. Kadow, A. W. Jackson, A. C. Goddard, "Free-space traveling-wave THz photomixers", *Proceedings of the 1999 IEEE Seventh International Conference on Terahertz Electronics*, Nara, Japan, (1999), p. 24
29. S. Matsuura, G. A. Blake, R. A. Wyss, J. C. Pearson, C. Kadow, A. W. Jackson, A. C. Gossard, "A traveling-wave THz photomixer based on angle-tuned phase matching", *Appl. Phys. Lett.* **74**, 2872, (1999).
30. E. K. Duerr, K. A. McIntosh, S. M. Duffy, S. D. Calawa, S. Verghese, C. Y. E. Tong, R. Kimberk, R. Blundell, "Demonstration of a 630-GHz photomixer used as a local oscillator", *Digests of the 1999 IEEE MTT-S International Microwave Symposium*, Anaheim, CA, (1999), p. 127.
31. T. Ishibashi, H. Fushimi, T. Furuta, H. Ito, "Uni-traveling-carrier photodiodes for electromagnetic wave generation", *Proceedings of the 1999 IEEE Seventh International Conference on Terahertz Electronics*, Nara, Japan, (1999), p. 36.
32. H. Ito, T. Furuta, S. Kodama, N. Watanabe, T. Ishibashi, "InP/InGaAs uni-traveling-carrier photodiode with 220 GHz bandwidth", *Electron. Lett.* **35**, 1556, (1999).
33. Y.-J. Chui, S. B. Fleischer, J. E. Bowers, "High-speed low-temperature-grown GaAs p–i–n traveling-wave photodetector", *IEEE Photonics Techol. Lett.* **10**, 1012, (1998).
34. Y. C. Lim R. A. Moore, "Properties of alternately charged coplanar parallel strips by conformal mappings", *IEEE Trans. Electron Devices* **15**, 173, (1968).
35. M. Kominami, D. M. Pozar, and D. H. Schaubert, "Dipole and slot elements and arrays on semi-infinite substrates", *IEEE Trans. Antennas Propag.* **33**, 600, (1985).
36. D. F. Filipovic, W. Y. Ali-Ahmad, G.M. Rebeiz, "Millimeter-wave douple-dipole antennas for high-gain integrated reflector illumination", *IEEE Trans. Microwave Theory Tech.* **40**, 962, (1992).
37. P. F. Goldsmith, "Quasi-optical techniques", *Proc. IEEE* **80**, 1729, (1992).
38. D.M. Pozar, *Microwave Engineering*, 2nd ed. (Wiley, New York, 1998).
39. C. A. Balanis, *Antenna Theory: Analysis and Design*, 2nd ed. (Wiley, New York, 1997).
40. K. C. Gupta, R. Garg, R. Chadha, *Computer-Aided Design of Microwave Circuits*, (Artech House, Dedham, MA, 1981).

41. D. B. Rutledge, M. S. Muha, "Imaging antenna arrays", *IEEE Trans. Antennas Propag*, **30**, 535, (1982).

42. G. M. Rebeiz, "Millimeter-wave and terahertz integrated circuit antennas", *Proc. IEEE*. **80**. 1748, (1992).

43. A. Skalare, T. De Graauw, H. van de Stadt, "A planar dipole array antenna with an elliptical lens", *Microwave Opt. Technol. Lett*. **4**, 9, (1991).

44. D. F. Filipovic, S. S. Gearhart, and G. M. Rebeiz, "Double-slot antennas on extended hemispherical and elliptical silicon dielectric lenses", *IEEE Trans. Microwave Theory Tech*. **41**, 1738, (1993).

45. D. F. Filipovic, *Analysis and Design of Dielectric-Lens Antennas and Planar Multiplier Circuits for Millimeter-Wave Applications*, PhD dissertation. University of Michigan, Ann Arbor, (1995).

46. D. Grischkowsky, S. Keiding, M. van Exter, C. Fattinger, "Far-infrared time-domain spectroscopy with terahertz beams of dielectrics and semiconductors", *J. Opt. Soc. Am. B* **7**, 2006, (1990).

47. T.-I. Jeon, D. Grischkowsky, "Nature of conduction in doped silicon", *Phys. Rev. Lett*. **78**, 1106, (1997).

48. S. Matsuura, unpublished data.

49. S. Verghese, K. A. McIntosh, S. D. Calawa, W. F. DiNatale, E. K. Duerr, K. A. Molvar, "Generation and detection of coherent terahertz waves using two photomixers", *Appl. Phys. Lett*. **73**, 3824, (1998).

50. S. Verghese, K. A. McIntosh, E. K. Duerr, S. M. Duffy, L. J. Mahoney, S. D. Calawa, "Terahertz spectroscopy of water vapor using a photomixer transceiver", *Proceedings of the 1999 IEEE Seventh Intl. Conf. on Terahertz Electronics*. Nara, Japan, (1999), p. 89.

51. N. Zamdmer, Q. Hu, K. A. McIntosh, S. Verghese, A. Forster, "On-chip frequency-domain submillimeter-wave transceiver", *Appl. Phys. Lett*. **75**, 3877, (1999).

Applications of Optically Generated Terahertz Pulses to Time Domain Ranging and Scattering

R. Alan Cheville, Matthew T. Reiten, Roger McGowan, and Daniel R. Grischkowsky

1 Introduction

This chapter focuses on the use of THz time-domain techniques for the measurement of time-resolved electromagnetic scattering. Electromagnetic scattering is a vast field owing to its application in a wide range of measurement techniques in addition to commercial and military radar. The vast majority of treatments look at scattering in the frequency domain. Owing to the high bandwidth, the phase coherence, and our ability to directly measure the electromagnetic field with subpicosecond resolution, optically generated THz-bandwidth pulses provide a valuable new method to investigate fundamental scattering mechanisms. This introductory section provides a brief background on electromagnetic scattering, and attempts to provide a perspective on the application of THz time-domain techniques in this broad field.

1.1 Perspective

The spectacular optical effects of the rainbow and the lesser-known glory have stimulated comment and explanation throughout much of recorded history [1] and provided early impetus to an understanding of electromagnetic scattering. A mathematical description of wave scattering from spherical particles was performed initially by Clebsch in 1861 [1], with the best-known work being that of Mie in 1908. Major technological impetus was given to studies of electromagnetic scattering by the development of sources of radio-frequency radiation in the early years of the 20th century. The adaptation of radio-frequency scattering for the detection and identification of objects at a distance received great attention during the Second World War [2]. The commercial and military importance of radio detection and ranging (RADAR) has led to intensive and ongoing technical and scientific investigation for applications ranging from biomedical imaging [3] to sensitive Doppler techniques for law enforcement and military target acquisition.

Outside the radio-frequency range, electromagnetic scattering encompasses a huge range of disciplines and applications, including light-scattering measurements of polymers and other macromolecules in chemistry [4]. Although both optical and radio-frequency scattering calculations begin from Maxwell's equations, work at optical and near-infrared frequencies rarely cites work on scattering in the radio and microwave spectral regions. This division is due to several factors, including the inability to directly measure phase information at optical frequencies, the difficulty of performing scattering measurements in the transition region between optical and radio frequencies, and different scientific goals, especially in the defense-oriented radar community.

1.2   Theory

Whether one is approaching scattering from the optical or the electronic point of view, the mathematical complexity is formidable. In fact, the scattering of electromagnetic radiation can be analytically calculated only in a small fraction of cases. The most familiar case is for simple objects with spherical or cylindrical geometry (Mie scattering). The Mie solution is obtained by a boundary method where the total field is broken into an incident and a scattered component and solved using the boundary conditions on the scatterer. The solution relies on excitation by a single-frequency plane wave, and complex incident phase fronts and pulse shapes are treated by Fourier expansion into the plane wave components. Analytical treatment of scattering is further complicated by the fact that the scattered radiation is typically aspect dependent – the scattered signal depends on the orientation of the scatterer relative to the k vectors of the incident and detected fields. The vast majority of scattering problems have no analytical solution, and are treated either by an approximate solution in the long- and short-wavelength limits, by numerical and computational techniques, or by experimental measurements.

The two regimes in which approximate solutions are found are when wavelengths are much larger (Rayleigh approximation) or much smaller (geometrical- or physical-optics approximation) than the scatterer. In the well-known Rayleigh approximation the scattered intensity varies as λ^{-4} and electrostatic approximations to the field can be used [4]. For wavelengths very small compared with features on the object, the geometrical-optics approximation is analogous to ray tracing. This regime also sees application of the physical-optics approximation based on the geometrical theory of diffraction [5], which treats the target as a series of flat surfaces.

These approximate solutions break down when the scatterer or, to use terminology from the radar community, the target, has dimensions or features on the order of a wavelength (the resonance region). To accurately determine the scattering in this case, numerical calculations are necessary. The recent progress in computational methods [6] is due largely to the increase in computing power, since numerical calculations become more computationally intensive as the size of the scatterer increases. The computational resources needed to accurately determine the scattered field typically scale with the size of the target expressed in wavelengths to the fourth power or more [6]. Even with the stunning Moore's Law increase in computational power in recent years, accurate scattering calculations for large targets or high frequencies are in many cases not feasible.

Numerical techniques include finite-difference time-domain (FDTD) techniques [7], which iteratively solve Maxwell's equations at discrete time and spatial points. Definitions of boundaries for the problem are critical as are correct definitions of each volume element, which must be small compared with a wavelength. The requirement of small volume elements and time steps leads to unreasonable computation times for targets large compared with a wavelength. The method of moments (MoM) [8] uses surface elements and calculates the interactions between each element simultaneously in the frequency domain. The element size must be small compared with the wavelength for accurate calculation, and the computational overhead scales as the sixth power of the target size. Each of these techniques is applicable in certain cases, and applications of the various computational techniques to problems of interest have generated a great deal of literature.

Predicting the scattering response either analytically or computationally from a given target, or forward scattering, is only part of the technical problem. The use of scattering as a remote-sensing tool is further dependent upon the ability to identify objects via a unique time or frequency dependence of the scattered radiation. This "signature" is due to many factors, illustrated in Fig. 1, determined by target composition, geometry, and environmental factors. The scattering is highly orientation dependent; the term "monostatic" is used to describe scattered radiation measured in the backscattering direction, and "bistatic" is used if the angle between the propagation vectors of the incident and scattered radiation is nonzero. In most scattering cases

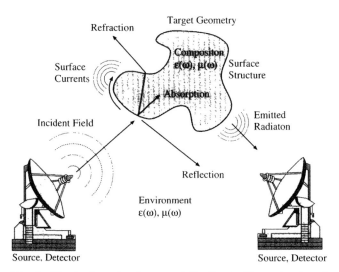

Fig. 1. Physical mechanisms contributing to the scattering signature $\varepsilon(\omega)$ and $\mu(\omega)$ are the frequency-dependent permittivity and permeability, respectively

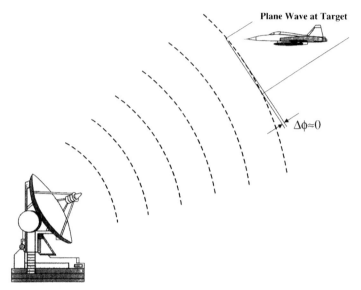

Fig. 2. Targets located a long distance from radar sources have phase fronts which can be approximated by a plane wave. The phase front curvature is measured by the phase variation, $\Delta\phi$, of the electric field over the total extent of the target

of interest the source is a long distance from the target, and the incident radiation is assumed to be a plane wave. The phase variations of the incident wavefront across the target are then less than a single cycle, so $\Delta\phi \ll 2\pi$, as illustrated in Fig. 2. A general algorithm for target identification would require computing scattering signatures at a variety of orientations, or aspects of all possible targets. The computational burden of this is formidable, and practical target identification often relies on libraries of known signatures and great importance is placed on identifying characteristic signatures which are aspect independent.

The simplification of this most general problem is highly desirable and relies on unique resonance frequencies exhibited by a particular target. Resonances are excited within and between scattering centers on the target; an example is the cylindrical cavity of a jet engine. Since the resonant frequencies are determined solely by the characteristics of the target itself, they are, to some extent, aspect independent. Scattering due to these resonances is called the "late-time response".

As discussed theoretically as early as the 1950s [9–11], a potentially simple way to identify complex targets is by the late-time response measured in the time domain after illumination by a short pulse rather than by a CW source. Advantages include the ability to isolate scattering mechanisms in time, the inherent broad bandwidth and phase coherence of pulsed excitation, and the ability to conveniently handle nonlinearities in calculations [12]. For target identification purposes, a short, broad-bandwidth pulse is initially specularly reflected from the target. However, the pulse also excites the target resonances which will radiate energy at later times. Temporal resolution permits the isolation of the late-time response. The use of broad-bandwidth pulses in conjunction with the temporal resolution of short-pulse radar gives insight into the physical characteristics of the target not obtainable with single-frequency CW sources, owing to the phase relationships between the scattered frequencies. The specular reflection and the late-time response can be understood through an analogy of hitting a bell with a hammer. The initial sharp "clang" is the specular reflection, while the "ringing" of the bell is the late-time response. The frequency of the emitted sound waves, as well as the damping time, gives information about physical attributes of the bell.

This ability to temporally isolate individual scattering mechanisms not only aids in target identification, but also permits a deeper understanding of the physical mechanisms of scattering. This potential insight into scattering processes has provided impetus for performing electromagnetic-scattering measurements in the time domain. Important factors for doing this are high bandwidth and phase coherence to achieve short excitation pulses, and the ability to directly measure the electromagnetic field with good time resolution.

1.3   Measurements

In both frequency- and time-domain target identification methods it is important to compare theoretical predictions with data from actual scattering experiments. Since the emphasis in THz impulse ranging is on experimental measurements of scattering, we briefly discuss experimental measurement techniques for scattering.

The most straightforward are outdoor ranges using actual targets [13]. The targets can be in motion or suspended on pylons or by cranes. This technique has the advantage that targets are measured as they will be detected by an actual radar system and effects due to nonideal targets such as condensation or ice buildup, weather, and the effects of paints can be determined. The drawbacks to these measurements are that they are time-consuming and expensive and that noncooperative (i. e. the enemy's) targets cannot in general be studied.

Indoor range measurements on full-size targets have also been attempted, however these can be quite cost prohibitive. In addition, for measurements to be comparable to those made in the field, the field incident on the target must be a plane wave, and compact range techniques [14] must be applied. A compact range uses an optical element of some type, typically a parabolic reflector, to generate a plane wave in a specified region where measurements are to be made. The region where the field approximates a plane wave is known as the "quiet zone", and typical reflector sizes required are two to three times the size of the quiet zone. Hence, for full-size targets very large reflectors are required, which must be designed with subwavelength accuracy across tens of meters. In addition, for CW or long-pulse measurements, anechoic shielding is required to prevent errors due to multiple reflections off the walls or other structural parts of the indoor range.

Since it is often impractical to obtain ranging data on full-scale targets, geometrically scaled targets are often used in conjunction with theoretical models and measurements from actual targets [15]. Such scale-model techniques have been used since the 1940s to obtain information about electromagnetic scattering [16]. The use of scale models to predict the response of full-size targets is valid owing to the scalability of Maxwell's equations. For model targets, the scattered electric and magnetic fields are linearly scaled compared with a real system, but the ratio, E/H, remains constant. Under these conditions measurements on scaled targets are directly comparable to those on full-size targets given several constraints [17], the most important of which are due to material properties. These constraints will be discussed in more detail in Sect. 3. Model measurements are particularly useful for the task of designing "stealthy" targets when the cost of an iterative design–model–measure cycle on a full-size target is prohibitive.

Before reviewing previous pulsed-scattering measurements, it is important to note that the distinction between pulsed and CW techniques is somewhat blurred. Many quasi-CW techniques use frequency-modulated pulses for EM scattering measurements to obtain the approximate position of the target through time of flight, and to concentrate more energy in the radar excitation pulse. Here we focus on what are typically known as ultrawide-band (UWB) radar pulses with bandwidths at least 25% of the center frequency. From an optics point of view, this means that the slowly-varying-amplitude approximation is invalid. The pulse cannot be considered an envelope whose temporal variation is slow compared with a carrier frequency.

The earliest UWB time-domain scattering measurements were performed in the 1960s and provided much impetus for the theoretical work. Here electronic pulse generation techniques utilizing high-speed electronics and pulse-forming lines covered bandwidths of several GHz . By the 1980s transient scattering work on indoor ranges was being used for both target measurement and antenna characterization. An overview with many references is given in [12]. Recent electronics-based time-domain scattering measurements have extended the frequency range to 50 GHz [18]. Extensive literature exists on measurements that use impulsive time-domain ranges obtained with electronic pulse generation methods to measure the impulse response of targets [19–22].

While electronically generated pulses have demonstrated features on the scale of tens of picoseconds, electromagnetic pulse generation using optically gated switches has demonstrated significantly faster rise times. Electromagnetic pulses contain observed features on timescales of hundreds of femtoseconds [23], leading to a two-order-of-magnitude increase in bandwidth. Although such work was not directly related to THz measurements of scattering, optically gated switches have been extensively developed for various radar applications. Photoconductive switches have demonstrated rise times of 10^{15} A/s, peak powers of over 100 MW, and lifetimes of over 2 million shots when optically gated with 40 μJ per MW of switched power [24].

Optoelectronic generation of extremely broad-bandwidth pulses has been used in the laboratory by several researchers for electromagnetic scattering measurements. Robertson has measured transient scattering from Fabry–Pérot interferometers, gratings, and cylinders using pulses with bandwidths extending from 15 to 140 GHz [25,26]. Optically gated horn antennas with a bandwidth extending from 5 to 70 GHz have been used to measure scattering from dielectric and metal spheres and aluminum strips [27–29]. Similar measurements have also been used to extract resonances from cylindrical cavities [30]. The bandwidth available for ranging measurements has been extended well beyond the range of conventional electronic techniques using micron-scale dipole antennas gated with femtosecond pulses [31,32]. Here bandwidths extend from 100 GHz to beyond 1 THz, with time resolutions less than 1 picosecond.

It should be noted that the extension of THz bandwidth pulses to ranging measurements is currently a laboratory application. Field use of frequencies in the millimeter and submillimeter regime is limited by strong absorption by atmospheric water vapor and oxygen. This attenuation is especially severe in inclement weather, as shown in Fig. 3.

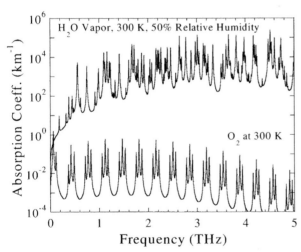

Fig. 3. Calculated power absorption coefficients of atmospheric water vapor and molecular oxygen. In addition the absorption coefficient of rain, depending on rain-fall rate, is on the order of 9 to 12 km^{-1}

1.4   Outline

The application of experimental methods based on THz time-domain spectroscopy (THz-TDS) for measurement of electromagnetic scattering in the laboratory is the topic of the rest of this chapter. Section 2 covers experimental methods. Calculations and theory are covered in Sect. 3. Section 4 presents measurements performed to date and discusses applications and current issues. The chapter concludes with a discussion of future directions and applications of THz electromagnetic scattering and of issues which need to be addressed in order for this technology to see wide application.

2   Experiment

As represented in the radar master equation in Sect. 3, a ranging system requires a source of radiation, a scatterer, and a method to detect the scattered radiation. This section provides an overview of THz impulse ranging, with specific focus on the details of THz radiation generation, propagation, scattering, and detection. The experimental methods are discussed in the context of THz time-domain spectroscopy, in which optoelectronic-based THz generation and detection techniques measure nearly-single-cycle transients of THz radiation whose electric field can be directly measured in the time domain with subpicosecond temporal resolution.

2.1   Overview of Experimental Configurations

A variety of experimental configurations can be used for THz impulse scattering. The basic configuration is a modification of the system used for THz time-domain spectroscopy [33,34], as shown in Fig. 4. Here deflecting mirrors steer the THz beam onto a target, and then steer scattered radiation to the THz detector. This configuration has the advantage that the deflecting mirrors can be removed for characterization of target materials via THz-TDS.

The THz source is placed at the focal distance of an off-axis parabolic mirror, which serves to collect and collimate the beam. The THz beam is deflected to the target by a flat mirror. Metal mirrors have nearly 100% reflectivity in the THz region of the spectrum [35]. For the scattering measurements presented here, the collimated THz beam propagates over a distance on the order of 50–100 cm and is incident upon a target, which then scatters the incident radiation over a 4π solid angle. The amplitude, phase, and polarization of the scattered field are dependent on the target composition and orientation, the angle at which the scattered radiation is observed, and the frequency and polarization of the incident radiation. Ranging measurements typically use two configurations. A monostatic ranging setup is one where the scattered radiation is detected along the same path as the incident radiation   i.e. the source and detector are co-located. Bistatic ranging is

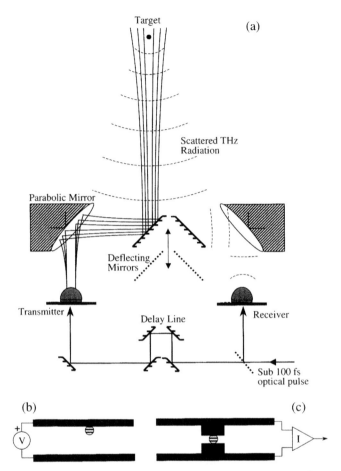

Fig. 4. (a) Schematic representation of experimental configuration for THz impulse ranging. The two-headed arrow near the deflecting mirrors represents the fact that these mirrors can be removed to aid system alignment. (b) Representation of the semi-insulating GaAs THz source chip. (c) Dipole antenna structure fabricated on silicon on sapphire, used in the THz detector

performed with a nonzero angle between the k vectors of the incident and detected scattered radiation. The configuration of Fig. 4a has a bistatic angle of approximately 11 degrees for a target distance of 67 cm.

The scattered radiation is collected by a second parabolic mirror, which directs it to the silicon lens attached to the receiver. The diameter of the parabolic mirror used to collect the radiation in the particular configuration described here is 63.5 mm, corresponding to a solid angle of 13×10^{-3} srad for a target 50 cm from the parabolic mirror. The actual solid angle collected is strongly frequency dependent, and at high frequencies the collection angle is smaller since the directivity of the THz detector is very high. From this it can be seen that for targets which scatter radiation isotropically only a small fraction of the power incident on the target will be collected by the detector.

The impulse range is contained in an airtight box, which is purged with dry air to minimize interference by water vapor. Seventeen strong water vapor absorption lines occur in the region of the spectrum from 0.1 to 2 THz with absorption coefficients as high as 1 cm^{-1} at 100% relative humidity and STP (see Fig. 3). With path lengths exceeding 1 m, water vapor levels of less than 30 parts per million can lead to signals which interfere with scattering measurements.

The goal of the THz impulse-ranging program at Oklahoma State University is to design and characterize a table-top impulse range to perform quantitative scattering measurements which can be compared with measurements obtained at facilities using radars at MHz and GHz frequencies. We now discuss in detail some experimental design parameters which permit such calibrated measurements to be performed.

2.2 Generation and Detection of Terahertz Electromagnetic Transients

A schematic of the THz ranging system is shown in Fig. 4a. The optical pulse train is generated from a mode-locked Ti:sapphire laser operating with a center wavelength of 820 nm with pulse widths of 50 fs. The pulse train is divided into two beams, one of which goes through a mechanical delay line to provide control over the relative timing. The beams are attenuated to approximately 10 mW before being focused on the THz source and detector antenna structures. Short-focal-length lenses are used to achieve a focused spot size of 5–10 μm diameter.

THz pulses are generated using the antenna structure shown in Fig. 4b. The antenna structure consists of two 10 μm wide coplanar transmission lines separated by 80 μm fabricated on semi-insulating GaAs, and DC biased in the range 60 to 100 V. Irradiating the metal–semiconductor interface (edge) of the positively biased line with focused ultrafast laser pulses produces synchronous bursts of THz radiation. The incident laser pulses create a large number of photocarriers in a region of extremely high (trap enhanced) electric field [36]. The consequent acceleration of the carriers generates the burst of radiation. The major fraction (proportional to $n^3 : 1$, where $n = 3.6$ in GaAs) of the laser-generated burst of THz radiation is emitted into the GaAs substrate in a cone normal to the interface [37]. The THz radiation is then collected and collimated by a crystalline silicon lens attached to the back surface of the chip. The lens structure used is a truncated 9.5 mm diameter silicon sphere, cut such that the ultrafast antenna is located at the focus of the lens. The resulting THz beam has been measured to have a Gaussian profile with a beam waist at the silicon lens [38]. Details of beam propagation are discussed in Sect. 2.3

The scattered THz beam is detected by a similar arrangement of a silicon lens attached to a photolithographically fabricated antenna structure. The antenna geometry of the THz detector is shown in Fig 4c. The antenna is fabricated on an ion-implanted silicon-on-sapphire (SOS) wafer. The 20 μm wide antenna structure is located in the middle of a 20 mm long coplanar transmission line consisting of two parallel 10 μm wide, 0.5 μm thick, 5 Ω/mm, aluminum lines. The separation of the lines, or size of the dipole, determines, to a large extent, the frequency response of the detector. A typical separation is 30 μm which provides an optimal compromise between responsiveness (large dipole) and bandwidth (small dipole).

The electric field of the incoming THz radiation is focused by the silicon lens on the 5 μm gap between the two arms of this receiving antenna. The electric field of the THz pulse induces a transient bias voltage across the antenna. When an optical pulse is incident on the gap between the dipole arms a current is produced which is proportional to the electric field of the THz pulse; this is measured using a low-noise current amplifier. The amplitude and time dependence of this transient voltage are obtained by measuring the collected charge (average current) versus the time delay between the THz pulse and the delayed optical-gating pulse. The detection process with gated integration can be considered as a subpicosecond boxcar integrator. Unlike the situation with purely electronic pulse generation techniques, there is no observable time jitter between the THz pulses and the optical sampling pulses gating the receiver, since the same optical pulse train is used both to generate the THz radiation and to gate the receiver.

A THz pulse measured with the system described above, using a flat mirror to reflect the pulse towards the detector, is shown in Fig. 5. The measured pulse shape can be reasonably well fit by a double Gaussian. The extended structure after the pulse was due to the effects of absorption and dispersion by atmospheric water vapor in the beam path. The pulse is measured using a lock-in amplifier connected to the output of the current amplifier and modulating the generated THz beam with an optical chopper. The amplitude spectrum of the incident pulse obtained by means of a numerical fast Fourier tranform is shown in Fig. 5b. The short temporal duration of the pulses results in a bandwidth that extends from approximately 100 GHz to 2 THz with an FWHM of 0.75 THz. The sharp absorption lines are due to atmospheric water vapor [39,40].

The subpicosecond optical pulse and correspondingly fast detector response permit THz impulse ranging to gate out unwanted reflections. These include reflections off table surfaces, target support structures, and the walls of the enclosure which result in a background signal in CW or quasi-CW measurements [18]. A background-free scattering scattering signal obtained is illustrated in Fig. 6. The scattered THz signal from a 0.5 mm diameter copper cylinder is shown in the upper trace, followed by a scan with the cylinder removed. Shown in the inset to Fig. 6 is the relative scattered signal

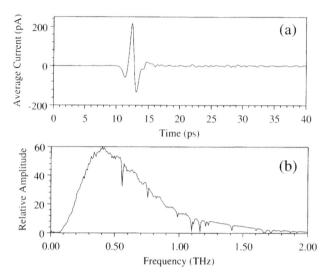

Fig. 5. (a) THz pulse measured on the impulse range shown in Fig. 4 using a large aluminum mirror as the target. (b) Numerical Fourier transform of pulse shown in (a) illustrating bandwidth of system. The sharp absorption lines are due to atmospheric water vapor

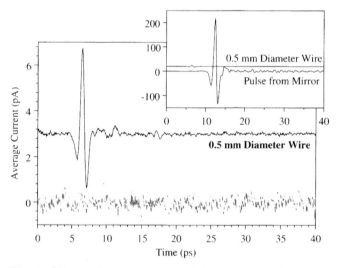

Fig. 6. Measured scattering signal from 0.5 mm diameter copper cylinder (*upper trace*) compared to measured background signal with cylinder removed (*lower trace*). The background signal has been expanded 10× for clarity. The inset shows the pulse scattered from the 0.5 mm cylinder in comparison with the pulse of Fig. 5a reflected from an aluminum mirror

strength from the 0.5 mm cylinder compared with a flat mirror. The background measurement of Fig. 6 appears unchanged if the incident THz pulse train is blocked.

The bandwidth of the impulse ranging system is dependent upon the dipole size used in the detector, which was 30 μm for these measurements. Smaller dipoles permit higher-frequency response with a decreased dynamic range, while larger dipoles give a bigger signal at the expense of bandwidth. A relative indication of the effect of dipole size on the usable bandwidth is given in Fig. 7a. Here the plotted bandwidth range is that of amplitude greater than 10% of peak. This gives a good "rule of thumb" range over which useful data can be collected. The horizontal bars at the top of Fig. 7a indicate the approximate useful range of each dipole, while the solid line and points show the theoretical dipole response. The high-frequency response of the 10 μm dipole is limited by the carrier lifetime to less than the theoretical bandwidth.

The size of the dipole used as a detector also determines the observed noise. since this is inversely proportional to the average resistance of the

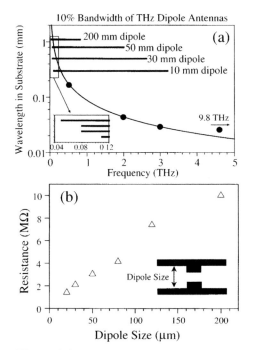

Fig. 7. (a) The bars show the approximate measured 10% bandwidth of dipole antenna structures used in THz impulse ranging. The curve shows the wavelength in the substrate at which interference limits the frequency response from a Hertzian dipole of the size indicated. (b) Measured open-circuit resistance of dipole antenna structures fabricated on SOS

receiver chip as seen by the current amplifier [41]. In the off state this resistance is typically $1\,M\Omega$ and drops to approximately $500\,\Omega$ for $0.6\,ps$ when the receiver is optically gated. The off-state resistance is to a large extent determined by the spacing between the transmission lines. Measured resistances are shown in Fig. 7b. For the measurement shown in Fig. 5a, the observed noise in front of the main pulse is $0.15\,pA$ RMS for a lock-in integration time of $100\,ms$. An identical noise value is obtained when the THz beam is completely blocked. The signal-to-noise ratio (S/N) in this 8.5 minute scan consisting of 2048 data channels is greater than 1000:1. Here we define the signal-to-noise ratio as the peak measured THz signal ($t = 12\,ps$) in pA divided by the RMS noise measured before the arrival of the THz pulse ($t = 5\,ps$).

The high signal-to-noise ratio demonstrated by this system is key to performing electromagnetic scattering measurements. The THz detectors used here have been shown capable of detecting THz radiation with an average power of $10\,nW$ with signal-to-noise ratios of approximately 10 000:1. The discrepancy between this optimal S/N and the measured S/N in ranging measurements is due to the nonideal beam coupling between the THz source and detector. This is discussed further in Sect 3.2. Calibration of the receiver dipole antenna structure has been performed [41] for a receiver of the same type as used in these ranging studies. The approximate calibration factor, which is alignment dependent, is such that a measured current of $1\,nA$ corresponds to a peak THz power level of $200\,\mu W$. The average power level is $10\,nW$. The results shown in Fig. 5 were obtained with an integration time of $100\,ms$, corresponding to an integration bandwidth of $1.6\,Hz$. Consequently, the demonstrated detection limit is $10^{-16}\,W$, for a signal-to-noise ratio of unity and a bandwidth of $1.6\,Hz$. This compares with detection limits of $10^{-13}\,W$ for liquid helium cooled bolometers, which are used as incoherent detectors in this frequency region. Good radio receivers often have detection limits as low as $10^{-19}\,W$ [41].

Another concern of importance to ranging measurements is the polarization of the electromagnetic field incident on the target, and the ability of the detector to distinguish the polarization of the scattered radiation. The source–detector combination has a polarization sensitivity of greater than 25:1 along the orientation of the dipole. The polarization of both the incident and the detected beams can be improved to extinction ratios of over 300:1 by the addition of wire grid polarizers. Simple rotation of the THz sources and polarizers allows both horizontal (H) and vertical (V) polarizations to be incident upon the target. Similarly, the detector can be configured to measure scattered waveforms with either H or V polarization. Thus THz ranging measurements can measure both the same polarization as the excitation beam (H source, H detector (HH) or, via rotation, VV) and cross-polarization terms (HV, VH).

2.3   Terahertz Beam Optics for Target Illumination

The role of the optics in the THz beam generated by the THz source is to produce, as closely as possible, a uniform plane wave at the target. This requires both a uniform intensity distribution and a planar phase front at the target. The THz beam generated by the source has been shown to approximate a Gaussian beam with a $1/e$ amplitude waist at the output face of the silicon lens [38,42]. We assume a Gaussian field distribution with a frequency-independent waist diameter of 5 mm over the bandwidth of the THz pulse. The broad-bandwidth THz pulse propagates out from the effective aperture of the THz source silicon lens, with each frequency component diverging owing to diffraction. The assumption of a Gaussian beam profile permits a description in terms of ray matrices [43] to determine the beam waist (intensity distribution) and phase front radius of curvature along the beam path.

As shown in Fig. 4 the output face of the silicon lens of the THz source is located at the focal plane of an off-axis parabolic mirror (OAPM). For these experiments, the mirror focal length f was 119 mm. For polarization sensitive measurements it is important that any polarizers be placed after the OAPM owing to cross-polarization terms generated in reflection [44]. Unlike the case in ray optics, which predicts a perfectly collimated beam, the THz beam has a second waist located one focal length from the parabolic mirror towards the target. The $1/e$ field radius as a function of distance for the optical configuration shown in Fig. 4a is illustrated in Fig. 8a. This second beam waist has frequency-dependent diameter, with higher frequencies having a smaller diameter than lower frequencies. The effect of this frequency-dependent focus is to provide an effective frequency-dependent aperture 238 mm from the THz source. The sequence of frequency-independent and frequency-dependent beam waists is repeated for each confocal optic added to the system.

Accurate scattering measurements require both field uniformity and planar phase fronts. The field uniformity is given by the $1/e$ field radius of the beam, which should be larger than the target. If w_o is the $1/e$ radius of the beam, the field variation over an aperture of radius $w_o/3$ is less than 10%, while the field variation is less than 25% over $w_o/2$. Additionally, the beam size should be frequency independent. As illustrated in Fig. 8a, the beam size is larger for lower frequencies, and the $1/e$ beam diameters for different frequencies begin to converge as the distance from the source increases. The waist diameter is nearly constant at frequencies greater than 500 GHz for target distances greater than 500 mm from the second focus. Since the beam diameter at the focus of the OAPM scales in proportion to wavelength, outside the Rayleigh range the divergence angles of the various frequency components reach the same value. Thus the optical configuration shown in Fig. 4 leads to uniform target illumination for frequencies above 0.5 THz.

The radius of curvature of the phase front, R, determines the phase error, $\Delta\phi$, across the target (see Fig. 1). An infinite radius of curvature ($R \rightarrow \infty$)

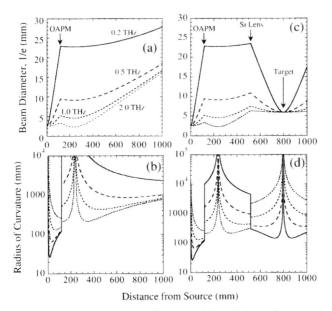

Fig. 8. THz electric-field 1/e radius (**a**) and phase front curvature (**b**), as a function of distance from THz source for the configuration shown in Fig. 4. Field 1/e radius (**c**) and phase front curvature (**d**) for the configuration shown in Fig. 9

is ideal. The deviation from a true plane wave is the phase variation, $\Delta\varphi$, across the extent of the target, expressed as a multiple of wavelengths, $N\lambda$. The phase variation in radians of a wavefront from a spherical source (outside the Rayleigh range) is

$$\Delta\phi = \frac{\pi N}{4r}(N\lambda), \tag{1}$$

where r is the distance from the source. A typical value for the quiet zone of a compact range (a spatial region approximating a plane wave) is $\Delta\phi < \pi/4$. Fig. 8b shows the calculated radius of curvature of the phase front R, for the configuration of Fig. 4. Owing to the logarithmic scale, this figure displays the absolute value of the curvature. The curvature becomes infinite at a focus, and has singularities at optical elements which change the beam from diverging (positive curvature) to converging (negative curvature). At the detector bandwidth limit of 2 THz and a distance from source to target of 1 m, the target size could be 72 wavelengths or just over 10 mm for phase variations $\leq \pi/4$. The amplitude variation over a 10 mm diameter target at 2 THz is approximately 10%.

Although the optical configuration illustrated in Fig. 4 produces nearly frequency-independent illumination above about 0.5 THz, there are discrepancies at lower frequencies. Here, to achieve uniform amplitude and minimal phase variations, the target must be several Rayleigh lengths from the frequency-dependent beam waist for all frequencies. The Rayleigh length at 0.2 THz is 1.2 m, leading to unrealistic range sizes. Extremely uniform phase fronts can be achieved with small range sizes across the entire bandwidth by addition of an additional phase-front-modifying optic in the THz beam path, as shown in Fig. 9. Addition of a 28 cm focal length, 50 mm diameter silicon lens located confocally with the OAPM generates a beam waist at the target position. The calculated waist diameters and radii of curvature of the THz beam for this configuration are shown in Figs. 8c,d respectively, at 0.2, 0.5, 1.0 and 2.0 THz. Note that at the focus of the silicon lens the beam waists have identical diameters for all frequencies. This is a consequence of the frequency-dependent beam waist at the previous focus. This pattern of alternating frequency-dependent and frequency-independent beam waists is repeated indefinitely for confocal focusing elements. For this configuration the target is placed in the quiet zone, which has a 1/e diameter of 12 mm with a phase variation $\Delta\phi = 0$ and is located 80 cm from the THz source.

In summary, the curvature of the phase front and diameter of the quiet zone are strongly dependent upon the optics used. For the configuration of Fig. 4, the size of the quiet zone increases with increasing source-to-target distance. For example, a 10 cm quiet zone ($> 300\lambda$ at 1 THz) requires a target

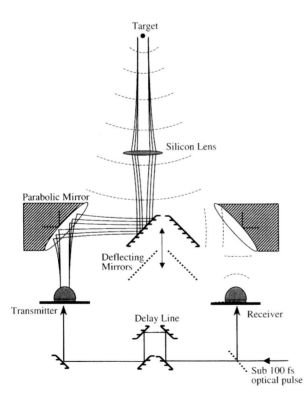

Fig. 9. Addition of a phase-front-modifying optic to the configuration of Fig. 4 to achieve true plane wave excitation at the target

distance of 3 m. Use of a lens in the configuration of Fig. 8 guarantees a planar phase front and a frequency-independent beam diameter, but places limits on the maximum permissible target size. Note that although the size of the quiet zone is on the order of 1 cm at 1 THz, this corresponds to a quiet zone of 1 m in the microwave X band (10 GHz). Investigations to date have used commercially available optics with diameters less than 50 mm. One practical advantage of the THz impulse range is the availability of commercial optics with much larger diameters. Additionally, owing to the very long wavelengths compared with typical optical tolerances, an optic with a surface flatness of $\lambda/2$ across the surface at optical frequencies has a surface flatness of nearly $\lambda/1000$ at 1 THz.

2.4   Targets

The THz beam is incident on an object of some type which scatters the radiation through the various processes illustrated in Fig. 2. The word "target" is borrowed from the radar literature to denote the object whose scattering properties are to be determined. The diameter and phase front curvature of the THz beam determine the permissible target size. Two types of targets are used – sample and reference – to borrow terms used in THz-TDS. The properties of the sample target are investigated through THz impulse ranging. The reference target has a known, frequency-dependent response and serves to characterize the impulse range. Characterization through actual measurement from a reference standard is used to deconvolve the system response from the measured scattering signatures. This concept is treated theoretically in Sect. 3.

Targets investigated to date have consisted mainly of simple geometrical shapes, including cylinders and spheres with calculable scattering, in order to characterize the THz impulse range. Cylinders which extend outside of the field distribution of the THz beam are treated theoretically as if they are of infinite length. Spheres must be contained in the quiet zone, which is approximately 10 mm diameter for the range configurations used.

Experimentally, targets must be at a known orientation relative to the incident k vector in both the θ and the ϕ directions (see Fig. 11 in Sect. 3). Alignment requirements are relaxed for simple target geometries: spherical targets are orientation independent, while cylindrical targets are independent of the θ orientation provided the k vector of the incident THz beam is orthogonal to the cylinder axis (see Fig. 11). Alignment in the ϕ direction is accomplished by attaching the target to a precision rotation stage. Alignment in the ϕ direction is accomplished using a kinematic mount.

For cylindrical targets which are of "infinite" extent, extending outside of the THz beam, mounting components can be placed below the THz beam and are "invisible" to the range. Targets such as spheres which must lie fully in the THz beam path require support mechanisms which perturb the scattered field as little as possible. Support mechanisms are borrowed from techniques used in the radar community. Spheres are typically supported with a styrofoam column shaped to minimize scatter in the direction of the detector. Styrofoam has a low refractive index, $n < 1.2$, at THz frequencies, corresponding to reflection of less than 9% of the field even for nonideal support columns. More complicated targets such as model aircraft can be supported on styrofoam pillars or suspended by thin dielectric fibers with a diameter smaller than a wavelength. Several support lines are required to keep the target from rotating during measurements and are attached to a frame much larger than the THz beam diameter which can be aligned in the θ and ϕ directions.

2.5   Scattered Radiation

Owing to the necessity of targets fitting within the quiet zone, the target typically intercepts a small fraction of the incident THz beam, and the scattered power is only a small fraction of the incident THz power. The scattered radiation from a particular target is a function of the direction and distance from the target to the point at which the scattered field is measured. These obviously depend upon the geometrical shape and composition of the target and, in general, must be calculated numerically for each particular target of interest. Complicated targets tend to have very rapid variations of scattered field strength and phase with variations in θ and ϕ. Geometrically simple targets have less rapid variations with scattered phase fronts whose curvature mimics the target shape; spherical waves for spheres, and cylindrical waves for cylinders. As a function of distance r from the target the field strength of the scattered radiation thus drops off as $r^{-1/2}$ for infinite cylindrical targets and as r^{-1} for spheres. The calculation of the scattered electric field for simple targets is addressed in detail in Sect. 3.

Optimal characterization of the scattered electric field would measure the field at a single point with infinite dynamic range. Practically, a dynamic range of 60 dB in power is sufficient. In cases where the spatial variation of the scattered electric field is rapid, small detector aperture solid angles (obtained by use of a detector with a small aperture and/or a large distance from the target) are required to avoid interference effects across the detector aperture. However small apertures lead to less power on the detector. There is therefore a trade-off between the detector size, the physical separation of target and detector, and the signal-to-noise ratio of the detector.

For the ranging system illustrated in Figs. 4 and 9, the scattered field propagates through space and is collected by an off-axis parabolic mirror, which focuses it on the detector. This provides a 13×10^{-3} srad solid angle of collection for a 500 mm target distance, an increase of 40 times over the detector silicon lens alone at the same distance. The OAPM additionally modifies the phase front of the scattered radiation. This is treated theoretically in Sect. 3.

2.6 Bistatic Range

Targets have a scattering signature which, in the most general case, is a function of frequency, the direction of the incident k vector, and the angles the scattered field is measured at, θ and ϕ. The impulse ranges shown in Figs. 4 and 9 measure the target response at a fixed bistatic angle determined by $\theta = 2\arctan(d/r)$, where d is the distance between the flat deflecting mirrors, and r is the distance from these mirrors to the target. For the configurations used, $\phi = 10° \pm 3°$ and $\theta = 0°$. To permit full angular characterization of targets a bistatic range, shown in Fig. 10, has been developed in which the angle ϕ is variable over the range 20° to 210°.

The THz beam, generated with a standard source, is collimated by a paraboloidal reflector with focal length $f = 119\,\text{mm}$. A planar deflection mirror placed 75 mm from the paraboloidal reflector steers the THz beam. The target is placed 400 mm from the deflection mirror and above the axis point of the rotating receiver arm. No subsequent focusing lens (see Fig. 9) is used, since the diameter of the beam at the target is 16.4 mm at the maximum frequency of 1.3 THz, leading to a negligible phase shift across the target, which results in a maximum temporal deviation from a plane wave of less than 10 fs.

The THz detector module is mounted on an optical rail which can rotate around the center of the target position. The gating optical beam for the detector is steered along the arm by a mirror attached to the arm as shown in Fig. 10. This permits adjustment of the angle ϕ over a large range, limited by the THz source position and the size of the range enclosure. The distance from

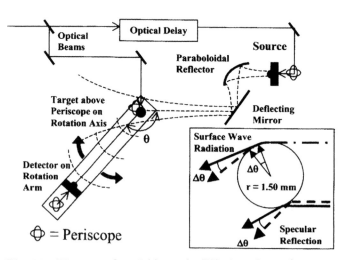

Fig. 10. Diagram of variable-angle THz impulse radar range. The target, the periscope beneath the target, and the detector rotation arm all share the same axis of rotation

the detector to the target can be adjusted from 100 to 1200 mm. This system uses only the silicon lens on the THz detector, giving a effective aperture size of approximately 7 mm. For a 50 cm target distance, the solid angle of the detector is then 50×10^{-6} srad, or over 300 times smaller than in the ranges shown in Figs. 4 and 9. The phase variation across the detector aperture from a point source is $\pi/6$ radians at 2 THz. from (1).

3   Theory

A complete description of the interaction of the THz pulse with the target is provided by Maxwell's equations in conjunction with the boundary conditions at the target in three-dimensional space. The total field consists of the sum of the incident, reflected, and transmitted fields. A general solution of Maxwell's equations does not exist, except for simple geometries. Although a general analytic solution is mathematically intractable, Maxwell's equations are linear and scalable in both space and time. THz impulse ranging makes use of this scaling property to shrink range sizes by the ratio of the wavelengths. The broad wavelength range of a THz pulse, 3 mm at 100 GHz to 150 µm at 2 THz, corresponds to a reduction in spatial scale by a factor of 10 to 200 over X-band radar frequencies, and radar signatures may be acquired from targets scaled geometrically by these amounts. The first part of this section discusses this scaling of Maxwell's equations and some material constraints important in scale ranging. Although the scattered field is determined by Maxwell's equations, any optical elements, in the incident beam path or used to increase the efficiency of the detector will change the spatial and temporal distribution of the incident or scattered electric field. The effect of these elements is described in terms of a system transfer function. The development of a frequency-domain transfer function description of THz impulse ranging is given in Sect. 3.2. This description permits the extraction of the target's response from the response of the impulse range.

The THz impulse range experimentally determines the electric field as a function of time. The description of the system response and the calculation of the target response are both done in the frequency domain. Although well known, the use of numerical Fourier transforms is reviewed in Sect. 3.3 since it determines the resolution of measurements.

To verify the accuracy of a THz impulse range, it is necessary to compare measurements of target response with calculated results. Section 3.4 focuses on numerical methods for calculation of the target response function. These calculations are limited to a relatively small class of objects of symmetrical geometry for which Maxwell's equations have a tractable analytical solution. Computational methods such as Finite Difference Time Domain (FDTD) are not discussed.

3.1   Scaling of Maxwell's Equations in Terahertz Impulse Ranging

The use of THz impulse ranging to investigate scattering from a variety of targets of practical interest relies on the ability to extract the response of a full-sized target from that of one scaled in size. The ability to scale results depends upon the linearity of Maxwell's equations. This is valid for all materials at the low power levels used in THz impulse ranging – 10 nW average power and 200 μW peak power (see Sect. 2.2). Maxwell's equations, represented in derivative form, for low-power EM fields are [45]

$$\nabla \cdot \boldsymbol{D} = \rho,$$
$$\nabla \cdot \boldsymbol{B} = 0,$$
$$\nabla \cdot \boldsymbol{E} = -\frac{\partial \boldsymbol{B}}{\partial t},$$
$$\nabla \cdot \boldsymbol{H} = \sigma \boldsymbol{E} + \frac{\partial \boldsymbol{D}}{\partial t}. \tag{2}$$

Here ρ is the charge density, and σ the conductivity. The relationships between the electric field $\boldsymbol{E}$ and the electric flux vector $\boldsymbol{D}$ and between the magnetic field $\boldsymbol{H}$ and the magnetic flux vector $\boldsymbol{B}$ are given by

$$\boldsymbol{D} = \varepsilon_0 \varepsilon_r \boldsymbol{E},$$
$$\boldsymbol{B} = \mu_0 \mu_r \boldsymbol{H}. \tag{3}$$

For the materials investigated to date, the permeability is that of free space, i.e. $\mu_r = 1$, and the relative permittivity, ε_r, is isotropic.

The linear scaling of Maxwell's equations means that both the electric and the magnetic fields, as well as time and space, can be scaled by separate factors. The spatial scaling factor L is defined as the ratio of the scale target to the full-size target, $L = x_{\text{scale}}/x_{\text{actual}}$ and is typically on the order of 10^{-2}. The temporal scaling factor $T = t_{\text{scale}}/t_{\text{actual}}$ relates time in the model frame to that on a full-size target. The frequency of the scale range f_{scale} to that used in an actual application f_{actual} are then related by $f_{\text{scale}}/f_{\text{actual}} = 1/T$. The electric and magnetic field strengths are scaled by factors such that $E_{\text{scale}} = \mathcal{E} E_{\text{actual}}$ and $H_{\text{scale}} = \mathcal{H} H_{\text{actual}}$.

To obtain a scaled measurement of an actual target by measuring the response of a scale model, the scaling properties of Maxwell's equations require one to adjust the material parameters of the scale model. The new material parameters, namely the permeability, permittivity, and conductivity as are given by [46]

$$\mu_{\text{scale}} = \mu_{\text{target}} \frac{T}{L} \frac{\mathcal{E}}{\mathcal{H}},$$
$$\varepsilon_{\text{scale}} = \varepsilon_{\text{target}} \frac{T}{L} \frac{\mathcal{H}}{\mathcal{E}},$$
$$\sigma_{\text{scale}} = \sigma_{\text{target}} \frac{1}{T} \frac{\mathcal{H}}{\mathcal{E}}. \tag{4}$$

Owing to the practical difficulty of finding or designing materials whose permeability and permittivity can be arbitrarily chosen over a broad frequency range, the measurements reported here scale both time and distance by an equal factor, $T = L = \kappa$. In addition, the scattered electric and magnetic field amplitudes are scaled linearly compared with a real system, but their ratio remains unity, $\mathcal{E}/\mathcal{H} = 1$. Under these conditions measurements on scaled targets are directly comparable to those on full-size targets given several constraints [17,46]. At a given frequency, the real part of the dielectric constant at THz frequencies, ε_R (THz), should be equal to that at the GHz frequencies actually employed, ε_R (GHz), i. e. ε_R (THz) $= \varepsilon_R$ (GHz). The loss due to absorption (the imaginary part of the dielectric constant, ε_1), and the conductivity σ should increase linearly with frequency. This is equivalent to maintaining a constant loss tangent $\tan \delta = (\sigma + \omega \varepsilon_1)/\omega \varepsilon_R$ [46].

For a large number of materials, ε_R changes by less than 1% when the frequency is increased from the GHz to the THz region [33]. Similarly, the imaginary part of the permittivity typically increases with frequency. This increase is quadratic for many materials, especially at higher frequencies, but can be approximated well by a linear increase over wide frequency ranges [33]. The natural frequency dependence of the permittivity leads to a limited ability to use the same dielectric materials in real and scaled targets at GHz and THz frequencies.

A potentially more serious constraint is found when scaling the conductivity σ for targets that support surface currents. Since σ_{model} must be equal to σ_{actual} multiplied by the scaling factor κ as specified in (4), physical limits on conductivity could nullify results at high frequencies for poor conductors. However most metals have effectively infinite conductivity over the GHz to low THz frequency range. As will be discussed in the next section, measurements on conducting targets [32] agree well with scattering theory based on the assumption of infinite conductivity. Physical limits on the conductivity may play a role at very high frequencies or for conducting targets that display resonances or surface currents which propagate over long distances.

3.2 Transfer Function Description of Terahertz Impulse Ranging System

The description of the THz impulse ranging system given here is based upon the radar master equation, traditionally used for determining radar scattering. This equation describes the measured scattered power on the receiver in watts per square meter coming from a known source [47]:

$$S = \frac{P_t G_t}{4\pi R^2} \times \sigma_{\mathrm{target}} \times \frac{1}{4\pi R^2} \times \frac{\lambda^2 G_r}{4\pi}. \tag{5}$$

Here the first term is the power on the target, determined by the transmitted power P_t and transmitter antenna gain G_t, assuming the incident wave is a spherical wave. The second term, σ_{target} is the scattering cross-section of the target. traditionally known as the radar cross section (RCS). Here the RCS is a function of the frequency and polarization of the incident radiation and of the spatial orientation of the target. The third term is due to scattering of the incident radiation over 4π steradians by the target. located a distance R from the transmitter and receiver. The final term relates the receiver aperture size to the wavelength and the receiver antenna gain G_r.

For the THz ranging system shown in Fig. 4 the transmitter. receiver and associated off-axis parabolic mirrors are in the quasi-optical limit owing to the large aperture sizes. For example. at 1 THz. $\lambda = 0.3\,$mm. the aperture of the silicon lens is approximately 20λ. and the mirror aperture is over 150λ. Assuming no ohmic loss in the antenna structures. the transmitter and receiver gains can be approximated by $G_t = G_r = 4\pi/\theta^2$. where θ is the divergence of the THz beam beyond the Rayleigh range. At 1 THz the gains G_t. G_r are approximately 9700. Owing to the quasi-optical nature of the THz impulse range and since both the transmitted and the received fields are strongly frequency dependent. the radar master equation can be rewritten as

$$E_{\mathrm{scat}}(\omega.\theta.\phi) = A_t(\omega)G_t(\omega) \times T(\omega.\theta.\phi) \times S(\omega.r.\theta.\phi)G_r(\omega). \tag{6}$$

Here $E_{\mathrm{scat}}(\omega.\theta.\phi)$ is the complex. frequency-dependent field amplitude. instead of the intensity as in (5). The angles θ and ϕ are defined for simple geometries in Fig. 11. and r corresponds to the distance from the target to the receiver. The frequency-dependent amplitude is determined from the numerical Fourier transform of the measured scattered pulse. discussed in Sect. 3.3.

The first term in (6). the field incident upon the target as a function of frequency. $E_{\mathrm{inc}}(\omega) = A_t(\omega)G_t(\omega)$. is the actual generated pulse $A_t(\omega)$ multiplied by a factor $G_t(\omega)$. which describes the effects of the source antenna. optics. and free-space propagation on the THz beam prior to the target. The second term in (6) is the response of the target to the incident THz pulse. given by $T(\omega.\theta.\phi)$. The propagation of scattered radiation to the detector is described by the third term. $S(\omega.r.\theta.\phi)$. which also takes into account the effect of the THz detector optics on the scattered phase front. The scattered electric field measured at the receiver is then $E_{\mathrm{sc}}(\omega) = E_{\mathrm{inc}}(\omega)T(\omega.\theta.\phi)S(\omega.r.\theta.\phi).G_r(\omega)$. where the term $G_r(\omega)$ is the gain- and frequency-dependent sensitivity of the THz detector.

In THz impulse ranging we wish to extract the response of the target. or measure $T(\omega.\theta.\phi)$. No method exists to fully characterize the THz source or detector independently. As a result any measurements show effects due to all of the terms in (6) and consequently the system response must be deconvolved from the target response. In THz impulse ranging we are able to bypass this deconvolution process by using the following procedure to isolate the target response. $T(\omega.\theta.\phi)$. This procedure is similar to that used in THz time-domain spectroscopy [33]. where two individual data sets are taken. The first is the response of a reference sample or target. $E_{\mathrm{ref}}(\omega)$. The second is the

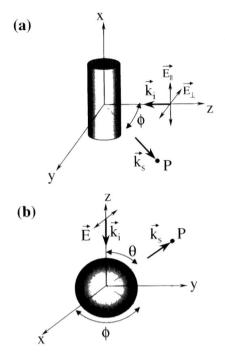

(a)

(b)

Fig. 11. Definition of k vectors of incident and scattered radiation for scattering calculations on (**a**) cylinders, and (**b**) spheres

response of the target object whose response is to be measured, $E_{target}(\omega)$. Taking the ratio of the measured scattered electric field of the target to that of the reference in the frequency domain is identical to doing a time-domain deconvolution [48]:

$$\frac{E_{target}(\omega)}{E_{ref}(\omega)} = \frac{T_{target}(\omega, \theta, \phi) S(\omega, r, \theta, \phi)}{T_{ref}(\omega, \theta, \phi) S(\omega, r, \theta, \phi)}. \tag{7}$$

Here the terms $E_{inc}(\omega)$ and $G_r(\omega)$ divide out of the ratio since they are independent of the target used. Two conditions need to be fulfilled to measure $T_{target}(\omega, \theta, \phi)$: the reference target needs to a have unity, frequency independent response; and the scattered phase fronts from both targets need to be identical so that $S(\omega, r, \theta, \phi)$ is the same for the reference and target.

Reference targets made of metal have a frequency-independent, 100% reflection across the THz spectrum measured in THz impulse ranging. By choosing a target large compared with the wavelength, such as a metal mirror, we obtain $T_{ref}(\phi, \omega, r) = 1$. Large sizes are required so that induced surface currents cannot propagate around the circumference of the target and contribute time delayed responses in the measurement window. In order to match $S(\omega, r, \theta, \phi)$ for both the target and the reference, the scattered phase fronts need to have identical spatial profiles. This can be accomplished for the investigation of simple geometrical objects by using a reference mirror whose shape is identical to that of the target: in other words, a spherical reference mirror for spherical targets or cylindrical mirror for cylindrical targets. When these conditions are fullfilled, the ratio of the target response to the reference response directly determines $T(\phi, \omega, r)$.

For more complicated targets with geometrical features several wavelengths in size, such as scale-model aircraft, the scattered phase front cannot be matched by a reference target with $T = 1$, and S will not be identical for the target and reference measurements. In this case it is necessary to measure the electric field either over a very small receiver aperture or at a very large distance. Here the phase front is constant across the detector aperture, i.e. it is a plane wave. In this limit the effect of any optical elements on the detector will be identical for both the target and the reference.

The propagation of the THz beam is also represented by S, for example the $1/r$ power decrease for a cylindrical phase front, in contrast to the inverse square law dependence for spherical phase fronts. In the limit of small detector aperture it is necessary to directly measure $E_{\mathrm{inc}}(\omega)$ at the target. This can be accomplished for a well-collimated THz beam by using a flat mirror as a reference target in the impulse range of Fig. 4, or by rotating the detector so that $\theta = 180°$ in the impulse range of Fig. 10.

Validation of the technique of THz impulse ranging requires a comparison of electromagnetic scattering measurements with well-established techniques in the microwave frequency range. Radar literature traditionally represents the scattering efficiency of a target by the RCS, which relates the incident power on a target to that measured by a far field detector. The working definition of the RCS is the ratio of the power scattered by a given target at a particular angle to the power scattered a perfectly reflecting target which radiates isotropically into a 4π solid angle. For a given measured scattered power on the detector from the actual target, the RCS is defined as the cross-sectional area of the perfect target required to provide the same scattered power. In terms of the scattered fields

$$\sigma = \lim_{R \to \infty} 4\pi R^2 \frac{\left|E_{\mathrm{scat}}^2\right|}{\left|E_{\mathrm{inc}}^2\right|} \tag{8}$$

where σ is the RCS, R is the distance to the target, and E_{scat} and E_{inc} are the scattered and incident fields respectively.

3.3   Calculation in Time and Frequency

The data from the THz impulse range corresponds to the time-dependent electric field. To compare theoretically the measured field with scattering predictions, it is necessary to convert the time data into a complex frequency-dependent amplitude through a Fourier transform. A numerical fast Fourier transform (FFT) [48] returns the complex amplitude of the electric field at discrete frequency points for both the target, $E_{\mathrm{target}}(\omega_n)$, and the reference, $E_{\mathrm{ref}}(\omega_n)$. The discrete time domain data consists of N points in time, spaced Δt seconds apart. N is traditionally a power of 2 for fast convergence of the

FFT algorithm, but the computational overhead for this is minimal compared with the calculation of the scattering coefficients (Sect. 3.4). A typical experimental value of N is 2048, with $\Delta t = 33\,\text{fs}$ (a delay line step of $5\,\mu\text{m}$). The FFT algorithm returns the complex amplitude at $N/2$ points, ω_n, in the frequency domain, evenly spaced over the frequency range $\omega_n = 0$ ($n = 1$) to $\omega_n = \pi/\Delta t$ ($n = N/2$). In the example above, 1024 frequency points are generated over the range 0–15 THz. Over the system bandwidth of 100 GHz to 2 THz, the temporal measurements correspond to approximately 130 discrete frequency points. By padding the time-domain data scans with zeros it is possible to interpolate in the frequency-domain and achieve an arbitrarily large number of frequency domain data points at the cost of increased computational overhead [48].

The target scattering coefficient in (7), $T(\omega_n, \theta, \phi)$ is calculated at each discrete frequency point ω_n by means of the numerical summations given in the following section at the measured scattering angle(s) θ and ϕ. In the case of a reference target with $T_{\text{ref}}(\phi, \omega, r) = 1$ and equivalent optical transfer functions $S(\omega, r, \theta, \phi)$ for both target and reference, the scattered field at the detector is calculated from

$$E_{\text{calc.}}(\omega_n) = T_{\text{target}}(\omega_n, \theta, \phi) E_{\text{ref}}(\omega). \tag{9}$$

To convert the calculated complex amplitude at the detector back to the time domain to perform direct comparison with the scattered pulse, an inverse numerical Fourier transform is performed on $E_{\text{calc.}}(\omega_n)$.

The scattered field has a rapid phase dependence, varying as e^{ikr}, where $k = 2.1 \times 10^4$ radians/m at 1 THz in free space (see (10), (11), and (15)). Thus determining the absolute phase requires a target positioning accuracy of $r < \pi/k$, or $r < 150\,\mu\text{m}$ over a distance of meters. Owing to the difficulty of obtaining repeatable submillimeter positioning accuracy between reference and target, the measured and calculated pulses are shifted in time to align the specularly reflected pulses.

Numerical calculations of the calculated pulse are prone to artifacts that appear non-causal. The numerical calculation of the scattering coefficient in the frequency domain includes all electric-field components scattered from the target. Since all scattering mechanisms are included in the frequency-domain description, portions of the calculated field have large phase delays, corresponding to temporally delayed features which occur at times longer than the measurement window of the reference pulse. The inverse FFT algorithm assumes a periodic function [48], causing scattering data falling outside the measurement window to appear to occur at earlier times. In other words if the measurement window is T seconds in duration and a scattering feature appears at $T + \Delta t$ then the numerical calculation will show this feature to occur at time Δt. This numerical artifact can be avoided by artificially extending the time-domain data with zeros. However this increases the number of frequency points required in the numerical calculation of $T(\omega_n, \theta, \phi)$ as well as computation time, which scales linearly with the number of points.

3.4   Calculation of Scattering Coefficients

In order to check the validity of THz impulse ranging it is necessary to compare measured scattering coefficients with theoretically predicted values. For spheres and cylinders the scattering coefficients can be calculated numerically directly from the Mie series [49]. The Mie series calculates scattered fields on the basis of boundary conditions for incident plane waves. Inherent in this calculation is the assumption that the target is excited by a linearly polarized plane wave, and that the total scattered field can be represented as a Fourier summation of single frequency sinusoids. Although the Mie series provides an exact solution for the scattered fields, a far-field approximation is used here, which assumes the distance from the detector to the target, r, is much greater than the target radius, a.

Two target types are used for range validation: infinite cylinders and spheres. The infinite-cylinder approximation is good when the target has a length which extends outside the incident field distribution, and the measurement window duration is short enough to prevent reflections of surface currents from the end of the cylinder being detected

3.4.1   Cylinders

The scattered electric field at a given frequency can be calculated for a cylindrical target of infinite axial extent at a given frequency ω_n, from [49]

$$E_{\parallel sc(\omega_n)} = E_{\parallel inc}(\omega_n)\sqrt{\frac{2}{\pi k_0 r}}e^{i(k_0 r - \pi/4)}T_\parallel(\phi). \tag{10}$$

$$H_{\perp sc(\omega_n)} = H_{\perp inc}(\omega_n)\sqrt{\frac{2}{\pi k_0 r}}e^{i(k_0 r - \pi/4)}T_\perp(\phi). \tag{11}$$

where $T(\phi)$ is the scattering coefficient, which depends upon the target composition, the polarization of the incident wave, either parallel ($\parallel$) or orthogonal ($\perp$) to the cylinder axis, and the angle ϕ between the incident and scattered k vectors. k_0 is the wave vector in free space. Here it is assumed the cylinder axis is orthogonal to the plane of the k vectors of the incident and scattered radiation. Note that the scattered radiation intensity for the cylinder falls off inversely with the distance, in comparison with the inverse square of the distance more commonly encountered with spherical wavefronts. The scattering coefficients calculated here, $T_\parallel(\phi)$ and $T_\perp(\phi)$ are calculated at a specific frequency, ω_n, and are related to the generalized scattering coefficient in (6), by: $T_\parallel(\phi), T_\perp(\phi) = T(\omega_n, \theta = 0, \phi)$.

The scattering coefficients are given by:

$$T_\parallel(\phi) = \Sigma_{n=0}^{\infty}(-1)^n \varepsilon_n A_n \cos(n\phi).$$

$$T_\perp(\phi) = \Sigma_{n=0}^{\infty}(-1)^n \varepsilon_n B_n \cos(n\phi). \tag{12}$$

which can be evaluated numerically, given A_n and B_n. The coefficient $\varepsilon_n = 1$ when $n = 0$ and is equal to 2 otherwise. The A and B coefficients depend on the composition of the cylinder – either dielectric or conducting. For a perfect conductor,

$$A_n = -\frac{J_n(k_0 a)}{H_n^{(1)}(k_0 a)},$$

$$B_n = -\frac{J_n'(k_0 a)}{H_n^{(1)\prime}(k_0 a)}. \tag{13}$$

Here J_n is the nth-order Bessel function, and H_n is the Hankel function. The primes in the expression for B_n represent derivatives of the functions with respect to $k_0 a$. For a dielectric cylinder, the coefficients A_n and B_n are given by

$$A_n = -\frac{(k_r/\mu_r)J_n(k_0 a)J_n'(k_r a) - (k_0/\mu_0)J_n(k_r a)J_n'(k_0 a)}{(k_r/\mu_r)H_n^{(1)}(k_0 a)J_n'(k_r a) - (k_0/\mu_0)H_n^{(1)\prime}(k_0 a)J_n(k_r a)},$$

$$B_n = -\frac{(k_r/\varepsilon_r)J_n(k_0 a)J_n'(k_r a) - (k_0/\varepsilon_0)J_n(k_r a)J_n'(k_0 a)}{(k_r/\varepsilon_r)H_n^{(1)}(k_0 a)J_n'(k_r a) - (k_0/\varepsilon_0)H_n^{(1)\prime}(k_0 a)J_n(k_r a)}. \tag{14}$$

Here, the permeabilities of free space and of the cylinder are μ_0 and μ_r respectively, and the permittivities of free space and of the cylinder are ε_0 and ε_r. The cylinder radius is a, and $k_r = [\omega^2 \mu_r \varepsilon_r]^{1/2}$ is the wave-vector in the cylinder. The wave vector in free space. k_0. is given by $[\omega^2 \mu_0 \varepsilon_0]^{1/2}$.

3.4.2   Spheres

The form of the scattered field for a spherical target is similar to that for a cylinder:

$$E_{sc}(\omega_n) = E_{inc}(\omega_n)\frac{e^{ik_0 r}}{k_0 r}T(\theta, \phi). \tag{15}$$

The power, however, falls off inversely as the square of the distance. The scattering coefficent $T(\theta, \phi)$ is now a function of the angles θ and ϕ, which are illustrated in Fig. 11. Here the incident radiation is assumed to be polarized along the x direction and incident along the z axis. The angle ϕ describes the relation of the plane formed by the k vectors of the incident and scattered radiation (the plane of incidence) to the polarization of the incident field. For $\phi = 0$ the incident polarization is in the plane of the incidence while for $\phi = \pi/2$ the polarization is orthogonal to the plane of incidence. The angle θ describes the angle between the k vectors of the incident and scattered fields.

The scattering coefficient in the far-field approximation is given by

$$T(\theta, \phi) = \cos\phi S_1(\theta)\hat{\theta} - \sin\phi S_2(\theta)\hat{\phi}. \tag{16}$$

This scattering coefficient shows that for $\phi \neq 0$ and $\phi \neq \pi/2$ the scattered field has components along two directions. Again, $T(\theta, \phi)$ is calculated at a specific frequency, ω_n, and is related to the generalized scattering coefficient in (6), by $T(\theta, \phi) = T(\omega_n, \theta, \phi)$. The coefficients S_1 and S_2 are given by [49]

$$S_1(\theta) = \Sigma_{n=1}^{\infty}(-i)^{n+1}\left(A_n \frac{P_n^1(\cos\theta)}{\sin\theta} + iB_n \frac{\partial}{\partial\theta}P_n^1(\cos\theta)\right),$$

$$S_2(\theta) = \Sigma_{n=1}^{\infty}(-i)^{n+1}\left(A_n \frac{\partial}{\partial\theta}P_n^1(\cos\theta) + iB_n \frac{P_n^1(\cos\theta)}{\sin\theta}\right). \quad (17)$$

where P_n^1 represents the Legendre function. For a perfectly conducting sphere, the coefficients A_n and B_n are given by

$$A_n = -(-i)^n \frac{2n+1}{n(n+1)} \frac{J_n(k_0a)}{H_n^{(1)}(k_0a)},$$

$$B_n = -(-i)^{n+1} \frac{2n+1}{n(n+1)} \frac{[k_0aJ_n(k_0a)]'}{\left[k_0aH_n^{(1)}(k_0a)\right]'}, \quad (18)$$

where again the primes represent derivatives with respect to k_0a. For a dielectric sphere the coefficients A_n and B_n are given by

$$A_n = -(-i)^n \frac{2n+1}{n(n+1)}$$
$$\times \frac{J_n(k_0a)[k_raJ_n(k_ra)]' - J_n(k_ra)[k_0aJ_n(k_0a)]'}{J_n(k_ra)\left[k_0aH_n^{(1)}(k_0a)\right]' - H_n^{(1)}(k_0a)[k_raJ_n(k_ra)]'}.$$

$$B_n = -(-i)^{n+1} \frac{2n+1}{n(n+1)}$$
$$\times \frac{J_n(k_0a)[k_raJ_n(k_ra)]' - n_r^2 J_n(k_ra)[k_0aJ_n(k_0a)]'}{H_n^{(1)}(k_0a)[k_raJ_n(k_ra)]' - n_r^2 J_n(k_ra)\left[k_0aH_n^{(1)}(k_0a)\right]'}. \quad (19)$$

The notation used is the same as used for the cylinder. $n_r = \varepsilon_r^{1/2}$ is the complex index of refraction equal to $2\pi/\lambda[n(\omega) + \alpha(\omega)\lambda/4\pi]$, and k_r is the complex wavevector equal to n_rk_0.

The Mie series calculation, for both the spherical and the cylindrical target, is expressed as an infinite summation. The number n of the terms, which are calculated determines the accuracy of the results. To achieve on accuracy on the order of four percent it is necessary to perform the summations of (12) and (17) to at least $n = 1.3k_0a$ [49]. For a target radius on the order of 2 mm and a frequency of 1 THz, k_0a is approximately 40. For the calculations performed here, the summations were taken to at least $1.5\,k_0a$. The summations can only be performed to a limited accuracy since numerical calculation techniques for the Bessel and Hankel functions have inherent numerical errors for large arguments. These errors must be monitored during calculations to avoid phase discrepancies in the predicted fields. Derivatives occurring in the A_n and B_n coefficients in (13), (14), (18), and (19) were calculated both numerically and through exact series calculation [50], with identical results.

4 Measurements

This section presents measurements on a range of targets analyzed using the formalism of Sect. 3. Sections 4.1 and 4.2 report measurements on cylindrical and spherical targets, respectively, used to test the calibration of the range. Section 4.3 discusses various fits of the data using both the Mie theory and a geometrical-optics theory. Variable-angle bistatic ranging is discussed in Sect. 4.4. The temporal resolution of the data permits direct observation of effects such as the Gouy phase shift, discussed in Sect. 4.5, while measurements on realistic and complex targets are presented in Sect. 4.6.

4.1 Conducting and Dielectric Cylinders

Initial target measurements and calibration of the range were performed using conducting cylinders. Cylinders are geometrically simple targets for which exact scattering solutions exist. Furthermore, the returned signal strengths are larger for a cylinder than for a sphere owing to the inverse square root dependence of the field on the distance from the target, as opposed to the inverse distance dependence for a sphere. In addition, the cross-sectional area of a long cylinder intercepts a larger fraction of the THz beam.

Scattering measurements were performed on copper cylinders 0.26 mm, 0.51 mm, and 1.02 mm in diameter which extended well beyond the extent of the THz beam to permit use of the infinite-cylinder approximation in analyzing the data. The experimental configuration in which the data was taken corresponded to that shown in Fig. 4. The wire targets were located 64 cm from the centers of the deflecting mirrors, with the angle between the incident wave and the measured scattered wave being less than $7°$.

Figure 12 shows the measured (circles) and calculated (line) pulse shapes in the time domain for three wire diameters. The calculated pulse shapes were obtained from (10)–(14). The cylinder axes were oriented perpendicular to the incident electric field vector. As seen in Fig. 12 there is good agreement between the calculated and measured responses (normalized in amplitude) in all cases.

The pulse that appears later in time (indicated by arrows in Fig. 12) is due to the "creeping wave". For a conducting cylinder with its axis oriented perpendicular to the incident field, the electric field of the THz radiation induces a pulse of current on the wire which propagates as a creeping wave around

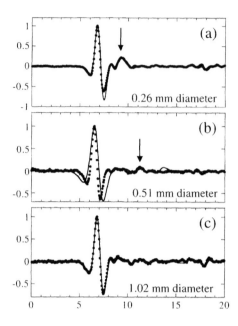

Fig. 12. Response of (a) 0.26, (b) 0.51, and (c) 1.02 mm diameter copper cylinders with axis oriented perpendicular to the incident field. The measured results are shown in dots and the calculated results by lines

the circumference of the wire [51]. The current pulse continually loses power to radiation as it propagates around the curved surface. As is evident from Fig. 12 the arrival of this secondary pulse is delayed in time in proportion to the wire diameter, and the pulse becomes strongly attenuated with increasing wire diameter. The other oscillations are due to residual water vapor in the enclosure. When the electric field vector is parallel to the cylinder axis, the generated pulse travels along the cylinder and is not seen by the detector. This is illustrated in Fig. 13, which shows the scattered pulse from a 0.51 mm diameter cylinder oriented both perpendicular (solid line) and parallel (dots) to the incident electric field. The reradiated pulse is identified by an arrow.

Similar results are obtained from dielectric cylinders, as shown in Fig. 14a, but the returned signature is much richer in features owing to the fact that the electromagnetic pulse can now penetrate the target. The target in this case is a 3.00 mm diameter finely ground alumina cylinder, for which the low-frequency real part of the index of refraction n is equal to 3.17. For these measurements the ranging system illustrated in Fig. 4 was utilized, and the reference target was a 30 mm diameter copper cylinder with a spectrally flat response. The reference pulse for this measurement has a frequency bandwidth from 0.1 THz to 1.5 THz. In contrast to the conducting cylinders, late-time resonances are observed as far as 205 ps after the first returned pulse.

A theoretical fit to the measured scattering data was performed using the Mie-scattering numerical model presented in Sect. 3. The absorption and

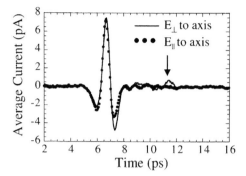

Fig. 13. Measured scattered THz pulse from 0.52 mm diameter copper cylinder. The dots represent the electric field parallel to the cylinder axis and the solid line the electric field orthogonal to the axis. The creeping-wave pulse, shown by the arrow, occurs only in the orthogonal orientation

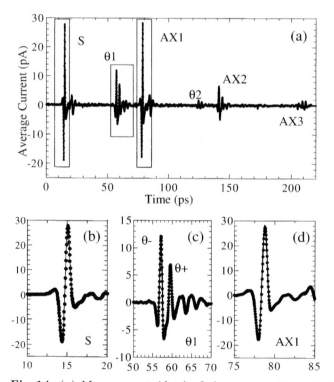

Fig. 14. (a) Measurement (dots) of electromagnetic scattering from a dielectric cylinder illustrating specular (S), surface ($\theta 1$ and $\theta 2$) and axial (AX1, AX2, and AX3) signatures. A fit using the Mie theory is shown as a solid line. The specular reflection is expanded in (b). (c) shows the surface wave, consisting of counterpropagating pulses $\theta +$ and $\theta -$. The first axial reflection is expanded in (d)

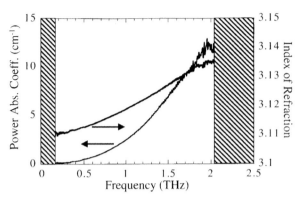

Fig. 15. THz-TDS measurement of absorption and index of refraction of the alumina used in the cylindrical and spherical targets

index of refraction of the alumina were measured using THz-TDS and are shown in Fig. 15; the values were used in the calculation of the complex wavevector k_r. Note that although crystalline Al_2O_3 (sapphire) is birefringent, with the ordinary and extraordinary indices being $n_o \cong 3.07$ and $n_e = 3.41$ respectively [33], the alumina used here was sintered from a powder and is not birefringent. The accuracy of the THz range is demonstrated by a comparison of the experimental data and the theoretical calculation directly in the time domain. The data (shown as dots) and the result of the numerical theory (solid line) are shown overlaid in Fig. 14a; the two graphs are virtually indistinguishable. The fit between the numerical results and the measured scattered field is shown enlarged in Figs. 14b–d for the first three features.

It should be noted that in fitting the data shown here and in all other comparisons between measurement and calculation, the only "floating" parameter was the amplitude of the calculated response. The amplitude was normalized to the specular reflection of the data using a factor of less than two. This normalization factor accounts for slight misalignments of the THz system, drifts in temperature, and time-dependent drifts of the excitation laser amplitude and pointing stability which occur during data acqusition. To obtain the fits shown here it was found that the eccentricity of the cylinder had to be kept to a minimum and the diameter should be constant over the illuminated area. The precision cylinder used was 3.001 ± 0.001 mm in diameter.

Similar accuracy is obtained in comparing the measured frequency-domain radar cross section (8) with the calculated value. The RCS is typically plotted in dB to illustrate the dynamic range of the measurement. The measured and numerically calculated RCSs are shown in Fig. 16, and has been expanded into three frequency regions for clarity. Owing to the finite duration of the time-domain data, the measurement is effectively windowed, which limits the

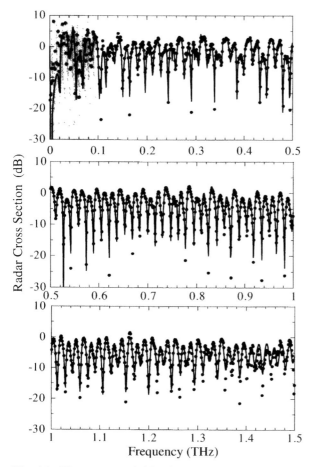

Fig. 16. The measured (dots) and calculated (lines) radar cross sections corresponding to the data shown in Fig. 14 for the alumina cylinder. The results are broken into three graphs for clarity

frequency resolution. For a temporal scan length of 220 ps, the corresponding full-width-at-half-maximum frequency resolution is 2.7 GHz. This limited resolution is accounted for in the theoretical fit to the data shown in Fig. 16. The discrepancy between the calculated RCS and the data at 1.16 THz is due to a strong water absorption line, owing to a relative-humidity difference of less than 0.5% or 150 parts per million of water vapor between the reference and sample scans.

4.2   Dielectric Spheres

Scattering from spheres is one of the classic problems in electromagnetics, owing to the large number of natural phenomena that can be understood through this scattering mechanism. The spectacular optical effects of the rainbow and the lesser known glory are due to scattering from water droplets in the atmosphere.

The use of ultrawide-band THz impulse ranging to measure scattering from dielectric spheres provides unique insight into the scattering from spheres, since it permits direct, time-dependent electric-field scattering measurements. The ability to temporally isolate the surface wave permits the first experimental verification of several surface-wave scattering theories.

The ranging system used to measure the scattered return from the dielectric spheres was of the kind shown in Fig. 9. The use of a 28 cm focal-length lens confocally with the off-axis parabolic mirror assures plane wave excitation at the target position. The $1/e$ amplitude diameter of the beam waist at the sphere was calculated to be 1.8 cm. The polarization of the incident field was in the plane of the page in Fig. 9. The target was an alumina sphere 6.36 mm (1/4 in) in diameter with a refractive index and absorption as shown in Fig. 15. The reference reflector in this configuration was a convex spherical metal mirror with a 12.5 mm radius of curvature. The time-dependent scattered electric field was measured with 0.3 ps resolution and a signal-to-noise ratio of approximately 30:1; the results are shown in Fig. 17. Comparing the magnitude of the scattered field (directly proportional to the average current measured) with that scattered from the cylindrical dielectric target, the field scattered from the sphere is seen to be approximately six times smaller. This is due to the decrease of the scattered field inversely with distance for the sphere (15) as opposed to the inverse square root decrease for a cylinder (10); also, the target cross section is smaller.

The measured scattered electric field shown in Fig. 17a consists of well-resolved, temporally separated pulses, each of which can be identified with a physical scattering mechanism. The first pulse, labeled "S" in Fig. 17a, shown in more detail in Fig. 17b, is due to specular reflection from the front surface of the sphere. The next feature, "θ", which is expanded in Fig. 17c is due to a surface wave. The pulse labeled "AX", shown in more detail in Fig. 17d is due to a back axial reflection; this is the portion of the incident pulse transmitted into the target and then reflected from the back surface of the sphere.

Numerical calculations of the scattering from the alumina sphere were made using the Mie scattering theory with the measured complex refractive index of the alumina. The predicted scattered pulses are shown by the solid lines in Fig. 17. As before, the only parameter which was varied during these fits was the amplitude of the calculated scattered pulse train.

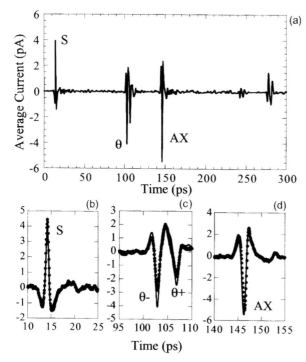

Fig. 17. (a) Measured specular (S), surface (θ), and axial (AX) features from a 6.35 mm diameter alumina sphere shown by dots. A fit using the Mie theory is shown as a solid line. The features are expanded in (b)–(d)

4.3   Data Analysis and the Geometrical Optics Model

The above analysis of the data was based on the Mie theory, which expands an incident plane wave into spherical or cylindrical wave functions to calculate exactly the internal and scattered fields. As mentioned previously although the solution is exact. it is expressed as an infinite summation of single-frequency waves and the visualization of the physics of electromagnetic scattering from such solutions is extremely difficult. This can be seen from Fig. 18, which shows the electric-field amplitude as a function of frequency obtained by a Fourier transformation of the data from the dielectric sphere of Fig. 17. Making measurements directly in the time domain using electromagnetic pulses can provide a simple and experimentally verifiable insight into the scattering process. Since. through the Fourier formalism, any pulse consists of a summation of an infinite series of single frequency waves, each with a fixed amplitude and phase, the observed data for a scattered pulse represents a superposition of Mie-theory scattering solutions with fixed phase and amplitude relationships. The Mie solutions are localized in energy

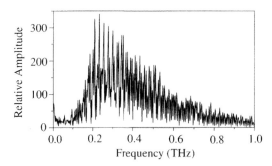

Fig. 18. Spectrum of time-domain measurement in Fig. 17 calculated using a numerical Fourier transformation

(or frequency) space, while the observed pulses are localized temporally and spatially.

The rapid oscillations of the electric-field amplitude as a function of frequency in Fig. 18 contrast with the measured scattered pulses in Fig. 17. The individual pulses contain continuously varying amplitude spectra which are qualitatively similar to if smaller than, the incident pulse represented in Fig. 5. The complexity and rapid oscillation of the spectrum in Fig. 18 arises from interference between the widely separated pulses shown in Fig. 17. To relate the spatially and temporally localized scattering mechanisms observed in the time domain to the well-developed Mie theory, each isolated pulse was Fourier transformed individually to remove the interference terms and obtain its frequency content.

The Fourier transforms of the individual pulses labeled "S". "θ". and "AX" in Fig. 17 are shown in Fig. 19 together with the amplitude spectrum of the measured reference pulse reduced in scale by 500 times. In contrast to the complex spectrum in Fig. 18. the frequency dependence of these pulses

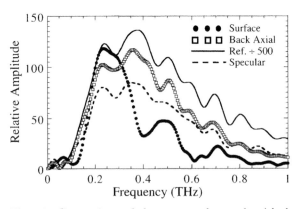

Fig. 19. Comparison of the measured specular (*dashed line*), surface (*dots*), and axial (*open squares*) pulses and the reference pulse scaled down 500× (*solid line*). The surface wave spectrum is strongly reshaped

varies slowly and, with the exception of the pulse labeled "θ", these spectra look very similar to the spectrum of the incident reference pulse. This is not surprising since their origin is due to simple reflection and refraction; the pulse labeled "S" is a specular reflection from the front surface of the sphere, and the pulse "AX" is the pulse reflected from the back surface of the sphere after making one round trip. Note that with increasing frequency the "S" spectrum from the front surface becomes larger than the "AX" spectrum, owing to the absorption of the alumina (Fig. 15). In contrast, the contribution from the pulse labeled "θ" shows a frequency-dependent oscillatory structure which can be predicted from the strong reshaping of this pulse. This pulse is due to a superposition of two "surface waves" which travel in opposite directions around the circumference of the sphere, and cannot be explained through simple ray tracing, or geometrical optics.

An explanation of this oscillatory structure and of the effect of the surface waves was first proposed by van de Hulst [52] through numerical approximations to the Mie series in order to explain the optical glory seen in water droplet clouds in back illumination. In this treatment the scattering coefficient $S(\omega, \theta)$, due to the surface wave is obtained from the Mie theory by applying the localization principle [52], considering only those terms corresponding to a ray striking at grazing incidence. The result is

$$S(\omega, \theta) = \frac{1}{2}C_2[J_0(u) + J_2(u)] + \frac{1}{2}C_1[J_0(u) - J_2(u)], \tag{20}$$

where $u = 2\pi a\theta/\lambda$, and C_1 and C_2 are constants. For water droplets, where $n \cong 1.33$, van de Hulst estimated the ratio C_1/C_2 to be in the range $-1/4$ to $-1/5$. The scattered electric-field amplitude for the surface wave contribution can be calculated from the measured spectrum of the reference pulse multiplied by the scattering coefficient and an empirical amplitude constant A, i.e. $|E_{\text{surf}}(\omega)| = |A \times E_{\text{ref}}(\omega) \times S(\omega, \theta)|$. The calculated spectrum due to the surface wave determined from (20) is plotted as a dashed line in addition to the data (dots) in Fig. 20; the best fit is obtained for $C_1/C_2 \cong -0.5$. Acceptable fits are obtained for C_1/C_2 in the range from -0.35 to -0.75. For comparison, the spectrum calculated from the full Mie theory is shown as a solid line in the figure. For this calculation, the scattered pulse was calculated via equations (15) to (19), isolated in time, and Fourier transformed. Despite the approximations made in this analysis, the qualitative agreement is good across the entire bandwidth. The absence of quantitative agreement shows that this simplification of the Mie theory for waves striking the surface is not completely valid and that other scattering mechanisms play a large role.

To help one understand the physical mechanisms of scattering from simple geometric objects, a "geometrical optics", or ray tracing, approximation and a "physical optics" approximation, which includes diffraction, are often used. These approximations are valid when the size of the scatterer is greater than the wavelength. This condition is often expressed in terms of the wavevector of the radiation in free space, k, multiplied by the target radius, a, as

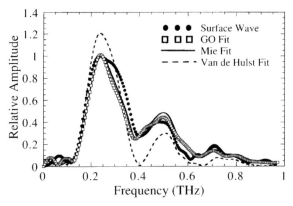

Fig. 20. Theoretical fits to the surface wave data using the Mie, van de Hulst, and geometrical optics models (GO) discussed in the text

$ka > 5$. A geometrical-optics picture of scattering from spheres and cylinders has been developed to describe the observed scattering from these objects [31,53,54]. The inherent assumption in this simple model is that the observed pulses propagate along well-defined paths which can be treated in a ray optics approximation. Thus each of the well-defined pulses observed for spheres and cylinders, as shown in Figs. 14 and 17, propagates on a different path through the target. This geometrical-optics model assumes that the propagation along a given path is described by complex wave-vector $k(\omega) = \omega n(\omega)/c$. The complex index of refraction $n(\omega)$ takes into account both amplitude attenuation and phase delay.

The possible paths taken by the pulses in this model are illustrated in Fig. 21. From this figure, we see that there are three main scattering mechanisms involved: the first pulse, labeled "S" in Figs. 14 and 17 is the specular reflection from the front surface of the target (cylinder or sphere). The pulses labeled "AX" (AX1, AX2, AX3 in Fig. 14, AX in Fig. 17) are contributions from the incident pulse coupled into the target at near-normal incidence which undergoes one or more reflections from the back surface of the target. The third contributing scattering mechanism is the surface, or creeping, wave. Here the THz pulse propagates along ray paths both through the material and along the surface. The THz pulse is weakly coupled to the surface owing to diffraction at the edge of the target when the plane of the surface and the k vector of the incident pulse are parallel. This feature is identified by "$\theta 1$" and "$\theta 2$" in Fig. 14, and by "θ" in Fig. 17. On the cylinder, the wave "$\theta 1$" makes a total rotation of π radians around the cylinder, while "$\theta 2$" makes a rotation of 3π.

In case of the surface wave, the incident plane wave couples to a weakly bound surface mode which begins to propagate around the circumference of the target. These external surface modes are treated explicitly and rigorously

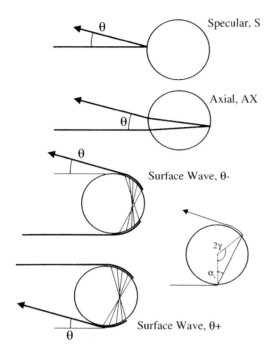

Fig. 21. Illustration of scattering mechanisms in the geometrical-optics model. The critical angle is α_c. The surface wave paths cutting through a chord of the cylinder or sphere are illustrative only, since an infinite number of equivalent paths exist

in the singularity expansion method, which is similar to the Mie theory expansion into an infinite series. In the time domain, we use a geometrical-optics model to provide insight into the scattering. Here, the coupling of the incident THz pulse into a surface wave is modeled by a frequency-independent constant A. This surface wave propagates along the surface of the target with an effective index of refraction n_{eff}. If the surface wave is assumed to propagate along a path coincident with the surface and to travel ϕ radians around the circumference, the total path length is ϕa, where a is the target radius. There are two energy loss mechanisms for this surface wave. Owing to the curvature of the surface, the surface wave is continuously radiating energy tangentially outward from the surface. This is taken into account by the imaginary component of n_{eff}. Energy can also enter the dielectric target and propagate at the critical angle $\alpha_c = \sin(1/n'(\omega))$, where $n(\omega) = n'(\omega) + in''(\omega)$ is the complex refractive index of the target (see Fig. 15). Since the imaginary part of the index is small, we ignore phase shifts in the surface coupling. Since there is no mechanism which distinguishes one point on the surface from any other, there are an infinite number of equivalent paths which the pulse may take, as illustrated in Fig. 21. The fraction of the surface wave coupled into the dielectric is given by a frequency-independent constant C_{12}. Inside the dielectric, the amplitude attenuation is determined by the complex index of refraction of the material, shown in Fig. 15 for alumina, where the complex index and power absorption coefficient are related by

$$a(\omega) = i\frac{4\pi}{\lambda}n''(\omega). \tag{21}$$

The length of propagation through the material is the length of the chord. $2a\sin\gamma$, which subtends an angle 2γ, where $\gamma = \pi/2 - \alpha_c$. The pulse, incident on the surface at the critical angle, is then coupled back into a surface wave with a frequency-independent coupling coefficient C_{21}.

Since the direction of the detected THz pulses is at an angle of θ to the incident radiation, determined by the geometry of the range shown in Fig. 4, two paths are available for this pulse. One, labeled as $\theta-$ travels counterclockwise around the cylinder, making a total angular rotation of $\pi - \theta$. The other, labeled $\theta+$, travels clockwise, making a total rotation of $\pi + \theta$. The different arrival times of these two pulses result in the "reshaping" of the pulses in Figs. 14 and 17. The total angle subtended by the surface path is given by the total angular rotation excluding the angle traversed through the cylinder, i.e. $\pi \pm \theta - 2\gamma$. The propagation of the surface wave is then described by

$$E_{\theta,1}(\theta,\omega) = A(C_{12}C_{21})E_{\text{ref}}(\omega)e^{i\eta\pi/2}$$

$$\times \left[e^{ik_s(\omega)a(\pi-2\gamma)}e^{ik_d(\omega)2a\sin\gamma}\left(e^{-ik_s(\omega)a\theta} + e^{ik_s(\omega)a\theta}\right)\right]. \tag{22}$$

Here $E_{\text{ref}}(\omega)$ is the complex amplitude of the incident pulse, and the two surface path lengths, $a(\pi \pm \theta)$, has been given explicitly. The wave-vectors of the radiation along the surface, $k_s(\omega)$, and through the dielectric, $k_d(\omega)$ are given by:

$$k_s(\omega) = \frac{\omega n_{\text{eff}}}{c},$$

$$k_d(\omega) = \frac{\omega n(\omega)}{c}, \tag{23}$$

where $n(\omega)$ is obtained from Fig. 15. The constant phase factor $\exp(i\eta\pi/2)$ is due to a topological Gouy phase shift and will be described in detail later. The factor η is 1 for a cylindrical target and 2 for a spherical target.

By floating the coupling constants A, C_{12}, and C_{21}, it is possible to fit the reshaped surface wave spectrum measured for the dielectric sphere, as shown in Fig. 20. Numerically fitting the measured amplitude spectrum of the surface wave to the geometrical model of (22) gives the fit shown by open squares in Fig. 20, with $n_{\text{eff}} = 0.98 + 0.08i$. Owing to the strongly reduced scattering from the sphere, the accuracy of the values is substantially less than that which can be obtained for a cylinder. The accuracy of the numerical fit is on the order of $\pm 5\%$ for the real part and $\pm 15\%$ for the imaginary part of n_{eff}. This fit compares quite favorably with the fit obtained from a numerical Mie theory analysis, also shown in Fig. 20. It can be seen that the surface wave spectrum, strongly reshaped compared with the specular and axial reflections, is due to the interference of the two pulses, $\theta+$ and $\theta-$, propagating different distances around the circumference of the spherical target. This geometrical-optics model provides insight into the scattered radiation patterns expected from dielectric spheres. Note that the geometrical-optics model, (22), can be written in the same form, within a constant, as the van de Hulst approximation to the Mie-scattering surface wave terms, (20), for $C_1/C_2 = -3$ and $n_{\text{eff}} = 1 + 0i$.

In the model discussed up to this point, A, C_{12}, C_{21}, and n_{eff} are not known and must be determined by a fit to the data. A, C_{12}, and C_{21} are treated as a single parameter which determines the amplitude of the surface wave relative to the incident pulse. Since the two surface waves $\theta+$ and $\theta-$ have different path lengths along the surface, n_{eff} determines their relative temporal separation and amplitude. A fit to the observed surface wave in the measurement on the dielectric cylinder (Fig. 14) is shown in Fig. 22a. Here $AC_{12}C_{21} = .025$ and $n_{\text{eff}} = 1.10 + i0.07$. Note that it is not possible to determine any frequency dependence of n_{eff} from this measurement.

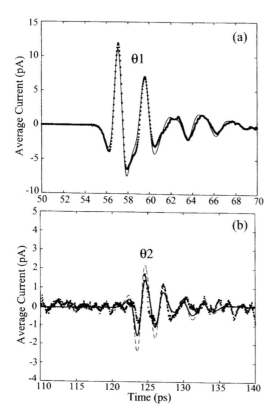

Fig. 22. Fit to single circumnavigation (a) and double circumnavigation (b) of a surface wave pulse on an alumina cylinder used to find the coupling constant $C_{12}C_{21}$

This model can be extended to measure the scattered pulse which travels an angle of $3\pi \pm \theta$ radians around the cylinder. The corresponding rays will travel three times through the cylinder, with a total angle of 6γ, with the remainder of the path being along the surface. This second surface wave is labeled $\theta 2$ in Fig. 14. Owing to the strong attenuation of the surface path compared with that in alumina ($\alpha = 29\,\mathrm{cm}^{-1}$ at 1 THz for the surface vs. $\alpha = 2.0\,\mathrm{cm}^{-1}$ at 1 THz in alumina) in this geometry, the rays which from two chords through the cylinder are not seen owing to the long surface path. The total surface path length for this second surface wave is $(3\pi - 6\gamma \pm \theta)a$. The frequency-dependent field is then

$$E_{\theta 2}(\theta, \omega) = A(C_{12}C_{21})^3 E_{\mathrm{ref}}(\omega)e^{i\eta(3\pi/2)}$$

$$\times \left[e^{ik_s(\omega)a(3\pi - 6\gamma)}e^{ik_d(\omega)6a\sin\gamma}\left(e^{-ik_s(\omega)a\theta} + e^{ik_s(\omega)a\theta}\right)\right]. (24)$$

Since the value of n_{eff} was determined by fitting (22) to the first surface wave, comparison of the second surface wave $\theta 2$ to the first surface wave $\theta 1$ permits a measurement of the magnitude of the coupling constant $C_{12}C_{21}$, as shown in Fig. 22b. The dashed line in the figure is the calculated result for $\theta 2$ assuming a coupling efficiency of $C_{12}C_{21} = 1$. To obtain the coupling efficiency, a best fit to the data was performed and is plotted as a bold solid line, corresponding to a total coupling efficiency of $C_{12}C_{21} = 0.89$. Note that most of the reduction of the $\theta 2$ signal is accounted for by the propagation effects of the wave on the surface and not through the alumina.

Other scattering phenomena may be predicted from this geometrical-optics model. For example, the scattered electric field of (22) shows a strong dependence on the bistatic angle θ. For backscattering ($\theta = 0$), the optical paths for all frequencies are equal, and constructive interference occurs at all frequencies. As θ changes, the relative optical path difference, ΔD, between the $\theta+$ and $\theta-$ pulses varies as $\Delta D = 2\theta a n_{\mathrm{eff}}$. In this case constructive interference occurs for $\Delta D = 2m\lambda$, or $\theta = m\lambda/a n_{\mathrm{eff}}$, and destructive interference for $\Delta D = (2m + 1)\lambda/2$ with $m = 0, 1, \ldots$. This interference is independent of the refractive index of the sphere, which is also the case for the frequency-dependent oscillations of (20).

This angle-dependent oscillatory structure is similar to the well-known optical glory. The glory is an enhancement of the scattered radiation in the backward direction and appears as concentric rings of colored light when suspended water droplets are illuminated from directly behind the observer. This phenomenon is observed much more rarely than its forward-scattering analogue, the rainbow. The relative rarity is due to the difficulty of obtaining an illumination source behind the observer: in fact, the first recorded observation of this phenomenon was around 1737 by a Spaniard, Ulloa, in the Andes mountains [1]. Its most common modern observation occurs around the shadow of an airplane flying over clouds.

As shown by Fig. 23, the amplitude of the scattered field varies as a function of angle for a given frequency. This variation is determined by con-

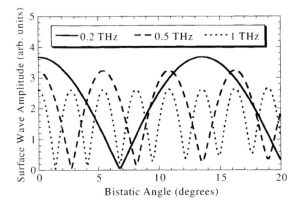

Fig. 23. Angular and frequency dependence calculated for aTHz glory

structive and destructive interference of the $\theta+$ and $\theta-$ waves, which leads to a "rainbow" variation of the broadband THz radiation as a function of angle. For an unpolarized incident field, this would result in a ring-like structure of the various frequency components of the THz radiation in the backscattering direction. Note that the frequency and angle dependence of the THz glory here is almost entirely due to the surface waves, unlike the situation for the optical glory, in which many competing physical effects contribute [55].

4.4 Angularly Resolved Scattering Measurements

The experimental technique of THz impulse ranging provides the ability to isolate individual scattered pulses in the time domain. The fact that the temporal separation between separate scattering mechanisms is long compared with the duration of the pulses permits investigation of details of the scattering mechanism through a geometrical-optics model. The measurements presented so far, however, are unable to resolve the frequency dependence of the surface propagation. The slight offset between the positions of the predicted and measured valleys at approximately 0.4 and 0.6 THz in Fig. 19 can be removed by using a frequency-dependent real part for n_{eff}; however, it is impossible to determine experimentally the exact form of this frequency dependence. The imaginary part of n_{eff} has been predicted to have a $\lambda^{2/3}$ dependence [55].

To determine the frequency dependence of surface wave propagation characteristics, it is necessary to perform measurements over a range of angles. Here, isolating the surface wave pulses in the time domain allows the direct measurement of surface wave propagation independent of other mechanisms such as radiation or coupling. The impulse ranging system used for these measurements, shown schematically in Fig. 10, was a modification of a fixed bistatic THz impulse ranging system. In this system the angle θ between the source and the detector can be adjusted from $21°$ to $180°$ about the axis of rotation as defined by the location of the target. The previous measurements were restricted to a fixed angle of approximately $10°$. Scattered THz radiation is detected by a receiver with an 8 mm diameter aperture placed 265 mm from the target. The 8 mm aperture corresponds to a phase front curvature at the receiver of less than $\lambda/30$ at the central wavelength and defines an angular resolution of $1.5°$. The absolute timing between the optical pulses in this free-space system is maintained between different measurement angles θ by steering the detector gating beam along the rotation arm axis with a periscope beneath the target.

The arrival of the incident THz pulse at the 3 mm diameter alumina cylinder used as the target establishes an absolute time reference. Unlike the case in previous measurements, which used a "reference reflector", this pulse was measured at $\theta = 180°$ with no target between the source and detector. The valid spectral range for the surface wave measurements was between 0.3 and 1.3 THz, determined by the signal-to-noise ratio of the surface wave reference spectrum $E_{\mathrm{ref}}(\omega)$. The time-dependent scattered electric field is shown in Fig. 24a, where the scattering mechanisms discussed in Sect. 4.3 are clearly visible: specular reflection (S), the surface wave (θ), and a wave propagating through the target (T), which corresponds to the axial reflection in the previous measurements. In this configuration, the absolute time of arrival of the surface wave radiation is measurable because the distance from the tangent of the cylinder to the detector is constant, as the rotation arm and the cylindrical target are axially aligned.

The energy path or Poynting vector of these waves is fundamentally different from that of surface waves measured at more acute angles ($\theta \cong 10°$). At small bistatic detection angles ($\theta < 36°$ for alumina), the observed surface wave takes a path through the dielectric at the critical angle and reemerges further along the surface to travel as a surface wave. Radiation which "cuts through" the cylinder and couples out to a surface wave suffers orders of magnitude less loss than does radiation which travels purely on the surface, for a given angular distance. In the present measurements, the "cut through" paths are not seen.

By measuring the surface wave radiation at oblique angles using the ranging system described above, the detected surface wave can be effectively isolated from paths propagating through the target dielectric. To measure surface wave propagation characteristics experimentally, we can vary θ such that the surface wave travels along path lengths differing by $d_{\mathrm{surf}} = r\Delta\theta$, where r is the radius of the target and $\Delta\theta$ is the change in θ between two measurements. Specular reflection (S) and surface wave (θ) measurements are shown in Fig. 24b over a range of θ from $150°$ to $120°$ in $10°$ increments.

To express the signal measured as a function of angle, we define $E_{\mathrm{rcv}}(\omega, \theta)$ as the signal detected at the receiver, $E_{\mathrm{inc}}(\omega, \theta)$ as the electric field of the surface wave pulse on the target surface, and a factor $R(\omega)$ describing the coupling in and out of radiation. The complex, frequency-dependent effective index of refraction n_{eff} is used to determine the wave vector of the surface

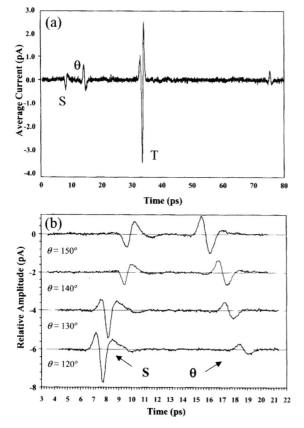

Fig. 24a,b. Measurement of specular (S), surface (θ), and traversal (T) waves on a dielectric cylinder measured at an angle of 150°. Details of the specular (S) and surface (θ) waves at bistatic angles of 150° (reference), 140°, 130° and 120°

wave tangentiel to the surface, $n_{\text{eff}} k_0$, where k_0 is the associated free-space wave vector. Here and in the following, r is the fixed radius of the cylinder and θ is the angular propagation of the surface wave measured from the point on the cylinder tangent to the wave vector of the incident plane wave. We have

$$E_{\text{rcv}}(\omega, \theta) = R(\omega)E_{\text{inc}}(\omega, \theta) = R(\omega)E_{\text{inc}}(\omega, 0)e^{in_{\text{eff}} k_0 r\theta}. \qquad (25)$$

To deconvolve the system response, we take the ratio of the signal spectrum to a reference spectrum to isolate the surface wave behavior from the coupling mechanisms.

$$\frac{E_{\text{sig}}(\omega, \theta_0 + \Delta\theta)}{E_{\text{ref}}(\omega, \theta_0)} = \frac{R(\omega)E_{\text{inc}}(\omega, 0)e^{in_{\text{eff}} k_0 r(\theta_0 + \Delta\theta)}}{R(\omega)E_{\text{inc}}(\omega, 0)e^{in_{\text{eff}} k_0 r\theta_0}}$$

$$= e^{-in_{\text{eff}}(\omega)k_0 r \Delta\theta} = e^{-ik_{\text{surf}}(\omega)d_{\text{surf}}}. \qquad (26)$$

Here E_{ref} is the reference spectrum, which is that of the isolated surface wave measured at $\theta_0 = 150°$ and E_{sig} is one of the corresponding signal spectra taken at the angles of 140°, 130°, and 120°. The use of a surface wave to provide the reference spectrum permits direct measurement of surface wave propagation, since any coupling mechanisms are divided out in (26). $\Delta\theta$ is the change in angle between the signal and reference spectra. This notation is formally equivalent to the notation used in Sect. 4.3 to describe a wave propagating along a surface path of length $d_{\mathrm{surf}} = r\,\Delta\theta$ with propagation constant $k_{\mathrm{surf}}(\omega) = n_{\mathrm{eff}}(\omega)k_0$.

Figure 25 shows the real part of the effective index, $n(\omega) = \mathrm{Re}(n_{\mathrm{eff}})$, and the field loss coefficient, $\alpha(\omega) = \mathrm{Im}(n_{\mathrm{eff}})k_0$, measured as a function of frequency. The measured values $n_{\mathrm{meas}}(\omega)$ and $\alpha_{\mathrm{meas}}(\omega)$ are shown as points and are the averaged values from the three angles to reduce noise. The individual measurements at each angle varied by no more than $\pm10\%$. These results agree well with results generated by Mie-theory numerical analysis (solid line) using parameters matching the experimental configuration. The values for $n_{\mathrm{Mie}}(\omega)$ and $\alpha_{\mathrm{Mie}}(\omega)$ used in Fig. 25 were determined following the same procedure as outlined above, but the scattering was determined numerically from the measured reference pulse. The real index in Fig. 25 is within 4% of that measured on the fixed-angle range in the valid spectral region and closely matches the Mie theory predictions. The index of refraction cannot be determined accurately below 0.3 THz since extracting the phase at low

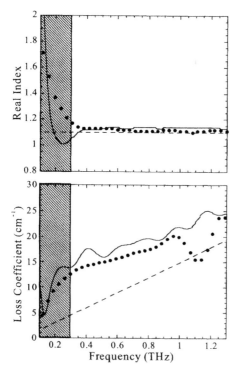

Fig. 25. *Top:* comparison of the measured real part of the effective index (*dots*) with the value calculated using Mie thoery (*solid line*) and the frequency-inependent value used previously (*dashed line*). *Bottom:* comparison of the loss coefficient measured from the effective index (*dots*) with the coefficient calculated using Mie theory (*solid line*) and the frequency-independent loss reported previously (*dashed line*). The measured values are averages for the three signal angles

frequencies has errors due to the 8 ps time window of the measurement and the decreased THz spectral content. The low index shows that the wave is weakly coupled to the surface. However, the loss component $\alpha_{meas}(\omega)$ shown in Fig. 25 is consistently higher (by approximately $9\,cm^{-1}$) than that of the surface wave which travels through the cylinder. The dashed line represents the constant n_{eff} determined by the measurements described in Sect. 4.3

4.5   Gouy Phase Shift

The THz impulse range permits direct observation of several phenomena which are explicitly treated in the Mie scattering theory but are not transparent from the mathematics. One of these phenomena is the Gouy effect, which can be observed for both cylindrical and spherical targets [54]. The Gouy effect is the well-known π phase shift that any spherical wavefront accumulates relative to a uniform plane wave as it propagates through a focus, which also applies to Gaussian beam propagation [56]. This Gouy phase shift is observed in focused single-cycle electromagnetic pulses [57] with direct observation of a π Gouy phase shift through a spherical focus. What is less well known is that the Gouy phase shift is geometrically dependent on the type of focus and that through a one-axis (cylindrical) focus, the phase shift is $\pi/2$, in contrast to the phase shift of π for a two-axis (spherical) focus. These phase shifts are required in interpreting the scattering results using the geometrical-optics model and were given explicitly in (22) and (24) [54].

The experimental data for the time-domain scattering of a THz impulse from the alumina sphere and cylinder are shown in Figs. 26a,b, respectively, for comparison. As before, the pulse labeled "S" in the figure is the specular reflection from the front surface of the sphere; the surface or creeping wave, located in time between S and AX1, is labeled "θ", and the pulses labeled "AX1 and "AX2" are the multiple orders of reflections between the front and back surfaces of the target. The reference pulses for both targets are shown as insets in the upper right corners. Here we focus our attention on the axial pulses, AX1, in Fig. 26. The Fresnel equations require that the specular reflection S is in phase with the reference pulse since the refractive index of the target ($n \cong 3.17$ for alumina) is higher than that of air. This is observed in both parts of Fig. 26. The reflection off the back surface, AX1. undergoes a π phase shift, however, since the far side of the interface has a lower index. This shift is evident in the results from the spherical target, Fig. 26a but not on the cylindrical, Fig. 26b.

The lack of the expected inversion of the cylindrical target is due to the Gouy phase shift [54]. This phase shift arises owing to focusing effects of the rays in the GO model refracting from the curved surfaces of the target. A ray diagram of the backscattered axial wave AX1 at the experimental detection angle of $\theta = 13°$ is shown in Fig. 27. For a spherical target, the rays will focus to a point owing to the spherical phase front induced by the target. Thus, through the Gouy phase shift, the ray bundle acquires a π phase shift at each

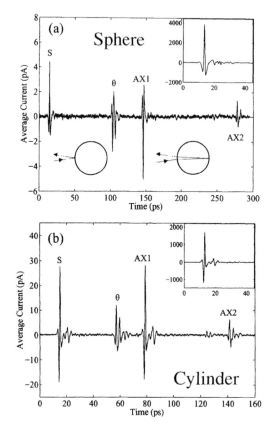

Fig. 26. (a) Measured scattering from a 6.346±0.001 mm alumina sphere. (b) Measured scattering from a 3.001 ± 0.001 mm diameter alumina cylinder. The insets show the reference pulses

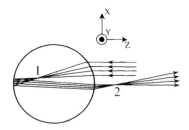

Fig. 27. Ray diagram of backscattered axial reflection AX1 at bistatic angles centered at $\theta = 13°$ showing the two Gouy phase focal regions

focus. Propagation through two focal points, one internal to the target and one external, lead to a total accumulated phase of $\pi + \pi = 2\pi$ from the Gouy effect. For the sphere, the accumulated phase shift is not observable except as a time shift of the returned pulse.

In the case of the cylinder, the spherical focal symmetry is broken. The focal regions are one-dimensional, i.e. the focused pulses have a line focus as with a cylindrical lens. Although it is not as well known in the optical literature, the accumulated phase shift through a one-axis focus or caustic is $\pi/2$ [58]. The scattered wave AX1 from the cylinder thus acquires a total accumulated phase shift from the Gouy effect of $\pi/2 + \pi/2 = \pi$, which accounts for the observed inversion of the scattered wave AX1 between the spherical and the cylindrical target in Fig. 26. This direct observation of the Gouy phase shift demonstrates the need to account for topologically dependent phase shifts when using a geometrical- or physical-optics approach to understanding scattering phenomena. The relative polarity between the specular reflection S and the backscattered axial reflection AX1 is a signature which can reveal the geometry of the scattering target in a phase sensitive system.

4.6   Realistic Targets

Characterization of a THz range using cylindrical and spherical targets illustrates the accuracy of the measurements which can be performed. As discussed in Sect. 3, the scalability of Maxwell's equations (given by (4)) permits scale measurements on realistic targets. If, for example, a radar measurement is required at some frequency f_0, the target can be characterized on a scaled THz range and the frequency of the scale range, f_S, is increased by a factor of N, i.e. $f_S = N f_0$. The size of the target is then scaled by a factor $1/N$. From (4) it can be seen that if both space (L) and time (T) are scaled by the same factor N, both the permittivity $\varepsilon(\omega)$ and permeability $\mu(\omega)$ at f_S must be equal to those at f_0. The conductivity $\sigma(\omega)$, however, must scale as N. This is especially important for targets that support surface currents, where physical limits on conductivity impose restraints.

The effect of finite conductivity at THz frequencies is currently an open question. Owing to the effectively infinite conductivity of many metals up to and through the THz frequency regime, this may not be a significant factor for many target characterization measurements. Previous measurements on conducting targets [32] agreed well with scattering theory based on the assumption of infinite conductivity. However, quantitative clarification of the importance of limited conductivity on the high-frequency response of metal models is required.

Preliminary measurements of the response of realistic targets with a more complex late time response have been made [32] to illustrate the scaling capabilities of THz impulse ranging. Measurements were made on metal scale-model aircraft with target/wavelength ratios of practical interest. Two 1/200 scale model aircraft, a MiG-29 Fulcrum and an F-117A Stealth, were measured using the ranging system shown in Fig. 4. The size of the model aircraft used were approximately 75 mm in length. The aircraft models were located 107 cm from the deflecting mirrors and oriented in various ways. Very fine nylon threads were used to suspend the model aircraft. The 1/200 scale determines the scaling ratio, $N = 200$. The THz impulse range bandwidth of 100 GHz to 2 THz and the target distance of approximately 1 m corresponds to measurements on a full-size target at 230 m distance over frequencies from

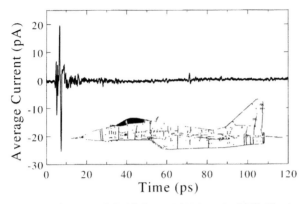

Fig. 28. Scattered field from 1/200 scale MiG-29 aircraft model in profile. The 1.07 meter target distance corresponds to a full size measurement at over 214 m distance

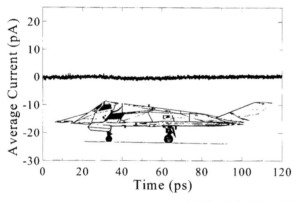

Fig. 29. Scattered field from model F-117A "Stealth" fighter aircraft in profile. measured under conditions identical to those used for the MiG-29 in Fig. 28

0.5 to 10 GHz. Measurements on the scale model aircraft are shown in Figs. 28 and 29. It should be noted that these measurements are qualitative. since both the MiG-29 and F-117 are constructed from composite materials. In addition. the F-117 makes use of radar-absorbing materials. which are not represented well by the metal scale model.

The time-domain response for the MiG-29, Fig. 28, has an extensive signal after the initial specular reflection. The model was oriented as shown in the figure. with the k vector of the incident THz radiation normal to the plane of the page. The peak signal of approximately 25 pA corresponds to a signal-to-noise ratio of approximately 150:1. In comparison to the relatively large signal for the MiG-29. with features extending over the 120 ps measurement window (corresponding to 24 ns for a full size aircraft), no signal is observable for the

F-117 model (Fig. 29). The lack of a scattered signal from the Stealth aircraft within our signal-to-noise ratio indicates the effectiveness of the aircraft's geometry, which is designed to reflect energy away from the detector, since no absorbing materials were incorporated in the model. The F-117 model, of the same physical size as the MiG-29 model, was oriented as shown in the inset to Fig. 29.

5 Summary and Future Directions

THz impulse ranging has demonstrated the ability to measure electromagnetic scattering of both specular and late-time responses from various targets. Optical gating of biased semiconductors results in the generation of temporally short, impulse-like electromagnetic pulses with THz bandwidth. The use of quasi-optical methods to collect the generated radiation permits these pulse trains to be collimated in near-diffraction-limited beams with a near-Gaussian profile [38]. The wavelength range of 3 mm to 150 µm permits the use of commercially available optics to generate near-planar phase fronts at the target.

The near-planar phase fronts and inherent dynamic range of the THz impulse ranging system have permitted measurements on geometrically simple targets which have shown excellent agreement with predicted returns based on numerical calculations of Mie scattering. Radar cross section measurements on dielectric cylinders have demonstrated over 20 dB of dynamic range over the range 400 GHz to 1.2 THz. This dynamic range is considerably lower than that obtained from RCS measurements made at single RF frequencies, in which dynamic ranges of over 50 dB are typically achieved. The lower dynamic range is partly due to fundamental differences between the impulse and single-frequency techniques. The impulse ranging system may be thought of as a fully open multichannel frequency system where the pulse width corresponds inversely to the bandwidth of the system, the measurement window (the length of the scan in time) determines the frequency separation between channels, and the temporal delay between data points determines the maximum frequency. It should be noted that the signal-to-noise ratio in the THz impulse range measurements compares favorably with that of other impulse ranging studies [19,22,25,27,30].

The most needed improvement to the THz impulse ranging system is to further increase the signal-to-noise ratio. An increase in the dynamic range of the system is critical for the ability to perform rapid scattering measurements at a range of bistatic angles. As in all areas, work in needed on higher-brightness THz sources since the average power in the THz beam is very low (10 nW average power, 0.1 fJ per pulse [41]). Other methods to improve dynamic range include signal averaging, and improving the efficiency of the optical system. Ranging systems based on real time digital sampling have an advantage in that many scans can be acquired in a short time to improve the signal-to-noise ratio by averaging [19.21]; however, the timing jitter of the electronics may degrade the temporal resolution. The adaptation of techniques used in THz imaging may permit real-time data acquistion with essentially no timing jitter, since in these techniques the same optical pulse is used to generate the THz radiation and gate the detector.

Issues of secondary importance needed to improve the utility of THz ranging are repeatable target positioning and alignment, optical systems to provide plane wave illumination over large target areas, accurate construction of small-scale models, and development of materials whose dielectric response at THz frequencies mimics that in the GHz frequency range. This last problem is exacerbated by the requirement that materials must have similar permittivity over a wide frequency range. Additionally issues on scaling the conductivity σ for targets that support surface currents need to be addressed along with fundamental investigations of the anomalous skin depth [46].

Further extension of this work beyond the simple targets and model aircraft measured so far has many possibilities based upon the ability to scale scattering measurement systems to a size which can easily fit on a typical optical table. These possibilities include construction of small-scale dioramas, multiple-path scattering measurements, and test beds for locating underground or hidden targets. Fundamental research would be needed to recreate the dielectric properties of soils in the frequency range of interest.

References

1. D. K. Lynch, S. N. Futterman, "Ulloa's observations of the glory, fogbow, and an unidentified phenomenon", *Appl. Opt.* **30**, 3538 (1991).
2. R. Buderi, *The invention that Changed the World: How a Small Group of Radar Pioneers Won the Second World War and Lauched a Technological Revolution* (Simon and Schuster, New York, 1996).
3. L. E. Larsen, J. H. Jacobi, *Medical Applications of Microwave Imaging* (IEEE Press, New York, 1985).
4. C. F. Bohren, D. R. Huffman, *Absorption and Scattering of Light by Small Particles*, (Wiley, New York, 1983).
5. V. A. Borovikov, *Geometrical Theory of Diffraction* (Institution of Electrical Engineers, London, 1994).
6. P. W. Barber, E. K. Miller, T. K. Sarkar, "Scattering by three-dimensional objects", *J. Opt. Soc. Am. A* **11**, 1380 (1994).
7. A. Taflove, *Computational electrodynamics: The Finite Difference Time Domain Method*, (Artech House, Boston, 1995).
8. C. A. Balanis, *Advanced Engineering Electromagnetics* (Wiley, New York, 1989).
9. E. M. Kennaugh, R. L. Cosgriff, "The use of impulse response in electromagnetic scattering problems", presented at Wescon IRE Convention, pp. 207–209 (1958), Los Angeles, CA.

10. E. M. Kennaugh, "The impulse response concept in single-body scattering", presented at the Second Interdisciplinary Conference on Electromagnetic Scattering, University of Massachusetts at Amherst (1965), published by Gordon and Bresch, New York, pp. 217–235.

11. E. M. Kennaugh, D. L. Moffatt, "Transient and impulse response approximations", *Proc. IEEE* **53**, 893 (1965).

12. E. K. Miller, J. A. Landt, "Direct time-domain techniques for transient radiation and scattering from wires", *Proc. IEEE* **68**, 1396 (1980).

13. M. O. White, "Radar cross-section: measurement, prediction, and control", *IEEE Electron. Commun. Eng. J.* 169, (1998).

14. R. C. Johnson, H. A. Ecker, R. A. Moore, "Compact range techniques and measurements", *IEEE Trans. Antennas and Propa.* **AP-17**, 568 (1969).

15. H. A. Corriher, "RCS Measurements on Scale Models" in *Techniques of Radar Reflectivity Measurement*, ed. by N. C. Currie (Artech House, Deham, MA, 1984), pp. 327–374.

16. G. Sinclair, E. C. Jordan, E. W. Vaughan, "Measurement of aircraft-antenna patterns using models", *Proc. IRE* **35**, 1451 (1947).

17. C. R. Schumacher, "Electrodynamic similitude and physical scale modeling of nondispersive targets", *J. Appl. Phys.* **62**, 2616 (1987).

18. M. A. Morgan, "Ultra-wideband impulse scattering measurements", *IEEE Trans. Antennas Propag.* **42**, 840–846 (1994).

19. M. A. Morgan, N. J. Walsh, "Ultra-Wide-Band Transient Electromagnetic Scattering Laboratory", *IEEE Trans. Antennas and Propa.* **39**, 1230 (1991).

20. M. A. Morgan, "Research in ultra-wideband impulse radar target identification", *Naval Res. Rev.* **44**, 28 (1992).

21. H. C. Strifors, G. C. Gaunaurd, B. Brusmark, and S. Abrahamson, "Transient interactions of an EM pulse with a dielectric spherical shell", *IEEE Trans. Antennas Propag.* **42**, 453 (1994).

22. H. C. Strifors, G. C. Guanaurd, S. Abrahamsson, and B. Brusmark, "Scattering of short EM-pulses by simple and complex targets in the combined time–frequency domain using impulse radar", presented at IEEE National Radar Conference (1993), Boston, MA, pp. 70–75.

23. N. Katzenellenbogen, D. Grischkowsky, "Efficient generation of 380 fs pulses of THz radiation by ultrafast laser pulse excitation of a biased metal– semiconductor interface", *Appl. Phys. Lett.* **58**, 222 (1991).

24. O. Zucker, I. A. McIntyre, "Ultra wideband signal synthesis using photoconductive switches", *Microwave J.* **34**, 60 (1992).

25. W. M. Robertson, G. V. Kopcsay, G. Arjavalingam, "Picosecond time-domain electromagnetic scattering from conducting cylinders", *IEEE Microwave Guided Wave Lett.* **1**, 379 (1991).

26. W. M. Robertson, G. Arjavalingam, G. V. Kopcsay, "Microwave diffraction and interference in reflection using transient radiation from optoelectronically pulsed antennas", *Appl. Phys. Lett.* **57**, 1958 (1990).

27. L. Carin and K. Agi, "Ultra-wideband transient microwave scattering measurements using optoelectronically switched antennas", *IEEE Trans. on Microwave Theory and Techniques* **41**, 250–254 (1993).

28. L. Carin, "Transient scattering measurements using photoconductively switched planar antennas", *Ultra-Wideband, Short-Pulse Electromagnetics*, ed. by H. L. Bertoni, L. Carin, L. B. Felsen (Plenum, New York, 1993) p.37.

29. D. Kralj, L. Carin, "Short-pulse scattering measurements from dielectric spheres using photoconductively switched antennas", *Appl. Phys. Lett.* **62**, 1301 (1993).

30. A. Rahman, K. Kralj, L. Carin, M. R. Melloch, J. M. Woodall, "Photoconductively switched antennas for measuring target resonances", *Appl. Phys. Lett.* **64**, 2178 (1994).

31. R. A. Cheville, R. W. McGowan, D. Grischkowsky, " Late time target response measured with THz impulse ranging", *IEEE Trans. Antennas Propag.* **AP-45**, 1518 (1997).

32. R. A. Cheville, D. Grischkowsky, "Time domain terahertz impulse ranging studies", *Appl. Phys. Lett.* **67**, 1960 (1995).

33. D. Grischkowsky, S. Keiding, M. van Exter, C. Fattinger, "Far-infrared time-domain spectroscopy with terahertz beams of dielectrics and semiconductors", *J. Opt. Soc. Am. B* **7**, 2006 (1990).

34. D. Grischkowsky, R. A. Cheville, "Limits and applications of THz time-domain spectroscopy", presented at National Science Foundation Forum on Optical Science and Engineering, San Diego, CA, (1995).

35. J. C. G. LeSurf, *Millimetre-Wave Optics, Devices, and Systems* (Adam Hilger, Bristol, 1990).

36. S. E. Ralph, D. Grischkowsky, "Trap-enhanced electric fields in semi-insulators: the role of electrical and optical carrier injection", *Appl. Phys. Lett.* **59**, 1972 (1991).

37. C. Fattinger, D. Grischkowsky, "Beams of terahertz electromagnetic pulses", *OSA Proc. on Picosec. Electron. Optolelctron.* **4**, 225, (1989).

38. P. U. Jepsen, S. R. Keiding, "Radiation patterns from lens-coupled terahertz antennas", *Opt. Lett.* **20**, 807 (1995).

39. M. van Exter, C. Fattinger, D. Grischkowsky, "Terahertz time-domain spectroscopy of water vapor", *Opt. Lett.* **14**, 1128 (1989).

40. S. D. Gasster, C. H. Townes, D. Goorvitch, F. P. J. Valero, "Foreign-gas collision broadening of the far-infrared spectrum of water vapor", *J. Opt. Soc. Am. B* **5**, 593 (1988).

41. M. van Exter, D. R. Grischkowsky, "Characterization of an optoelectronic terahertz beam system", *IEEE Trans. Microwave Theory Tech.* **38**, 1684, (1990).

42. P. U. Jepsen, R. H. Jacobsen, S. R. Keiding, "Generation and detection of terahertz pulses from biased semiconductor antennas", *J. Opt. Soc. Am. B* **13**, 2424 (1996).

43. J. T. Verdeyen, *Laser Electronics* (Prentice Hall, Englewood Cliffs, 1995).

44. M. J. Gans, "Cross polarization in reflector type beam waveguides and antennas", *Bell Syst. Tech J.* **55**, 289 (1976).

45. S. Ramo, J. R. Whinnery, T. Van Duzer, *Fields and Waves in Communications Electronics* (Wiley, New York, 1984).

46. R. W. P. King, G. S. Smith, *Antennas in Matter* (MIT Press, Cambridge, 1981).

47. D. R. Wehner, *High Resolution Radar* (Artech House, Norwood, MA, 1995).

48. W. H. Press, S. A. Teukolsky, W. T. Vetterling, B. P. Flannery, *Numerical Recipes in Fortran*, 2nd ed. (Cambridge University Press, New York, 1993).

49. G. T. Ruck, D. E. Barrick, W. D. Stuart, C. K. Krichbaum, *Radar Cross Section Handbook* (Plenum, New York, 1970).

50. M. Abramowitz, I. A. Stegun, *Handbook of Mathematical Functions* (Dover, New York, 1970).

51. L. B. Felsen, N. Marcuvitz, *Radiation and Scattering of Waves* (IEEE Press, New York, 1973).

52. H. C. van de Hulst, *Light Scattering by Small Particles*: (Wiley, New York, 1957) p. 365.

53. R. A. Cheville, R. W. McGowan, D. Grischkowsky, "Time resolved measurements which isolate the mechanisms responsible for terahertz glory scattering from dielectric spheres", *Phys. Rev. Lett.* **80**, 269 (1998).

54. R. W. McGowan, R. A. Cheville, D. Grischkowsky, "Direct observation of the Gouy phase shift in THz impulse ranging". *Appl. Phys. Lett.* **76**, 670 (2000).

55. V. Khare, H. M. Nussenzveig, "Theory of the Glory", *Physical Review Lett.*, **38**, 1279 (1977).

56. A. E. Siegman, *Lasers* (University Science Books. Mill Valley, 1986).

57. S. Feng, H. G. Winful, R. W. Hellwarth, "Gouy shift and temporal reshaping of focused single-cycle electromagnetic pulses", *Opt. Lett.* **23**, 385 (1998).

58. R. G. Kouyoumjian, L. Peters Jr., D. T. Thomas, "A modified geometrical optics method for scattering by dielectric bodies", *IEEE Trans. Antennas Propag.* **11**, 690 (1963).

Bio-medical Applications of THz Imaging

Martin Koch

Abstract. We present examples of the application of THz time-domain imaging in the life-sciences and medical diagnosis. Since this new technique will have to compete with established techniques, we start with a comparative discussion of modern imaging systems followed by a brief description of a THz imaging setup. We then give three examples in the life sciences where THz imaging is potentially useful. First, we describe an examination of wood density variations due to tree rings and show that THz imaging can be a valuable tool for dendrochronologists. Second, we describe a study of water transport processes in plants. We investigated the slow rehydration process of drought-stressed plants and the fast water transport that occurs in mimosa after mechanical stimulus. Third, we describe an investigation of histopathological samples including the larynx of a pig and a human liver containing metastases, to explore the potential of THz imaging for medical diagnosis. In addition, we present the first continuous-wave THz imaging spectrometer employing a two-color external-cavity laser diode. This new THz imaging system is less expensive and more compact than traditional time-domain spectrometers.

1 Introduction

1.1 Some General Remarks Regarding Biomedical Imaging

Only a few weeks after Wilhelm Conrad Röntgen discovered x-rays in 1895, he took the first medical image with this new kind of radiation. Since then, x-ray examination has become an invaluable tool in medical diagnosis. However, despite its fascinating possibilities, this technique has some shortcomings and drawbacks. The most obvious one is that x-rays are ionizing and thus harmful to living beings, as they can, for example cause cancer. Furthermore, the spatial resolution of x-ray images is limited not by the very short wavelength of the radiation of 0.1 pm to 10 nm but by the fundamental effect of Rayleigh scattering. Even with modern techniques, the resolution is hardly better than 50 µm. The absorption of x-rays is proportional to the thickness and density of the tissue investigated and to the cube of the atomic number of the elements it contains. As an unfortunate consequence soft tissue and parenchyma (organ tissue) are hard to distinguish owing to low contrast.

Since these disadvantages are inherent in investigations with x-rays they also apply to computer tomography (CT). Here, three-dimensional resolution is obtained by use of x-ray emitters and detectors that quickly orbit the body under investigation. Information is again drawn from the transmitted intensity. The spatial resolution of CT is slightly poorer than that of two-dimensional x-ray imaging. Furthermore, CT suffers from low contrast: the application of contrast media is often required.

If high contrast is required, nuclear magnetic resonance imaging (NMRI), where microwave transitions between nuclear spin levels are used, is the tool of choice. The transition frequency varies with the external magnetic field, leading to a spin splitting of the nuclear states. By using magnetic field gradients, distinct spatial positions can be addressed. The signal strength depends on the longitudinal relaxation time of the microwave transition. Since this has strikingly different values for different tissues (e.g. owing to different water contents), high contrast is achieved. However, the spatial resolution is only about 0.5 mm.

It is worth mentioning positron emission tomography (PET), since it may soon develop into a routine technique for early cancer detection. Here, glucose that contains radioactive isotopes which emit positrons is injected into the patient. The glucose molecules accumulate in regions of high metabolic activity. The annihilation radiation of positrons which is emitted when they encounter an electron on their flight through the body, is detected. Since this radiation typically originates within one or two millimeters from the emission point of the positron, the spatial resolution of PET is quite poor.

Sonography is an imaging technique which uses longitudinal sound waves instead of transverse electromagnetic waves. Here an ultrasound generator emits sound bursts with frequencies between 2 and 30 MHz. Owing to impedance mismatching between different types of tissue, echoes are reflected back and detected at the position of the emitter. From the echo pattern, cross-sectional images that show details of organs are obtained. The spatial resolution and penetration depth depend on the frequency used. While low-frequency ultrasound waves at 2 MHz can penetrate several tens of centimeters of body tissue but lead to a resolution of approximately 400 μm, waves at 30 MHz achieve a resolution of 25 μm with a penetration depth limited to a few centimeters.

A related optical technique that analyzes backscattered light is optical coherence tomography (OCT). This rather new technique performs cross-sectional imaging by measuring the time delay and magnitude of optical echoes at different transverse positions. Here, the spatial resolution is 1–15 μm and thus one to two orders higher than in sonography. The penetration depth of 2–3 mm, however, is much less.

While all the above-mentioned techniques can be performed in vivo because of their large penetration depth, several other techniques require small samples that have been removed from a larger specimen. The most prominent of these techniques is optical microscopy, which has a resolution of several hundred nanometers. However, it often requires the usage of stains to enhance contrast. Modern microscopy methods include laser scanning microscopy and two-photon scanning microscopy, both of which yield three-dimensional imaging with a resolution comparable to that of optical microscopy. A very high resolution of one or a few nanometers is achieved in electron microscopy, but this requires rather involved sample preparation. The spatial resolution of all these microscopy methods is sufficient to differentiate details within single cells. Less spatial resolution is achieved as the wavelength becomes longer. Spatially resolved Fourier transform infrared (FTIR) spectroscopy in the near infrared (120–380 THz) and mid-infrared (12–120 THz) is suitable for mapping out the chemical composition of biological and medical samples with a resolution between one and a few tens of microns (depending on the frequency).

Terahertz imaging at far-infrared frequencies between 0.3 and 10 THz will naturally have a resolution of a few hundred microns. As in FTIR spectroscopy it is possible to map out the chemical composition of samples if they are not too thick. Furthermore, impulsive THz imaging as described in this chapter shares some of the advantages inherent in sonography and OCT owing to the short coherence length of the source. The potential of this relatively young technique for bio-medical imaging has not yet been fully explored. The examples described in this chapter represent a starting point towards more extensive and systematic measurements.

1.2   The Closure of the Terahertz Gap

The so-called "THz gap" was closed only recently. Until then, investigations in the frequency range between several hundred GHz and 10 THz had been very difficult. This was due to a lack of effective, flexible, and inexpensive far-infrared sources and detectors. As described elsewhere in this volume, huge steps to close this gap were accomplished in the late 1980s by groups at AT&T Bell Laboratories and at IBM's T. J. Watson Research Laboratory. The newly developed technique relied on the gating of biased photoconducting dipole antennas [1–5]. Shortly thereafter, alternative generation and detection schemes using optical rectification and electro-optic sampling were demonstrated [6–11].

The spectral width of the THz pulses generated by these techniques, either optoelectronically or via optical rectification, makes them an ideal tool for far-infrared spectroscopy. During the last decade numerous investigations using these techniques have been reported on semiconductors [12–16], superconductors [17–19], gases [20–22], liquids [23–26], and biomolecules [27].

Exactly one hundred years after the first x-ray photography by Röntgen, Nuss and coworkers took the first THz image [28]. They developed the photoconducting antenna emitter–receiver scheme further to the point where two-dimensional images of objects a few centimeters in diameter could be accumulated in a reasonable time. The first THz images taken were of a leaf freshly cut from a tree and a semiconductor integrated-circuit package

[28]. This work has led to the development of the first commercially available THz time-domain imaging system [29]. Furthermore, new dimensions in THz imaging have been explored since its first demonstration in 1995. Some effort has been made to improve the spatial resolution by means of near-field imaging [30–36] and resolutions down to $\lambda/6$ have been reported. The internal structure and composition of objects can be visualized using THz tomography [37,38]. This technique is closely related to OCT. Finally, imaging with enhanced depth resolution using phase-shift interferometry was recently demonstrated [39]. In addition, there is a desire for smaller and less expensive systems.

THz imaging has a high potential for practical applications such as packaging inspection, quality control, and chemical composition analysis [28,38,40,41]. As we show here, it may also hold considerable potential for biomedical diagnosis.

2   Description of the Terahertz Imaging Setup

Since an extensive description of time-domain THz spectroscopy and THz imaging can be found in other chapters of this volume (e.g. the chapters by Mittleman, Cheville et al., and Jianga and Zhang) and many review articles [28,32,40,42] we give only a rather brief discussion here.

For much of the work presented here, we used a standard time-domain THz imaging setup with photoconductive dipole antennas for THz generation and detection (Fig. 1a). The time profile of a typical THz pulse as detected by the system is plotted in Fig. 1b. A Fourier transformation of this waveform reveals the broad spectrum of such pulses, which extends from 0.1 to 3 THz (Fig. 1c). The THz pulses are guided by off-axis parabolic mirrors to form an intermediate focus of submillimeter diameter, through which the samples are scanned in a raster pattern. At each sample point, a complete transmitted waveform is acquired and subsequently Fourier transformed. The transmitted intensity integrated over a given spectral window (e.g. 0.75 to 1.2 THz, indicated by the hatched area in Fig. 1c) is translated into a grayscale value to obtain a pixel. The time required to generate a complete two-dimensional THz image is typically 10 minutes, depending on the sample size and scan resolution.

In addition, we present the first continuous-wave THz imaging system. This new imaging scheme has advantages over time-domain imaging when high spectral resolution is required. This results from the much higher spectral power density of the CW radiation and from the fact that the spectral resolution of a CW system depends merely on the frequency stability of the two laser modes used for photomixing. In contrast, the power of a THz pulse is distributed over its entire spectral width. Furthermore, time-domain imaging in a distinct narrow frequency interval requires a time-consuming increase of

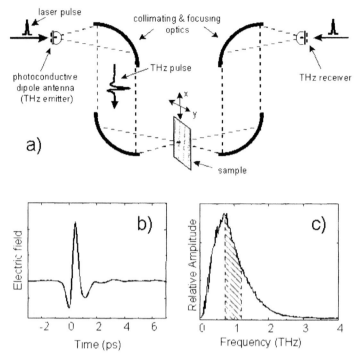

Fig. 1. (a) THz imaging setup consisting of THz emitter and receiver antennas, and an intermediate focus through which the sample is scanned in a raster pattern. (b) Typical THz pulse (electric field). (c) Spectrum of the THz pulse. The hatched area indicate a typical spectral integration window

scanning range and resolution, together with additional effort in the Fourier transformation and spectral integration.

The CW THz imaging setup is shown in Fig. 2. Continuous-wave radiation is generated via photomixing [43,44], a technique described in the chapter by Duffy and coworkers in this volume. We use a recently developed external-cavity laser diode which allows two-line emission. In the original concept tunability was achieved by insertion of a liquid crystal array, used as an electronically controllable aperture [45]. Here, we use a V-shaped mirror as a frequency selective element for simplicity (Fig. 2) [46]. The frequency spacing of the two laser modes can be adjusted by moving the V-shaped mirror vertically. The two-line laser emission, with a total optical power of 29 mW is focused onto the 5 μm wide gap of a 50 V DC-biased stripline dipole antenna. The dipole has a length of 400 μm and was deposited on a low-temperature-grown (LTG) GaAs substrate [47]. The epitaxial layer was grown at 300° C and annealed at 600° C for 10 minutes to achieve short carrier lifetimes below 1 ps, which are a prerequisite for photomixing. For detection of the THz intensity, a standard bolometer is used. Again, an image is acquired by two-

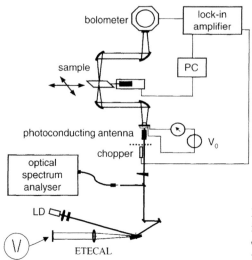

Fig. 2. Setup of a CW THz imaging system based on a two-mode external-cavity laser diode, a photomixer, and a bolometer

dimensional scanning of the sample through the intermediate focus. Early results obtained with this system are shown below.

3   Dendrochronology: Density Mapping of Wood

While the macroscopic density of wood is a key parameter for the wood and paper industry and is routinely measured in wood-processing plants, the microscopic density fluctuations of wood are interesting from a scientific point of view. In particular, density inhomogeneities associated with tree rings are of high interest and are the core of a field of research called dendrochronology. From the density profiles of these rings, dendrochronologists obtain information about varying climate conditions and forest fires over the centuries.

Over the years various methods have been developed to determine the wood density with sub-year ring resolution. In addition to simple gravimetric methods [48,49], density determination techniques include the "hardness feeler" [50], which determines the pressure involved in boring with a needle into a section of wood, and the acid corrosion method [51], in which wood is subjected to acid erosion and the resulting surface is investigated using an optical microscope. More advanced methods include radiodensitometry with x-rays [52,53] and sandblasting [54], in which the resulting height profile of the surface after damage is directly related to the density. The density of wood has also been determined using microwaves at 10 GHz [55]. However, the spatial resolution associated with this long-wavelength radiation is not sufficient to resolve tree rings.

In the following, we show that THz imaging can be an interesting alternative to these traditional techniques. For demonstration, we investigated a sample of beech wood (*fagus silvatica*). Because of the high sensitivity of THz radiation to water, the sample, which is $14 \times 14 \times 1.7\,\text{mm}^3$ in size, was predried in an oven. However, after storage under ambient conditions, moisture in the air was reabsorbed by the wood so that the final water concentration in the measured samples was approximately 12% by weight.

Figure 3a shows a THz image of a cross section of the beech sample. Tree rings are visible as vertical structures. The darker areas represent late wood, which is less transparent owing to its greater density. Early wood, on the other hand, has larger cell lumina and thinner cell walls, and consequently is more transparent and appears grey to white. In addition, radial rays can be observed as horizontal structures.

To relate this THz transmission image to the true wood density, the transmission data have to be calibrated against a density obtained by the gravimetric-volumetric method. Such a calibration has to be performed only once for any given plant species, as long as all experimental parameters stay constant. Part of the sample was cut along the y-direction into thin slices. The volume and weight of these slices were then determined. Fig. 3b shows the density obtained over the position range from $x = 6$ to 15 mm and $y = 0$ to 4.5 mm. Several peaks are clearly observed. These peaks correspond to regions of late wood, which are marked in Fig. 3a by the set of arrows. Since reflection losses at the two surfaces of the sample are estimated to be below 3% (we have determined an average refractive index of 1.4 in the THz range) they have been neglected here, and the transmission data can be transformed into the optical thickness αd according to Beer's law:

$$\int_{0.75\,\text{THz}}^{1.2\,\text{THz}} d\upsilon\, \alpha(\upsilon)d = -2\ln\left[\int_{0.75\,\text{THz}}^{1.2\,\text{THz}} d\upsilon E(\upsilon) \Big/ \int_{0.75\,\text{THz}}^{1.2\,\text{THz}} d\upsilon E_0(\upsilon)\right], \qquad (1)$$

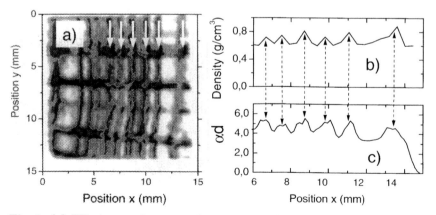

Fig. 3. (a) THz image of a piece of beech wood. (b) Density profile obtained by the gravimetric–volumetric method. (c) Absorption profile extracted from a defined area of the THz image shown on the *left* (see text)

where d is the sample thickness of 1.7 mm, α is the absorption coefficient at frequency v, and E and E_0 represent the transmitted and incident THz field amplitudes, respectively. To make them comparable to the density data obtained from the slices, the absorption data extracted for the sample area under investigation were integrated over the y coordinate. The resulting absorption data are plotted in Fig. 3c. A reliable correlation between the density data and the THz data is apparent. A quantitative calibration between absorption and density can be obtained by extracting several data points from each plot and plotting the values for the density ρ versus the corresponding absorption data (dots in Fig. 4a). The data points in Fig. 4a can be approximated by a line (solid line) obeying

$$\rho = 0.145 \frac{\text{g}}{\text{cm}^3} \alpha d. \tag{2}$$

This dependence can be used to directly transform the transmission image of Fig. 3a into a density map of this sample. The result is shown in Fig. 4b. While white and light-gray regions represent low-density regions of early wood, the darker areas represent the high-density regions of late wood (for a color reproduction see [58]). Overall, the density in this particular sample varies from 0.4 to $1.3\,\text{g/cm}^3$.

The above results show that THz imaging provides an interesting alternative method for determining the density of wood samples with sub-tree-ring resolution. In contrast to the sandblasting method, the samples can be reused for further examinations as THz imaging is nondestructive. Furthermore, THz radiation is not harmful to human beings at the field strengths used here. Hence, the investigation with THz waves has advantages over direct scanning with x-rays. Moreover, because for its high sensitivity to water, THz radiation could be used to reveal water deposits in wood and other biological specimens, a task which cannot be performed by radiodensitometry or the

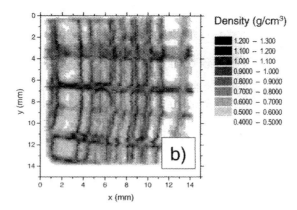

Fig. 4. (a) Density plotted versus absorption. (b) Density map of the beech sample

sandblasting method. In the next section, we shall describe an investigation making use of this fact and show that THz imaging allows us to observe the water transport in living plants.

4   Plant Physiology:
Monitoring the Water Flow in Plants

The physical mechanisms responsible for long-distance water transport in higher plants are still controversial. Since capillary forces alone cannot be responsible for the transport of water from the roots to the leaves in tall trees with heights exceeding 100 m, it was assumed that water transpiration from tree leaves effectively pulls water up from the ground. This so-called cohesion theory introduced by Dixon and Joly [57] in 1895, was widely accepted for nearly one century. The theory was based on a variety of experiments which indicated that transpiration was very important for water transport in plants [58–60]. Recent experimental evidence, however, has led to a reexamination of the cohesion theory owing to the incompatibility of data from two competing techniques, one of which uses xylem pressure probes [61] and the other of which uses pressure chambers [62]. A discussion of the current controversies related to the cohesion theory can be found in [63,64].

It is therefore obvious that more accurate experimental data on water flow rates in plants are needed. One technique that has recently proven to be useful is nuclear magnetic resonance (NMR) [65,66]. Kuchenbrod et al. [67] investigated maize plants under controlled transpiration conditions, showing that the water flow velocity can be as high as 2 mm/s. However, as mentioned earlier, NMR is quite expensive and needs to be supplemented by other methods which are portable and can be used on large plants.

As we shall show in the following, THz imaging represents an interesting alternative method for monitoring water transport in plants. This method is based on the high sensitivity of THz radiation to water mentioned earlier. The absorption spectrum of water around 1 THz is plotted in the left part of Fig. 5. It is featureless, and the absorption strength increases steadily with frequency in the frequency window plotted. Although it is widely accepted that collective translational motions of the water molecules are the origin of this featureless absorption, the detailed mechanisms involved are still controversial [23–25,68]. At 1 THz the absorption coefficient is 235 cm^{-1}. Using this value we can calculate the attenuation of the THz signal as a function of the water film thickness (for simplicity we neglect losses due to reflection). The result is shown in the right part of Fig. 5. From this plot we can conclude that a biological sample should ideally contain an amount of water equivalent to a water film of 100 to 200 μm thickness. Fortunately enough, this criterion is nicely fulfilled by most leaves.

In most cases, a THz investigation of a plant must be limited to a single leaf. Hence it is not possible to gather information on flow rates in different

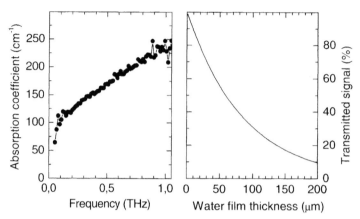

Fig. 5. *Left:* absorption spectrum of water. *Right:* attenuation of the THz signal by a water film as a function of the film thickness, at $\nu = 1$ THz

sections of the plant. However, one can investigate the dynamics of water uptake and thus answer the question of how long the rehydration process takes in drought-stressed plants. In addition, THz measurements can give an estimate of how much water is taken up.

Mittleman and coworkers [40] were the first to study water transport in a living plant. In a proof-of-principle experiment, they monitored the water uptake in a house plant (coleus) after subjecting the plant to water stress. The overall transmission through a leaf was found to drop by more than 10% within eight hours after watering [69]. This drastic reduction of the transmitted signal is due to the water which is taken up during the rehydration process. A close inspection reveals that the rehydration process can be divided into two regimes [69]: the THz transmission shows an initial fast drop within the first hour (referred to in the following as the fast rehydration regime (FRR)), and a subsequent nearly linear decrease (referred to in the following as the slow rehydration regime (SRR)), which was still not complete at the end of the measurement, i.e. after nearly eight hours.

Here we present THz transmission data on the rehydration process in clematis. In agreement with the data obtained by Mittleman et al., we observed two time regimes. However, the initial fast rehydration regime was completed after 15 minutes in our measurements, i.e. it was four times as fast as that observed for coleus. We also give an estimate of the amount of water that was taken up by the leaf.

In addition, we studied the change in the water content of leaf joints in mimosa after the plant had experienced an external stimulus. Since we observed a fast decrease of the THz signal transmitted through this section of the plant, we can conclude that the net amount of water in the leaf joint increases. This is an important piece of information which cannot be obtained in a simple way otherwise.

4.1 Clematis

The clematis had not been watered for a period of about a week prior to the experiments. To scan a living leaf through the THz focus, we clamped it gently in a special holder. During the measurements, the plants were illuminated with white light from mercury halide lamps to trigger adequate transpiration. To decrease the data acquisition time, we did not record a complete image but restricted ourselves to a line scan. We could further reduce experimental noise by increasing the number of waveforms averaged per data point. The acquisition time for one line scan was approximately five minutes.

The spatially resolved transmitted intensity through a clematis leaf, directly before watering and 144 minutes after watering, is shown in Fig. 6. Here, the integration range is 0.2 to 0.9 THz and the data are normalized by assuming that transmission through air (from $x = 0$ mm to 10 mm, data points not shown in this expanded view) is 100%. Losses due to reflection have been ignored. The two pronounced transmission minima at $x = 21$ mm and 26 mm correspond to two of the vascular bundles containing aggregates of xylem vessels. After watering, the overall transmission drops significantly as water flows into the leaf. This is most obvious at the positions of the minima.

Figure 7 shows the reduction of the overall transmission as a function of the time after watering. Here, the data plotted in Fig. 6 have been averaged from $x = 14$ mm to $x = 28$ mm and subsequently normalized to the value of transmission before watering. Within the first 15 minutes, the transmitted THz intensity quickly drops by 2.5% (the FRR). Subsequently, the THz transmission decreases further at a reduced rate (the SRR). These observations are in qualitative agreement with those of Mittleman and coworkers [69]. However, for clematis, the duration of the FRR is much shorter than for coleus.

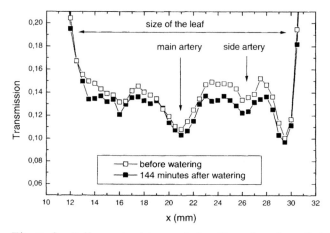

Fig. 6. Spatially resolved transmission through a clematis leaf

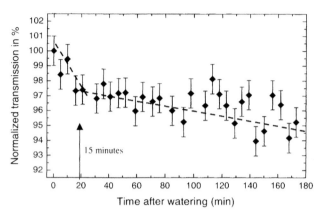

Fig. 7. Overall transmission through the leaf as a function of time after watering

Several interesting questions arise from these findings. Why is the rehydration process divided into two regimes? Is this behavior a typical one that can be observed in every plant? Is the duration of the FRR characteristic of a particular species or does it vary within a single plant; for example, does it depend on the distance of the leaf under investigation from the ground? How do the rehydration dynamics depend on environmental conditions such as humidity, illumination intensity, and temperature? We are looking forward to addressing all these questions in future measurements.

Nevertheless, one can extract further information from the drop in transmission observed here. It is possible to estimate crudely the amount of water that is taken up by the leaf. Before watering, the averaged transmitted THz intensity is 13.43% (see Fig. 6). Using an average absorption coefficient of $165 \, \text{cm}^{-1}$ (this value corresponds to 0.5 THz (see Fig. 5)) and ignoring losses due to reflection, this attenuation corresponds to a water film $121.6 \, \mu\text{m}$ thick. This is a reasonable value since for clematis the thickness of a leaf typically varies between 180 and $220 \, \mu\text{m}$ (only in the main arteries does it reach $500 \, \mu\text{m}$). Three hours after watering the transmitted THz intensity has dropped to 12.80%. With the above value of the absorption coefficient, this corresponds to a water film with a thickness of $124.6 \, \mu\text{m}$. Hence the amount of water increases by 2.5% within three hours after watering. Although the average absorption coefficient is somewhat uncertain (since we are averaging over the measured THz spectral range), we can conclude that the amount of water in the leaf has increased by somewhere between 2 and 3%. We would like to stress that it is extremely difficult to obtain this information from living plants by any other method of measurement used today.

4.2   Mimosa

The rehydration process studied above persists for many hours. There are, however, plants which show water transport phenomena with much faster dynamics. One example is mimosa (*Mimosa pudica*), a plant that originates from the tropical part of the American continent.

A schematic drawing of this sensitive plant is shown in Fig. 8. Upon mechanical stimulus, the plant responds with rapid movements of the leaves. This is due to water translocations in the leaf joints, the pulvini. These pulvini have a core of vascular tissue which contains vessels. The vascular tissue is surrounded by a bulky cortex of thin-walled parenchyma cells, which are the basic plant tissue. It is commonly accepted that these cortex cells cause movements by either losing or gaining internal cell pressure, or turgor [70,71].

As shown in the right part of the schematic drawing, the primary pulvini respond with downward motions, whereas the tertiary pulvini respond with upward movements of the leaflets so that the leaf folds up. What remained unknown was whether there was a net increase or decrease of pulvinar water during the movement cycle.

To settle this question, we measured the THz transmission through a single pair of tertiary pulvini. One leaf was clamped gently between two transparent plastic sheets. This prevented the leaf from closing in response to the stimulus. Hence, the joint remained in place and we could avoid any changes in signal amplitude due to a variation in sample geometry. We cut holes of few millimeters in diameter in the plastic exactly at the position of the THz focus to prevent multiple reflections of the THz pulses between the plastic sheets.

The THz transmission, integrated from 0.1 to 1.0 THz through a tertiary pulvinus as a function of time is plotted in Fig. 9. At time zero, the plant experiences a mechanical stimulus and the THz transmission shows a sudden drop by 6%. Subsequently, the transmitted signal recovers within a few

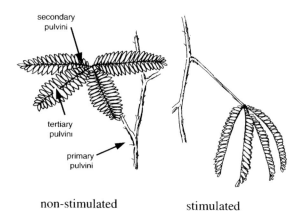

secondary pulvini

tertiary pulvini

primary pulvini

non-stimulated   stimulated

Fig. 8. Schematic drawing of *Mimosa pudica* before and after mechanical stimulation

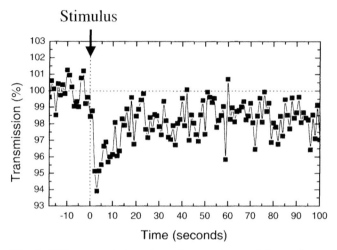

Fig. 9. THz transmission through a tertiary pulvinus after mechanical stimulus of the plant

seconds to 98% of the initial value. This fast recovery is followed by a much slower recovery which is typically completed only after 30 minutes. After this time the leaf appears nonstimulated again.

From the observed decrease of the THz transmission, we can conclude that there must be a net increase of turgor in the tertiary pulvini. However, it is difficult to obtain reliable quantitative values, since the spot size of the THz radiation is somewhat larger than the entire leaflet joint. Therefore a portion of the THz beam passes the joint and does not interact with it. This is particularly true for the low-frequency components, which form a larger focal spot. It is possible that recently developed THz near-field-imaging techniques could be used to obtain more quantitative results, however, in studies of this type. From these results it is clear that THz imaging shows great promise as a noninvasive tool for probing varying water concentrations in different sections of plants.

5   Medical Imaging on Histopathological Samples

The potential of THz imaging for medical diagnostics is currently, being intensively investigated. The groups of both Zhang and Arnone have been able to identify different types of tissue in "fresh" untreated samples [72,73], Mittleman at al. have shown that THz imaging may be useful for diagnosis of burns [32,40]. Detection of caries and of skin cancer has been demonstrated by Arnone et al. [73,74].

However, much more work is needed to fully explore the potential of THz imaging for medical diagnosis. Here, we present an initial investigation of some samples that have been treated by the standard procedures for histopathological examination. We investigated the larynx of a pig and a human liver with metastases. In order to fix the protein structure, the samples were put in a formalin solution for several hours. Subsequently, they were subjected to alcohol and xylol treatment to remove water. Finally, they were embedded in paraffin wax. For histopathological examination, such paraffin wax blocks are normally cut into thin slices with a thickness of a few micrometers, which are then stained and scrutinized with an optical microscope. We, in contrast, needed much thicker slices to obtain an appropriate absorption, since THz radiation can penetrate several millimeters of dehydrated tissue (note that the penetration depth for water-containing "fresh" samples is only a few hundred micrometers). A handsaw was used to cut these thick slices, with dimensions of $67 \times 42 \times 7\,\mathrm{mm}$ (larynx) and $50 \times 40 \times 4\,\mathrm{mm}$ (liver) from larger blocks. Optical images are shown in Figs. 10a and 13a.

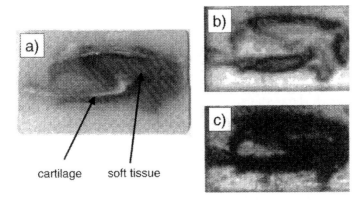

Fig. 10. (a) Optical image of a larynx sample taken from a pig. (b) THz transmission image of this sample obtained using a frequency window from 1.0 to 1.5 THz. (c) Image obtained using a window from 0.2 to 0.5 THz

5.1   Time-Domain Imaging

The larynx contains regions of cartilage and regions of soft tissue. A distinction between these tissue types is difficult on the basis of the optical image in Fig. 10a. The following results demonstrate that, in this case, different types of tissue can be differentiated by means of a comparison of images obtained using different integrated frequency windows. Figure 10b shows THz transmission image in the frequency window from 1.0 to 1.5 THz. The dark areas correspond to regions of low transmission through the biological tissue. All tissue in the sample seems to absorb the THz radiation equally well. Hence, a distinction between the two different tissue types cannot be made in this frequency window. The paraffin wax that surrounds the larynx is nonpolar and appears bright owing to its negligible absorption.

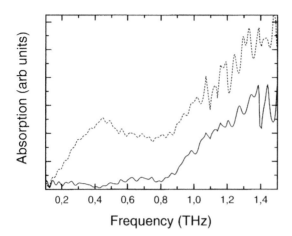

Fig. 11. Absorption of soft tissue (*solid line*) and cartilage (*dashed line*)

If the spectral integration is performed from 0.2 to 0.5 THz the transmission image looks entirely different (Fig. 10c). Now the dark areas correspond to cartilage structures. In this frequency window, the absorption of cartilage is obviously much stronger than that of soft tissue. This is confirmed by Fig. 11, which shows the absorption spectra of the two tissue types. At 0.45 THz the absorption of cartilage exceeds that of soft tissue by a factor of 20. Hence, these very different types of tissue can be identified perfectly by comparing two THz images obtained using different frequency windows.

However, it should also be noted that images taken using different spectral windows have, by their nature, different spatial resolutions. To demonstrate this effect, we scanned a razor blade through the focus and determined how rapidly the signal decreased form 90% to 10% for different integration windows. The result is shown in Fig. 12. While the drop in signal occurs within

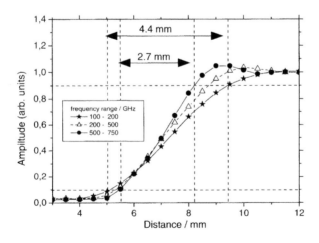

Fig. 12. THz line scans across the edge of a razor blade

4.4 mm for a frequency window from 0.1 to 0.2 THz the edge is much steeper for an integration range from 0.5 to 0.75 THz (note that the latter data are somewhat modulated by diffraction effects). Hence, for a fair comparison of images obtained using different frequency ranges a numerical treatment may prove necessary. Investigations of how to deal with this effect are in progress.

The next example shows that there are also situations where two types of tissue can be differentiated within a single frequency window. Figure 13a shows an optical image of a liver sample containing several tumors, obtained using an integration window from 0.2 to 0.5 THz. The cancerous areas are somewhat brighter than the regular liver tissue. In the THz image shown in Fig. 13b they appear as dark patches of reduced transmission. Now the question arises as to whether the higher THz absorption of the cancerous material is due to variations in the density or in the chemical composition of the tissue. However, since the density is believed to be rather similar for metastases and regular tissue, we suspect that the contrast is due to a different chemical content. Further investigations to clarify this point are currently under way.

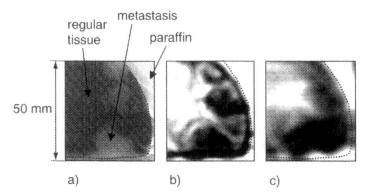

Fig. 13. Images of a liver sample containing tumors. (a) optical image, (b) THz transmission image obtained using a frequency window from 0.2 to 0.5 THz using pulsed radiation (interpolated to contour lines), (c) THz transmission image at 230 GHz obtained using CW radiation (interpolated to contour lines)

5.2 Continuous-Wave Imaging

To demonstrate CW THz imaging, we used the same liver sample that was used in the investigation above. Figure 13c shows a cw image acquired at a frequency of 230 GHz, which is the resonance frequency of the photoconducting dipole antenna used. Again, the tumors appear as dark patches of reduced transmission and are clearly distinguishable from regular tissue. While the areas of paraffin wax and regular tissue exhibit a maximum transmission of approximately 60% (normalized to the free-space transmission), the transmission of the cancerous regions decreases drastically, down to values of 18%.

The level of detail is not yet as good as that of the time-domain transmission image shown in Fig. 13b. The cw frequency of 230 GHz, located at the lower end of the integration interval used for the pulsed-THz-radiation image, is responsible for the reduced spatial resolution. In addition, in this first demonstration, the signal-to-noise ratio was only 75, which diminishes the quality of the image. Currently we are working towards an improved signal-to-noise ratio by optimizing the LTG GaAs emitter antennas.

6   Conclusion

In this chapter, we have discussed a few THz-imaging experiments which lead to the conclusion that imaging in this frequency range can be complementary to other, conventional imaging techniques. THz imaging is ready to be a valuable tool for dendrochonology. It will likewise be of great value for plant physiology and might help to settle the ongoing basic controversies about water transport in plants.

While these two problems are rather academic ones, diagnostic applications in medical science could have considerable impact and may therefore have great market potential. One big advantage could be that, in contrast to x-ray imaging, CT, and PET, all of which use ionizing radiation, THz waves are safe and nondestructive. However, at present medical THz imaging is still in its infancy, i. e. in a state in which the initial data are being collected. In this early stage of research the main question is whether different types of tissue can be differentiated and unambiguously identified. Of particular importance is the distinction between healthy and diseased tissue. However, even if this can be accomplished, such a distinction is no guarantee that THz imaging will be useful as a diagnostic tool, since in many cases this can be also accomplished with established and/or much cheaper techniques. The future of THz imaging in medicine depends crucially on the question of whether diagnostic problems can be identified to which it is a superior solution.

Acknowledgments

It is my pleasure to thank my coworkers, without whom the results described above, would not have been possible. The work on wood was done at Bell Labs in cooperation with S. Hunsche and M. C. Nuss. The work on plants was done in cooperation with J. Fromm of the Technical University of Munich. The measurements on clematis were performed by I. Libon and N. E. Hecker in J. Feldmann's group at the University of Munich. The measurements on mimosa and on the medical samples were done at the Technical University of Braunschweig by P. Knobloch, K. Schmalstieg and T. Kleine-Ostmann. The histo-pathological samples were provided by E. Rehberg, M. Sperling and K. Donhuijsen. The two color laser for CW THz generation was provided by the group of M. Hofmann in Bochum. The LTG GaAs were fabricated by C. Hein and K. Pierz of the Physikalisch-Technische Bundesanstalt in Braunschweig.

References

1. C. Fattinger, D. Grischkowsky, "Terahertz beams", *Appl. Phys. Lett.* **54**, 490 (1989).
2. M. van Exter, D. Grischkowsky, "Characterization of an optoelectronic terahertz beam system", *IEEE Trans. Microwave Theory Tech.* **38**, 1684 (1990).
3. M. Tani, S. Matsuura, K. Sakai,, S. Nakashima, "Emission characteristics of photoconductive antennas based on low-temperature-grown GaAs and semi-insulating GaAs", *Appl. Opt.* **36**, 7853 (1997).
4. P. R. Smith, D. H. Auston, M. C. Nuss, "Subpicosecond photoconducting dipole antennas", *IEEE J. Quantum Electron.* **24**, 255 (1988).
5. P. Uhd Jepsen, R. H. Jacobsen, S. R. Keiding, "Generation, detection of terahertz pulses from biased semiconductor antennas", *J. Opt. Soc. Am. B*, **13**, 2424 (1996).
6. A. Rice, Y. Jin, X. F. Ma, X.-C. Zhang, D. Bliss, J. Larkin, M. Alexander, "Terahertz optical rectification from ⟨110⟩ zinc-blende crystals", *Appl. Phys. Lett.* **64**, 1324 (1994).
7. A. Bonvalet, M. Joffre, J. L. Martin, A. Migus, "Generation of ultrabroadband femtosecond pulses in the mid-infrared by optical rectification of 15 fs light pulses at 100 MHz repetition rate", *Appl. Phys. Lett.* **67**, 2907 (1995).
8. A. Nahata, A.S. Weling, T.F. Heinz. "A wideband coherent terahertz spectroscopy system using optical rectification and electro-optic sampling", *Appl. Phys. Lett.* **69**, 2321 (1996).
9. Q. Wu, M. Litz, X.-C. Zhang, "Broadband detection capability of ZnTe electro-optic sampling field detectors", *Appl. Phys. Lett.* **68**, 2924 (1996)
10. Q. Wu, X.-C. Zhang, "Free-space electro-optics sampling of mid-infrared pulses", *Appl. Phys. Lett.* **71**, 1285 (1997).
11. P. Uhd Jepsen, C. Winnewisser, M. Schall, V. Schyja, S. R. Keiding, H. Helm, "Detection of THz pulses by phase retardation in lithium tantalate", *Phys. Rev.* E53 R3052 (1996).
12. P. N. Saeta, J.F. Frederici, B. I. Greene, D. R. Dykaar, "Intervalley scattering in GaAs and InP probed by pulsed far-infrared transmission spectroscopy", *Appl. Phys. Lett.* **60**, 1477 (1992).
13. J. Zielbauer, M. Wegener, "Ultrafast optical pump THz-probe spectroscopy on silicon", *Appl. Phys. Lett.* **68**, 1223 (1996).
14. D. M. Mittleman, J. Cunningham, M. C. Nuss, M. Geva, "Noncontact semiconductor wafer characterization with the terahertz Hall effect", *Appl. Phys. Lett.* **71**, 16 (1997).
15. T.-I. Jeon, D. Grischkowsky, "Nature of conduction in doped silicon", *Phys. Rev. Lett.* **78**, 1106 (1997).
16. R. Kersting, R. Bratschitsch, G. Strasser, K. Unterrainer, J. N. Heyman, "Sampling a terahertz dipole transition with subcycle time resolution", *Opt. Lett.* **25**, 272 (2000).
17. M. C. Nuss, P.M. Mankiewich, M. L. O'Malley, E. H. Westerwick, P. B. Littlewood, "Dynamic conductivity and coherence peak in YBa2Cu3O7 superconductors", *Phys. Rev. Lett.* **66**, 3305 (1991).

18. M. Hangyo, S. Tomozawa, Y. Murakami, M. Tonouchi, M. Tani, Z. Wang, K. Sakai, S. Nakashima, "Terahertz radiation from superconducting $Yba_2Cu_3O_{7-\delta}$ thin films excited by femtosecond optical pulses", *Appl. Phys. Lett.* **69**, 2122 (1996).

19. Wilke, M. Khazan, C.T. Rieck, P. Kuzel, T. Kaiser, C. Jaeckel, H. Kurz, "Terahertz surface resistance of high temperature superconducting thin films", *J. Appl. Phys.* **87**, 2984 (2000).

20. H. Harde, N. Katzenellenbogen, D. Grischkowsky, "Terahertz coherent transients from methyl chloride vapor", *J. Opt. Soc. Am. B* **11**, 1018 (1994).

21. D. M. Mittleman, R. H. Jacobsen, R. Neelamani, R. G. Baraniuk, M. C. Nuss, "Gas sensing using terahertz time-domain spectroscopy", *Appl. Phys. B* **67**, 379 (1998).

22. G. Mouret, W.Chen, D. Boucher, R. Bocquet, P. Mounaix, D. Lippens, "Gas filter correlation instrument for air monitoring at submillimeter wavelengths", *Opt. Lett.* **24**, 351 (1999).

23. L. Thrane, R.H. Jacobsen, P. Uhd Jepsen, S.R. Keiding, "THz reflection spectroscopy of liquid water", *Chem. Phys. Lett.* **240**, 330 (1995).

24. J.T. Kindt, C.A. Schmuttenmaer, "Far-infrared dielectric properties of polar liquids probed by femtosecond terahertz spectroscopy", *J. Phys. Chem.* **100**, 10373 (1996).

25. C. Rønne, P.-O. Åstrand, S. R. Keiding, "THz spectroscopy of liquid H_2O and D_2O", *Phys. Rev. Lett.* **82**, 2888 (1999).

26. E. Knoesel, M. Bonn, J. Shan, T. F. Heinz, "Charge transport and carrier dynamics in liquids probed by THz time-domain spectroscopy", *Phys. Rev. Lett.* **86**, 340 (2001).

27. M.Walter, B. Fischer, M. Schall, H. Helm, P. Uhd Jepsen, "Far-infrared vibrational spectra of aa-trans, 9-cis and 13-cis retinal measured by THz time-domain spectroscopy", *Chem. Phys. Lett.* **332**, 389 (2000).

28. B. B. Hu, M. C. Nuss, "Imaging with terahertz waves", *Opt. Lett.* **20**, 1716 (1995).

29. J. V. Rudd, D. Zimdars, M Warmuth, "Compact fiber-pigtailed terahertz imaging system", in Proc. SPIE volume 3934, 27 (2000).

30. R. Merz, F. Keilmann, R. J. Haug, K. Ploog, "Nonequilibrium edge-state transport resolved by far-infrared microscopy", *Phys. Rev. Lett.*, **70**, 651 (1993).

31. S. Hunsche, M. Koch, I. Brener, M. C. Nuss, "THz near-field imaging", *Opt. Commun.* **150**, 22 (1998).

32. D. M. Mittleman, M. Gupta, R. Neelamani, R. G. Baraniuk, J. V. Rudd, M. Koch, "Recent Advances in Terahertz Imaging", *Appl. Phys. B* **68**, 1085 (1999).

33. K. Wynne, D. A. Jaroszynski, "Superluminal terahertz pulses", *Opt. Lett.* **24**, 25 (1999).

34. Z. Jiang, X. G. Xu, X.-C. Zhang, "Improvement of terahertz imaging with a dynamic subtraction technique", *Appl. Opt.* **39**, 2982 (2000).

35. O. Mitrofanov, I. Brener, M. C. Wanke, R. R. Ruel, J. D. Wynn, A. J. Bruce, J. Federici, "Near-field microscope probe for far infrared time domain measurements", *Appl. Phys. Lett.* **77**, 591 (2000).

36. Q. Chen, Z. Jiang, G. X. Xu, X.-C. Zhang, "Near-field terahertz imaging with a dynamic aperture", *Opt. Lett.* **25**, 1122 (2000).

37. D. M. Mittleman, S. Hunsche, L. Boivin, M. C. Nuss, "T-ray tomography", *Opt. Lett.*, **22**, 904 (1997).

38. S. Hunsche, D. M. Mittleman, M. Koch, M. C. Nuss, "New dimensions in t-ray imaging", *IEICE Trans. Electron.* **E81-C**, 269 (1998).

39. J. L. Johnson, T. D. Dorney, D. Mittleman, "Enhanced depth resolution in terahertz imaging using phase-shift interferometry", *Appl. Phys. Lett.* **78**, 835 (2001).

40. D. M. Mittleman, R. H. Jacobsen, M. C. Nuss, "T-ray imaging", *IEEE J. Sel. Top. Quantum Electron.* **2**, 679 (1996).

41. M. Brucherseifer, P. Haring Bolivar, H. Klingenberg, H. Kurz, "Angle-dependent THz tomography - characterization of thin ceramic oxide films for fuel cell applications", *Appl. Phys. B* **72**, 361 (2001).

42. M. Herrmann, M. Tani, K. Sakai, "Display Modes in Time-Resolved Terahertz Imaging", *Jpn. J. Appl. Phys., P. 1* **39**, 6254 (2000).

43. K. A. McIntosh, E. R. Brown, K. B. Nichols, O. B. McMahon, W. F. DiNatale, T. M. Lyszczarz, "Terahertz photomixing with diode lasers in low-temperature-grown GaAs", *Appl. Phys. Lett.* **67**, 3844 (1995).

44. S. Verghese, K. A. McIntosh, E. R. Brown, " Optical and terahertz power limits in the low-temperature-grown GaAs photomixers", *Appl. Phys. Lett.* **71**, 2743 (1997).

45. J. Struckmeier, A. Euteneuer, B. Smarsly, M. Breede, M. Born, M. Hofmann, L. Hildebrand, J. Sacher, "Electronically tunable external-cavity laser diode", *Opt. Lett.* **24**, 1573 (1999).

46. P. Gu, F. Chang, M. Tani, K. Sakai, C.-L. Pan, "Generation of coherent cw-terahertz radiation using a tunable dual-wavelength external cavity laser diode", *Jpn. J. Appl. Phys., P.2* **38**, L1246 (1999).

47. K. A. McIntosh, K. B. Nichols, S. Verghese, E. R. Brown, "Investigation of ultrashort photocarrier relaxation times in low-temperature-grown GaAs", *Appl. Phys. Lett.* **70**, 354 (1997).

48. O. Lenz, E. Schär, F. H. Schweingruber, "Methodische Probleme bei der radiographischen-densitometrischen Bestimmung der Dichte und Jahrringbreiten von Holz", *Holzforschung* **30** 114 (1976).

49. J.R. Davis, J. Ilic, P. Wells, "Moisture content in drying wood using direct scanning gamma-ray densitometry", *Wood Fiber Sci.* **25**, 153 (1993).

50. H. Mayer-Wegelin, "Der Härtetaster: Ein neues Gerät zur Untersuchung von Jahrringbau und Holzgefüge", *Allg. Forst- und Jagdzeitung* **122**, 12 (1950).

51. J. Kisser, I. Lehnert, "Ein neues Verfahren zur Bestimmung des Früh- und Spätholzanteils bei Nadelhölzern", *Mitt. Österr. Ges. Holzforschung*, Wien (1951).

52. H. Polge, "Fifteen years of wood radiation densitometry", *Wood Sci. and Technol.* **12**, 187 (1978).

53. D. J. Cown, B. C. Clement, "A wood densitometer using direct scanning with X-rays", *Wood Sci. and Technol.* **17**,91 (1983).

54. G. Lesnino, "The laser–sandblasting method: a new method for the qualitative annual ring analysis of conifers", *Wood Sci. and Technol.* **28**, 159 (1994).

55. B. Goy, P. Martin, J. M. Leban, "The measurement of wood density by microwave sensor", *Holz Roh- Werkstoff.* *50*, 163 (1992).

56. M. Koch, S. Hunsche, P. Schumacher, M. C. Nuss, J. Feldmann, J. Fromm, "THz imaging: a new method for density mapping of wood", *Wood Sci. and Technol.* **32**, 421 (1998).

57. H. H. Dixon, J. Joly, *Phil. Trans. R.. Soc. London B* **186**, 563 (1895).

58. G. Huber, "Die Gefäßleitung" in Encyclopedia of Plant Physiology, Vol. 3, W. Ruhland (ed.), 541 (Springer, Berlin, 1956).

59. J. Böhm. Capillarität und Saftsteigen, Ber. Dtsch. Bot. Ges. 11, 203 (1893).

60. M. H. Zimmermann, Xylem Structure and the Ascent of Sap (Springer, Berlin, 1963).

61. Balling, U. Zimmermann, K.-H. Bchner, O. L. Lange, "Direct measurement of negative pressure in artificial-biological systems", Naturwissenschaften 75, 409 (1988).

62. P. F. Scholander, H. T. Hammel, E. D. Bradstreet, E. A. Hemmingsen, "Sap pressures in vascular plants", Science 148, 339 (1965).

63. J. A. Milburn, "Sap ascent in vascular plants: challengers to the cohesion theory ignore the significance of immature xylem and the recycling of mnch water". Ann. Bot. 78, 399 (1996).

64. M. T. Tyree, "The cohesion–tension theory of sap ascent: current controversies", J. Exp. Bot. 48, 1753 (1997).

65. H. Van As, T. J. Schaafsma, Biophys. J. 45, 469 (1984).

66. G. A. Johnson, J. Brown, P. J. Kramer, "Magnetic resonance microscopy of change in water content in stems of transpiring plants", Proc. Nat. Acad. Sci. 84, 2752 (1987).

67. E. Kuchenbrod, M. Landeck, F. Thrmer, A. Haase, U. Zimmermann, "Measurement of water flow in the xylem vessels of intact maize plants using flow-sensitive NMR imaging", Bot. Acta 109, 184 (1996).

68. P. L. Silvestrelli, M. Bernasconi, M. Parrinello, "Ab initio infrared spectrum of liquid water". Chem. Phys. Lett 277, 478 (1997).

69. D. M. Mittleman, private communication.

70. E. Brcke. Müllers Arch. Anat. Physiol. 443 (1848).

71. J. Fromm, W. Eschrich, "Seismonastic movements in Mimosa", in "The Pulvinus: Motor Organ for Leaf Movement", ed. By R. L. Satter, H. L. Gorton, T. C. Vogelmann, (American Society of Plant Physiologists, Rockville, MD, 1990).

72. P. Y. Han, G. C. Cho, X.-C. Zhang, "Time-domain transillumination of biological tissues with terahertz pulses", Opt. Lett. 25, 242 (2000).

73. D. D. Arnone, C. M. Ciesla, A. Corchia, S. Egusa, M.Pepper, J. M. Chamberlain, C. Bezant, E. H. Linfield. "Applications of THz technology to medical imaging", Proc. SPIE, volume 3828, 209 (1999).

74. D. Arnone, C. Ciesla, M. Pepper. "Terahertz imaging comes into view", Phys. World, April, p. 36 (2000).

75. R. M. Woodward, B. Cole, V. P. Wallace, D. D. Arnone, R. Pye, E. H. Linfield, M.Pepper, A. G. Davis, "Terahertz imaging of in-vitro basal cell carcinoma samples", in Conference of Lasers and Electro-Optics 2001, Technical digest (Baltimore, 2001), p. 329

Electronic Sources and Detectors
for Wideband Sensing in the Terahertz Regime

Daniel W. van der Weide

Abstract. Broadband sensing and spectroscopy using both reflection and transmission, can now be performed in the 1–1000 GHz regime with pulsed terahertz circuits. The use of stable, coherent driving waveforms from phase-locked microwave synthesizers – or a new single-sideband-modulator frequency translation technique – in combination with these NLTLs enables measurement of both instantaneously wideband spectra and single absorption lines with Hertz-level precision, something is not readily done with more conventional optoelectronic THz techniques.
We have taken two approaches to coherent measurements: (1) use of a conventional source/detector arrangement with sampling detectors, and (2) spatially combining the freely propagating beams from two matched picosecond pulse generators. The latter method employs a dual-source interferometer (DSI), modulates each harmonic of one source with a precisely offset harmonic from the other source, both sources being driven with stable phase-locked synthesizers, and the resultant beat frequency can be low enough for detection by a standard bolometer. Possible methods of detection at room temperature for the DSI include antenna-coupled Schottky diodes. Finally, we have recently introduced a reflectometer based on serrodyne modulation of a linearized delay line, using a technique that is process-compatible with pulse generator circuits.

1 Introduction

With the greater acceptance of terahertz techniques for spectroscopic and sensing applications, there is a growing need for integrated circuit (IC) sources and detectors in this regime. While a hybrid optoelectronic system might incorporate laser sources and photoconductive switches [4–9] or photomixers [10,9,11], this chapter will focus on fully electronic systems. Two primary classes of electronic THz ICs are continuous-wave (CW) and pulsed systems. While much important work has been done on frequency multipliers and detectors for astrophysics [11–20], these are primarily narrowband systems, usually operated at cryogenic temperatures. The solid-state electronic sources reaching toward THz frequencies are narrowband tunable CW sources [21,22], low-order frequency-multiplied sources [15] and wideband sources (pulsed, using high-order frequency multiplication).

Here we discuss wideband generation and detection systems that incorporate circuits generating over 100 harmonics of the drive frequency [23–26]. The applications of such devices extend beyond sensing to communication, working in an unregulated portion of the electromagnetic spectrum. We have performed measurements on quantum dots [1], gas absorption spectra [2] and reflection spectra of energetic materials [3] with such instruments, which use phase-locked microwave sources to drive (sub)picosecond GaAs nonlinear transmission lines (NLTLs). While our focus here will be on low-power uses, new vacuum sources aimed at achieving high power are under development [27], while solid-state amplifiers continue to improve in power and frequency performance [28–30].

The applications of wideband systems include multi-species gas sensing to monitor automotive and factory emissions, particularly in the presence of particulates (e.g. soot from diesels) that scatter near-infrared radiation used in today's near-infrared (NIR) gas sensors. Other applications include measuring reflections with portable, inexpensive equipment for security and location of buried objects. Finally, in order to make them economical and compact, all of these electronic techniques would depend on IC-compatible coherent measurement capabilities[31,32]. For wavelengths longer than the NIR, Fourier transform infrared (FTIR) spectroscopy is the dominant method for measuring absorption spectra [33]. While FTIR spectroscopy has been used in the millimeter-wave regime with specialized equipment [34], the relative weakness of black-body sources below $10\,\mathrm{cm}^{-1}$ ($300\,\mathrm{GHz}$) requires prohibitively long scans and averaging times.

As an alternative to FTIR spectroscopy, we can employ a dual-source interferometer (DSI) that is not only compact but also free of the need for a coherent detector so that an incoherent (e.g. photoconductive or bolometric) power detection mechanism can be used [35–37]. This eliminates the noise of a sampling detector and establishes a baseline for future fully integrated implementations that would use sampling detectors. In the first section of this chapter we describe and analyze the DSI, and demonstrate its use in measuring gas absorption spectra and electron cyclotron resonance.

We then discuss measurements of reflection spectra of energetic materials using a more conventional sampling detector, and finally review a new means of generating coherent signals from a single source. This approach will play a role in future handheld electronic THz systems.

To generate wideband (discrete) spectral energy in the 1-1000 GHz regime, we can use nonlinear transmission-line (NLTL) pulse generators coupled to wideband planar antennas [24,25,38–42]. The GaAs IC NLTLs used in the work described here consist of series inductors (or sections of high-impedance transmission line) with varactor diodes periodically placed as shunt elements (Fig. 1). A fast (~ 0.5–$2\,\mathrm{ps}$) voltage step develops on this structure at room temperature from a sinusoidal input because the propagation velocity u is modulated by the diode capacitance: $u(V) = 1/\sqrt{LC(V)}$, where L is the line inductance and $C(V)$) is the sum of the diode capacitance and the parasitic line capacitance [23,25]. Limitations of the NLTL arise from its periodic cutoff frequency, waveguide dispersion, interconnect metallization losses, and

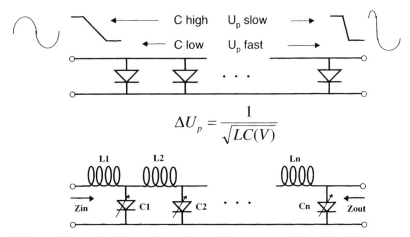

$$\Delta U_p = \frac{1}{\sqrt{LC(V)}}$$

Fig. 1. Schematic illustration of a nonlinear tranmission line. *Top:* arrangement of diodes along a high-impedance transmission line and illustration of waveform compression due to voltage-dependent propagation velocity U_p. *Bottom:* equivalent-circuit representation of NLTL, indicating its impedance-transforming properties

resistive losses in the diodes. Improvements in NLTL design have resulted in subpicosecond pulses at room temperature [26].

The first electronic THz systems used resonant or broadband (bow-tie) antennas driven directly by NLTLs (Fig. 2) for generation of freely propagating pulses. These pulses could be focused and collected in a manner exactly analogous to what was done in the optoelectronic THz systems covered in

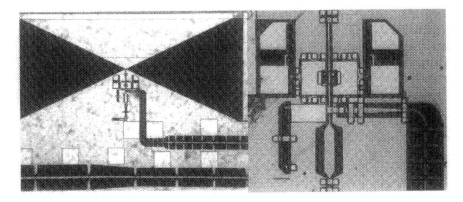

Fig. 2. *Left:* GaAs NLTL integrated with planar bow-tie antenna. The circuit shown includes a diode sampling bridge – note for four bond pads near antenna apex. *Right:* close-up view of diode sampling bridge integrated with NLTL. The sampler drive signal enters from the right; the signal to be sampled travels on the vertically oriented line; the four bond pads are sampler outputs

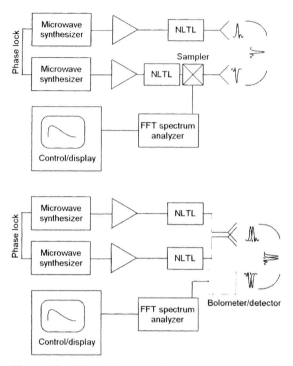

Fig. 3. *Top:* block diagram of conventional source/detector electronic THz system with a sampling detector, which coherently downconverts the detected pulse to baseband. *Bottom:* dual-source interferometer with copropagating pulses from each source – the resultant beat frequency spectrum is low enough to fall within the response bandwidth of a composite bolometer

detail elsewhere in this volume, using substrate lenses and reflective optics. For coherent detection, the pulse from the receiver's NLTL was used to drive a sampler, essentially a broadband frequency mixer (Fig. 2), whose temporal output could be sent to an oscilloscope or a Fast Fourier Transform (FFT) spectrum analyzer for display in the frequency domain (Fig. 3).

2   Dual-Source Interferometer

In contrast to the first THz systems, the dual-source spectrometer does not use a coherent detector. In this spectrometer, we spatially combine the freely propagating beams from two coherent NLTL pulse generators, resulting in a compact, reliable, and easy-to-implement microwave and millimeter-wave source for spectroscopy (Figs. 3 and 4). Because the DSI modulates each harmonic of one source with a precisely offset harmonic from the other source–both sources being driven with stable phase-locked synthesizers–the resultant beat frequency can be low enough for detection by a standard composite

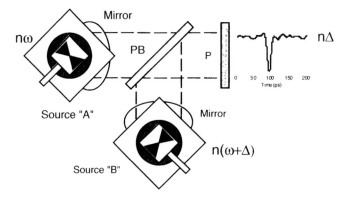

PB = polarizing beamsplitter
P = wire grid polarizer

Fig. 4. Configuration of dual-source interferometer. Each source/antenna combination is at the focus of a paraboloidal mirror and radiates a polarized beam, which is transmitted (source A) or reflected (source B) by the polarizing beam splitter (PB). One source is driven at a frequency ω or $\omega + \Delta$. The output polarizer (P) selects half the power of each beam. The output waveform is detected by a bolometer [2,3,35–37]. Harmonics of order n emerge from NLTL sources; the harmonic spacing is $n\Delta$ at the difference frequency. ©IEEE

bolometer. Possible methods of detection at room temperature for the DSI include antenna-coupled Schottky diodes [15].

3   Description of DSI

In the dual-source interferometer described here, the output of each NLTL feeds an integrated bow-tie antenna mounted at the focus of a hyperhemispherical high-resistivity silicon lens [12,43,44,55,46]. Each of these lenses, in turn, is mounted at the focus of an off-axis paraboloidal mirror. The beam collimated by the mirrors is either transmitted (source A) or reflected (source B) by a wire-grid polarizing beam-splitter (Fig. 4). Each beam then contributes equally to the final, linearly polarized beam after passing through a final wire-grid polarizer mounted at 45° to the beam-splitter. The prototype system is small ($\sim$ 170 mm long, 120 mm wide, and 80 mm high), but we have already fabricated a power combiner driving one antenna on the same substrate, making the whole system extremely compact [37]. We also note that other workers have described similar ideas using laser-triggered photoconductive switches [8,47,48], but one advantage of the current system is its all-electronic approach, which offers precise amplitude (0.01 dB) and frequency (1 Hz) control of the fundamental excitation.

Each source is fed by a 100–900 mW sine-wave generated by one of two microwave synthesizers, which share a common 10 MHz timebase. The output of one synthesizer is offset by $\Delta f \ll f_0 (\Delta f \sim 100\,\text{Hz}, f_0 = 3\text{--}10\,\text{GHz})$. and this offset is used as a trigger for a FFT spectrum analyzer or digital lock-in amplifier. While the synthesizers and wideband power amplifiers used in the present demonstration are expensive, they could be replaced by an inexpensive single source combined with an NLTL-based frequency translator [31,32], which we describe below.

4   Analysis of Dual-Source Interferometer

Here we analyze the viability of a detection method based on combining two waveforms with slightly offset frequencies at a power sensor such as a bolometer. Assume first that we have two periodic waveforms $v(t)$ and $w(t)$ of the same frequency $f = 1/T$, so that

$$
\begin{aligned}
v(t+T) &= v(t). \\
w(t+T) &= w(t).
\end{aligned}
\tag{1}
$$

We combine the waveforms and excite a detector with the result. The detector is slow enough that it can only measure the average power of the radiation on it. If one of the waveforms is shifted with respect to the other by a, we obtain

$$
\begin{aligned}
P(a) &= \frac{1}{T} \int_0^T [v(t) + w(t+a)]^2 \, dt \\
&= \frac{1}{T} v^2(t) \, dt + \frac{1}{T} \int_0^T w^2(t+a) \, dt + \frac{2}{T} \int_0^T v(t)w(t+a) \, dt \\
&= \frac{1}{T} \int_0^T [v^2(t) + w^2(t)] \, dt + \frac{2}{T} \int_0^T v(t)w(t+a) \, dt
\end{aligned}
\tag{2}
$$

as the detected power. Clearly, only the last term of the above formula is dependent on the relative shift a between the two waveforms; the first term is merely a DC offset. In the subsequent calculations we shall focus only on the shift-dependent term, which we shall call $2\tilde{P}(a)$. Since the waveforms $v(t)$ and $w(t)$ are real functions, we have

$$
\tilde{P}(a) = \frac{1}{T} \int_0^T v(t)w(t+a) \, dt = \frac{1}{T} \int_0^T v^*(t)w(t+a) \, dt,
\tag{3}
$$

where the asterisk denotes complex conjugation. Since $v(t)$ and $w(t)$ are periodic, we can write them as Fourier series:

$$
v(t) = \sum_{m=-\infty}^{\infty} v_m \exp\left(\frac{2\pi j m}{T} t\right).
\tag{4}
$$

$$
w(t) = \sum_{n=-\infty}^{\infty} w_n \exp\left(\frac{2\pi j n}{T} t\right).
\tag{5}
$$

This yields

$$\int_0^T v^*(t)w(t+a)\,dt$$

$$= \int_0^T \sum_{m=-\infty}^{\infty} v_m * \exp\left(-\frac{2\pi i m}{T}t\right) \sum_{n=-\infty}^{\infty} w_n \exp\left(\frac{2\pi i n}{T}(t+a)\right) dt$$

$$= \sum_{m=-\infty}^{\infty} v_m^* \sum_{n=-\infty}^{\infty} w_n \int_0^T \exp\left(-\frac{2\pi i m}{T}t\right) \exp\left(\frac{2\pi i n}{T}(t+a)\right) dt$$

$$= \sum_{m=-\infty}^{\infty} \sum_{n=-\infty}^{\infty} v_m * w_n 2\pi T \delta_{mn} \exp\left(\frac{2\pi i n}{T}a\right)$$

$$= 2\pi T \sum_{n=-\infty}^{\infty} v_n^* w_n \exp\left(\frac{2\pi i n}{T}a\right) = 2\pi T \sum_{n=-\infty}^{\infty} v_n^* w_n \exp\left(\frac{2\pi i n}{T}a\right), \quad (6)$$

or

$$\tilde{P}(a) = \frac{1}{T}\int_0^T v(t)w(t+a)\,dt$$

$$= \frac{1}{T}\int_0^T v^*(t)w(t+a)\,dt$$

$$= 2\pi \sum_{n=-\infty}^{\infty} v_n^* w_n \exp\left(\frac{2\pi i n}{T}a\right). \quad (7)$$

If the relative shift a between the two waveforms is varied (slowly compared with the frequency $f = 1/T$), the detected average power will vary as well, according to the above equation. In particular, if $a = Tf't$, where $f' \ll f$ is the offset frequency, then

$$\tilde{P}(t) = 2\pi \sum_{n=-\infty}^{\infty} v_n^* w_n \exp(2\pi i f' n t) = \sum_{n=-\infty}^{\infty} \tilde{P}_n \exp(2\pi i f' n t). \quad (8)$$

That is, the detected power $\tilde{P}$ varies in time with a frequency f', such that $\tilde{P}(t+1)/f' = \tilde{P}(t)$, and the harmonics of the original waveforms $v(t)$ and $w(t)$ are encoded in its spectral composition as in the above formula. The attenuation of the nth harmonic,

$$\left|\frac{v_n'}{v_n}\right| = \left|\frac{w_m'}{w_n}\right|, \quad (9)$$

where v_n' and v_n are the amplitudes of the signal with and without the sample present in the beam path, can be easily recovered from the spectral analysis of the detected power:

$$\left|\frac{v'_n}{v_n}\right| = \left|\frac{w'_m}{w_n}\right| = \sqrt{\left|\frac{\tilde{P}'_n}{\tilde{P}_n}\right|} = \sqrt{\left|\frac{P'_n}{P_n}\right|}. \tag{10}$$

where P'_n and P_n are the nth harmonics of the detected power signal with and without the sample in the beam path.

5   DSI Results from Dual-Source Interferometer

As shown in Fig. 5, the DSI can illuminate a single-pass gas absorption cell using off-axis paraboloidal mirrors to collimate the radiation and collect it for detection by a bolometer. This is useful for low-pressure gas spectroscopy, in this case of a simple rotor molecule, OCS.

Using a digital lock-in amplifier, we could measure a single harmonic produced by the DSI, in this case at 133.785 980 GHz, the 20th harmonic of the fundamental, 6.689 299 GHz. To measure the data shown in Fig. 5, we used a third source to provide the 10 MHz timebase for the microwave synthesizers and then varied this 10 MHz signal to continuously sweep the harmonics of the DSI. This enabled the extremely narrow linewidth of 0.1 torr OCS to be resolved, with an equivalent Q ($f_0/\Delta f$) of $> 8 \times 10^7$. In Fig. 6 we show the

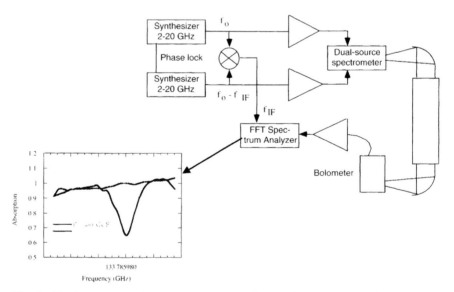

Fig. 5. Dual-source spectrometer system configured for gas absorption measurements. The inset shows data from 0.1 torr OCS compared with vacuum obtained from a 76 cm long cell, uncorrected for background power slope. The horizontal scale is 2 MHz per large division [2]. ©IEEE

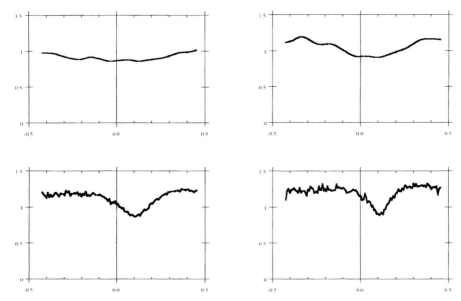

Fig. 6. Pressure broadening of OCS line measured by DSI. Center frequency 133.78597 GHz (20th harmonic of 6.6893 GHz), sweep width 11.8 MHz. Pressure, from left to right, top to bottom: 0.5, 0.3, 0.2 and 0.1 torr [3]. ©SPIE

effect of pressure broadening on this line, which was expected. Likewise, the system can be applied for excitation of low-dimensional electron systems [1]. One simple illustration of this system's utility is the measurement of cyclotron resonance absorption in a two-dimensional electron gas (2DEG), as shown in Fig. 7. This is advantageous in that the entire spectrum is captured at once, as opposed to the traditional method of using CW sources and tuning the magnetic field at the sample, to change its energy levels.

6   Reflection Spectroscopy

Increasingly sophisticated weapons and explosives require increasingly sophisticated detection technologies. Nonmetallic varieties of these threats are especially important because they elude familiar metal-detecting portals, so they have motivated the development of a multipronged approach to detection, including residue sniffing and computerized tomography. These techniques, however, have significant drawbacks, suffering from invasiveness, slowness, unfamiliarity to the public, and significant potential for false negatives [49].

Threats such as these, however, appear to be readily detectable and even identifiable using wideband signals in the sub-THz regime (1–500 GHz) [50,51], below the point of significant water vapor absorption. Traditional equipment for generating and detecting these frequencies has, however, been

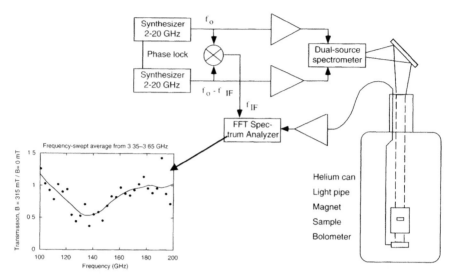

Fig. 7. Experimental setup for spectroscopy of a two-dimensional electron gas. The output of the dual-source interferometer is guided to the 2DEG sample through a light pipe in a sealed can of liquid He, pumped to 20 mtorr ($T = 1.8$ K). The signal from the bolometer is amplified and Fourier transformed as shown. *Inset:* broadband transmittance through 2DEG at fixed 315 mT magnetic field. The apparent scatter in the data is actually the effect of standing waves. ©IEEE

difficult to use, bulky and expensive. We propose applying all-electronic and monolithically integrated technology to generate and detect these wideband signals for the purpose of imaging the reflection spectra of plastic weapons and explosives carried by human subjects.

Many of the concepts we employ here are being pursued at lower frequencies for target detection at higher resolutions than are allowed by traditional narrowband radar. Ultrawideband (UWB), carrier-free, impulse, or baseband radar has been rapidly gaining popularity in applications where complex, elusive targets are the norm [52]. UWB radar has benefited from very recent advances in semiconductor technology that enable the production of subnanosecond pulses with peak powers of over 1 MW but having average powers in the milliwatt regime [53]. By contrast, the NLTL essentially trades power for speed, producing picosecond or even subpicosecond pulses with peak powers less than 1 W and average powers in the low microwatt regime. These power levels are nonionizing and biologically inconsequential, but because we can employ coherent detection, rejecting noise outside the frequencies of interest, we can still measure useful spectra with them.

As described elsewhere in this volume, baseband picosecond and subpicosecond pulses of freely propagating radiation, usually generated and coherently detected with photoconductive switches and ultrafast lasers [6,54,55], have been useful for wideband coherent spectroscopy of materials, liquids, and gases in the THz regime. Such systems have even been used for what could be called scale-model UWB radar [9]. These highly versatile beams of ultrashort electromagnetic pulses can be treated quasi-optically: they can be diffracted and focused with mirrors and lenses, and the resultant effects can be readily observed in the time-domain waveform at the detector. Consequently, such beams are singularly useful for spectroscopy in a difficult-to-access spectral regime, and reports of spectroscopic imaging with such optoelectronic systems have generated much interest [56].

Using system concepts identical to those of optoelectronic THz systems but with electronic components (Fig. 8), we have configured an NLTL–antenna generator and coherent sampling detector in a focused reflection spectroscopy system using off-axis ellipsoidal mirrors.

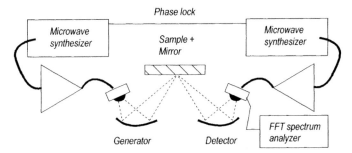

Fig. 8. System diagram for electronic THz reflection spectroscopy [3]. © SPIE

Wideband pulsed (as opposed to single-wavelength) imaging has the chief advantage of flexibility: if weapons change composition over the years, a single-wavelength or narrowband source may no longer detect the new composition, but having a broad range of frequencies maximizes the opportunity to detect the new threat's signature.

One example of this advantage is shown in the data we present here. C-4 consists primarily of RDX with a plasticizing binder. As shown in Fig. 9, the wideband reflection spectra out to 450 GHz show a clear similarity between C-4 and RDX; both are distinct from HMX, demonstrating specificity in the dielectric response of these targets. The contrast that we measure is not fully understood, but probably arises from the granularity of the materials, their dielectric constants and orientation of the sample, as distinct from rotational absorption of chemical bonds.

These spectra each show the average and standard deviation of eight data sets obtained using 50 harmonics of a 7.75 GHz fundamental, with the sample being occasionally repositioned to simulate a security screening application. The presence of standing waves in the highly reflective arrangement resulted in sample-to-background ratios exceeding unity at some frequencies.

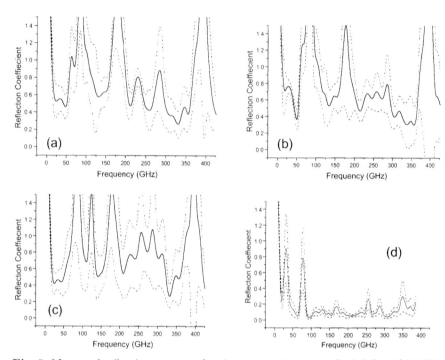

Fig. 9. Measured reflection spectra of various energetic materials: (**a**) C-4. (**b**) RDX and (**c**) HMX. These can be compared with the measured reflection spectrum from a human hand (**d**). In each figure, the dashed lines represent one standard deviation [3]. © SPIE

We have measured reflection spectra from several potential threats, such as HMX, RDX, C-4, TNT, napthalene (which has a similar chemical structure to some explosives), ammonium nitrate solids and slurries, and gasoline, as well as from common objects such keys, other metal objects, wallets, credit cards, and notebooks. All show unique and repeatable signatures.

The new information gained from these results could enable progress toward developing multipixel arrays of sources and detectors for use in new screening technologies. Our current focus is on building inexpensive "single pixel" screening technology by pursuing advances in the microfabrication of the devices described above, integrating more functions on a chip, and reducing the cost of these systems.

7   Coherent Signal Generation
with Scanned Delay Lines

In this section, we review a new frequency translator based on an NLTL phase shifter. Rather than forming shock waves on the NLTL by using excitation with a large signal, we have used its voltage-variable delay together with both amplitude and phase linearization to modulate the phase of a small 0.5–3.0 GHz microwave signal, with a performance that can be scaled with integration of the circuit. The resultant single sideband modulator exhibits > 45 dBc carrier and spurious suppression. This new approach has significant applications in both instrumentation and sensing, particularly because it offers a clear path toward complete integration of a coherent measurement system.

This is a new solution that could enable complete integration of wideband network analyzers, directly addressing the need for instruments to characterize 100 GHz devices, circuits, and systems, as well as the growing opportunities for sensors in this regime. Combining an NLTL phase shifter with serrodyne (sawtooth) modulation results in a frequency translator that can use an inexpensive–ultimately integrated–microwave source to coherently convert a wideband microwave signal directly to baseband. This circuit, coupled with improved directional sampling circuits, could enable high-performance, inexpensive, field-capable 100 GHz vector network analysis, as well as several other new military and commercial applications which would benefit from a monolithic coherent generation/detection system.

As shown in Fig. 10, a coherent electronic measurement system need not rely on phase-locked sources, but instead could use a modulated delay line to impart a (single-sideband) frequency shift onto one arm of an interferometer. When the delay line is serrodyne modulated, additional sidebands are generated owing to the flyback transient; these can be minimized by sufficiently short flyback times or triangle-wave modulation with appropriate Fourier transform reversals at baseband.

We have built a demonstration phase shifter on a brass block with 30 abrupt-junction diodes ($C_{j0} = 2\,\mathrm{pF}$) spaced at intervals of 4.6 mm in the first section and with 20 abrupt-junction diodes ($C_{j0} = 0.8\,\mathrm{pF}$) spaced at intervals

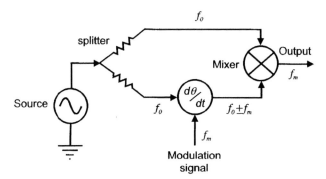

Fig. 10. Block diagram of a frequency translator application. The phase shifter is serrodyne modulated at f_{m} and the output of the mixer is a sinusoid at frequency f_{m} [32]. © IEEE

of 0.6 mm in the second section. This circuit was able to achieve > 2 ns delay, yet was usable beyond 3 GHz. By comparison, single GaAs IC NLTLs can achieve ∼ 160 ps delays [26] with higher usable frequencies; they can also be cascaded for additional delays.

We implemented a phase linearization and amplitude compensation scheme using a gain-controlled amplifier (HP IVA 14208) for the amplitude and a 12-bit digital-to-analog converter for phase. We measured the beat note between the original microwave signal and its frequency-shifted version using both serrodyne and triangle-wave modulation. Both methods gave nearly pure sinusoidal results, but the serrodyne version exhibited a flyback transient. We could reconstruct the output sinusoid without this transient by using triangle-wave modulation time-reversing second cycle of the output.

In spite of the cleaner time-domain appearance of the triangle-wave modulated output, when the system was driven at 0 dBm the serrodyne version exhibited a record > 45 dB carrier and harmonic suppression compared with ∼ 35 dB for the triangle wave. Figure 11 shows a downconverted spectrum obtained with serrodyne modulation. In the future, we expect to address the shortcomings of the triangle-wave modulation with more careful waveform reconstruction.

Modulated NLTL frequency translators are ultimately candidates for integration with NLTL pulse generators and diode sampling bridges, enabling for the first time the foundations of a complete monolithic electronic THz

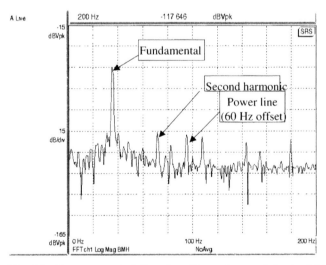

Fig. 11. Downconverted (baseband) spectrum of single-sideband modulator. The fundamental is at f_m while harmonics and spurious contributions due to the line frequency and the sawtooth retrace transient are evident [3]. © SPIE

system. We have extended this approach to building a 1-3 GHz reflectometer with performance comparable to a commercial network analyzer [31,32].

8   Conclusions

We have described and demonstrated a new technique for absorption spectroscopy using a new, all-electronic THz interferometer having no moving parts; rather, its phase is controlled electronically via a frequency offset between its two identical sources. We have also discussed a more conventional coherent source/detector arrangement, but used for reflection spectroscopy of energetic materials. Finally, we have presented a new means for eliminating the precision sources used in these systems in order to produce a very compact, and low-cost system. While this technique lays a promising foundation for inexpensive coherent microwave instrumentation, it can also be extended to other microwave and millimeter-wave sensors, such as handheld reflectometers operating in the THz regime for applications such as de-mining, and sensing gases, nonmetallic weapons, and explosives.

Acknowledgments

This work has been supported by a Ford Motor Company University Research Award, Defense Advanced Research Projects Agency, Office of Naval Research, Young Investigator Program, National Science Foundation, Presidential Early Career Award for Scientists and Engineers (ECS-0196021), Army Research Office, and Federal Aviation Administration. Thanks are due to J. Murakowski, V. Agrawal, P. Akkaraekthalin, and S. Kee for some of the measurements and modeling, to M. Choi, A. Wangsanata, J. Truitt, and B. Rosner for current work on the THz system, and F. Keilmann, R. Blick, and K. McNesby for helpful discussions.

References

1. R. H. Blick, D. W. van der Weide, R. J. Haug, K. Eberl, "Complex broadband millimeter wave response of a double quantum dot: Rabi oscillations in an artificial molecule", *Phys. Rev. Lett.* **81**, 689 (1998).
2. D. W. van der Weide, J. Murakowski, F. Keilmann, "Gas-absorption spectroscopy with electronic terahertz techniques", *IEEE Trans. on Microwave Theory Tech.* **48**, 740 (2000).
3. D. W. van der Weide, J. Murakowski, F. Keilmann, "Spectroscopy with electronic terahertz techniques", *Proc. SPIE*, **3828**, 276 (1999).
4. C. Fattinger, D. Grischkowsky, "Terahertz beams", *Appl. Phys. Lett.*, **54**, 490 (1989).
5. M. van. Exter, D. Grischkowsky, "Optical and electronic properties of doped silicon from 0.1 to 2 THz", *Appl. Phys. Lett.*, **56**, 1694 (1990).

6. D. Grischkowsky, S. Keiding, M. van. Exter, C. Fattinger, "Far-infrared time-domain spectroscopy with terahertz beams of dielectrics and semiconductors", *J. Opt. Soc. Am. B.* **7**. 2006 (1990).

7. N. Katzenellenbogen, D. Grischkowsky, "Efficient generation of 380 fs pulses of THz radiation by ultrafast laser pulse excitation of a biased metal-semiconductor interface", *Appl. Phys. Lett.*. **58**, 222 (1991).

8. S. E. Ralph, D. Grischkowsky, "THz spectroscopy and source characterization by optoelectronic interferometry", *Appl. Phys. Lett.*, **60**, 1070 (1992).

9. R. A. Cheville, D. Grischkowsky, "Time domain terahertz impulse ranging studies", *Appl. Phys. Lett.*. **67**, 1960 (1995).

10. S. Verghese, K. A. McIntosh, E. K. Duerr, "Generation and detection of THz waves with photomixers", *Proc. SPIE* **3828**. 118 (1999).

11. S. M. Duffy, S. Verghese, A. McIntosh, A. Jackson, A. C. Gossard, S. Matsuura, "Accurate modeling of dual dipole and slot elements used with photomixers for coherent terahertz output power". *IEEE Trans. Microwave Theory Tech.* **49**. 1032 (2001).

12. B. Kormanyos, G. Rebeiz, "A 26 to 220 GHz harmonic mixer-receiver". *Microwave J.*, **36**, 103-108 (1993).

13. A. Skalare, W. R. McGrath, B. Bumble, H. G. LeDuc, P. J. Burke, A. A. Verheijen, D. E. Prober, "A heterodyne receiver at 533 GHz using a diffusion-cooled superconducting hot electron bolometer mixer", *IEEE Trans. Appl. Supercond.* **5**. 2236. (1995).

14. G. Chattopadhyay, F. Rice, D. Miller, H. G. LeDuc, J. Zmuidzinas, "A 530-GHz balanced mixer". *IEEE Microwave Guided Wave Lett.*, **9**, 467 (1999).

15. T. W. Crowe, J. L. Hesler, R. M. Weikle, S. H. Jones, "GaAs devices and circuits for terahertz applications". *Infrared Physics and Technology*, **40**, **3**, 175-189 (June 1999).

16. J. Kawamura, C. Jian, D. Miller, J. Kooi, J. Zmuidzinas, B. Bumble, H. G. LeDuc, J. A. Stern, "Low-noise submillimeter-wave NbTiN superconducting tunnel junction mixers". *Appl. Phys. Lett.*. **75**, 4013. (1999).

17. H. Kazemi, S. T. G. Wootton, N. J. Cronin, S. R. Davies, R. E. Miles, R. D. Pollard, J. M. Chamberlain, D. P. Steenson, J. W. Bowen, "Active micromachined integrated terahertz circuits". *Int. J. Infrared Millim. Waves* **20**. 967 (1999).

18. G. Chattopadhyay, D. Miller, H. G. LeDuc, J. Zmuidzinas, "A dual-polarized quasi-optical SIS mixer at 550 GHz", *IEEE Trans. Microwave Theory Tech.* **48**, 1680 (2000).

19. J. W. Kooi, J. Kawamura, J. Chen, G. Chattopadhyay, J. R. Pardo, J. Zmuidzinas, T. G. Phillips, B. Bumble, J. Stern, H. G. LeDuc, "A low noise NbTiN-based 850 GHz SIS receiver for the Caltech Submillimeter Observatory", *Int. J. Infrared Millim. Waves*, **21**, 1357 (2000).

20. I. Siddiqi, A. Verevkin, D. E. Prober, A. Skalare, B. S. Karasik, W. R. McGrath, P. Echternach, H. G. LeDuc, "Noise and conversion efficiency of aluminum superconducting hot-electron bolometer mixer". *IEEE Trans. Appl. Supercond.* **11**. 958, (2001).

21. H. Eisele, "Two-terminal devices as fundamental solid state terahertz oscillators", *Proc. SPIE* **3828**, 70 (1999).

22. H. Eisele, A. Rydberg, G. Haddad, "Recent advances in the performance of InP Gunn devices and GaAs TUNNETT diodes for the 100–300-GHz frequency range and above", *IEEE Trans. Microwave Theory Tech.* **48**, 626 (2000).

23. M. J. W. Rodwell, M. Kamegawa, R. Yu, M. Case, E. Carman, K. S. Gi-
    boney, "GaAs nonlinear transmission lines for picosecond pulse generation and
    millimeter-wave sampling", *IEEE Trans. Microwave Theory Tech.* **39**, 1194
    (1991).

24. Y. Konishi, M. Kamegawa, M. Case, Y. Ruai, S. T. Allen, M. J. W. Rodwell,
    "A broadband free-space millimeter-wave vector transmission measurement sys-
    tem", *IEEE Trans. Microwave Theory Tech.* **42**, 1131 (1994).

25. M. J. W. Rodwell, S. T. Allen, R. Y. Yu, M. G. Case, U. Bhattacharya, M.
    Reddy, E. Carman, M. Kamegawa, Y. Konishi, J. Pusl, R. Pullela, J. Esch,
    "Active and nonlinear wave propagation devices in ultrafast electronics and
    optoelectronics (and prolog)", *Proc. IEEE*, **82**, 1035 (1994).

26. D. W. van der Weide, "Delta-doped Schottky diode nonlinear transmission lines
    for 480-fs, 3.5-V transients", *Appl. Phys. Lett.* **65**, 881 (1994).

27. J. H. Booske, C.L. Kory, D. Gallagher, V. Heinen, K. Kreischer, D. W. van
    der Weide, S. Limbach, P. Gustafson, W.-J. Lee, S. Gallagher, and K. Jain,
    "Terahertz-regime, micro-VEDs: evaluation of micromachined TWT concep-
    tual designs", IEEE Conf on Pulsed Power Plasma Science, p. 161, Las Vegas,
    NV, 2001.

28. Q. Lee, S. C. Martin, D. Mensa, R. P. Smith, J. Guthrie, M. J. W. Rodwell,
    "Submicron transferred-substrate heterojunction bipolar transistors", *IEEE
    Electron Device Lett.* **20**, 396 (1999).

29. M. Rodwell, Q. Lee, D. Mensa, J. Guthrie, S. C. Martin, R. P. Smith, R. Pullela,
    B. Agarwal, S. Jaganathan, T. Mathew, S. Long, "Transferred-substrate HBT
    integrated circuits", *Solid State Electronics*, **43**, 1489 (1999).

30. L. S. McCarthy, I. P. Smorchkova, X. Huili, P. Kozodoy, P. Fini, J. Limb, D. L.
    Pulfrey, J. S. Speck, M. J. W. Rodwell, S. P. DenBaars, U. K. Mishra, "GaN
    HBT: toward an RF device", *IEEE Trans. Electron Devices* **48**, 543 (2001).

31. P. Akkaraekthalin, S. Kee, D. W. van der Weide, "Distributed broadband fre-
    quency translator", in *1998 IEEE MTT-S International Microwave Symposium
    Digest* (IEEE, New York, 1998), vol. 3, p. 1431.

32. P. Akkaraekthalin, S. Kee, D. W. van der Weide, "Distributed broadband
    frequency translator and its use in a 1–3 GHz coherent reflectometer", *IEEE
    Trans. Microwave Theory Tech.*, **46**, 2244 (1998).

33. G. W. Chantry, *Long-Wave Optics* (Academic Press, London, 1984), vol.1.

34. M. N. Afsar, "Dielectric measurements of millimeter-wave materials", *IEEE
    Trans. Microwave Theory Tech.* **32**, 1598 (1984).

35. D. W. van der Weide, R. H. Blick, F. Keilmann, R. J. Haug, "Electronic
    picosecond-pulse interferometer probing the millimeter-wave response of a
    quantum-dot system", presented at Quantum Optoelectronics, Dana Point, CA
    (1995).

36. D. W. van der Weide, F. Keilmann, "Picosecond dual-source interferometer
    extending Fourer-transform spectrometer to microwave regime", in *1996 IEEE
    MTT-S International Microwave Symposium Digest* (IEEE, New York, 1996),
    vol. 3, p. 1731.

37. D. W. van der Weide, F. Keilmann, "Coherent periodically pulsed radiation
    spectrometer." US Patent 5,748,309, (1998).

38. D. W. van der Weide, J. S. Bostak, B. A. Auld, D. M. Bloom, "All-electronic
    free-space pulse generation and detection", *Electron. Lett.* **27**, 1412 (1991).

39. Y. Konishi, M. Kamegawa, M. Case, R. Yu, M. J. W. Rodwell, R. A. York, "Picosecond electrical spectroscopy using monolithic GaAs circuits", *Appl. Phys. Lett.*, **61**, 2829, (1992).

40. D. W. van der Weide, J. S. Bostak, B. A. Auld, D. M. Bloom, "All-electronic generation of 880 fs, 3.5 V shockwaves and their application to a 3 THz free-space signal generation system", *Appl. Phys. Lett.* **62**, 22 (1993).

41. J. S. Bostak, "All-electronic generation and detection of terahertz free-space radiation with applications to free-space terahertz spectroscopy", PhD dissertation, Stanford University (1994), p. 90.

42. J. S. Bostak, D. W. van der Weide, D. M. Bloom, B. A. Auld, E. Özbay, "All-electronic terahertz spectroscopy system with terahertz free-space pulses", *J. Opt. Soc. Am. B*, **11**, 2561 (1994).

43. G. M. Rebeiz, "Millimeter-wave and terahertz integrated circuit antennas", *Proc. IEEE* **80**. 1748 (1992).

44. J. P. Rice, E. N. Grossman, D. A. Rudman, "Antenna-coupled high-T_c airbridge microbolometer on silicon", *Appl. Phys. Lett.* **65**, 773 (1994).

45. D. W. van der Weide, "Planar antennas for all-electronic terahertz systems", *J. Opt.Soc. Am. B* **11**, 2553 (1994).

46. V. M. Lubecke, K. Mizuno, G. M. Rebeiz, "Micromachining for terahertz applications", *IEEE Trans. Microwave Theory Tech.* **46**, 1821 (1998).

47. B. I. Greene, J. F. Federici, D. R. Dykaar, R. R. Jones, P. H. Bucksbaum, "Interferometric characterization of 160 fs far-infrared light pulses", *Appl. Phys. Lett.* **59**. 893 (1991).

48. C. Karadi, S. Jauhar, L. P. Kouwenhoven, K. Wald. J. Orenstein, P. L. McEuen. "Dynamic response of a quantum point contact", *J. Opt. Soc. Am. B* **11**, 2566 (1994).

49. P. Mann, "TWA disaster reopens tough security issues", in *Aviation Week & Space Technology*, **145**, 23–27 (July 22, 1996).

50. V. Agrawal, T. Bork, S. Kee, D. W. van der Weide, "Electronic THz reflection spectroscopy for detecting energetic materials", *Proceedings of the 6th International Conference on Terahertz Electronics* (Leeds, UK, 1998), pp. 34–38.

51. D. M. Sheen, D. L. McMakin. T. E. Hall, "Combined illumination cylindrical millimeter-wave imaging technique for concealed weapon detection", *Proc. SPIE* **4032**, 52 (2000).

52. D. Herskovitz, "Wide, wider, widest", *Microwave J.*, **38**, 26 (1995).

53. C. E. Baum. L. Carin, A. P. Stone, *Ultra-Wideband, Short-Pulse Electromagnetics 3* (Plenum Press, New York, 1997).

54. M. van. Exter, "Terahertz time-domain spectroscopy of water vapor", *Opt. Lett.*, **14**, 1128 (1989).

55. M. C. Nuss, K. W. Goossen, J. P. Gordon, P. M. Mankiewich, M. L. O'Malley, M. Bhusan, "Terahertz time-domain measurement of the conductivity and superconducting band gap in niobium", *J. Appl. Phys.*, **70**, 2238 (1991).

56. B. B. Hu, M. C. Nuss, "Imaging with terahertz waves", *Opt. Lett.* **20**, 1716 (1995).

Index

Springer Series in
OPTICAL SCIENCES

Springer Series in
OPTICAL SCIENCES

Printing (Computer to Film): Saladruck Berlin
Binding: Stürtz AG, Würzburg

Printed by Printforce, the Netherlands